To Charlie Towns
with Best Regards
Brad Moore

Профессору Ч. Таунсу
с наилучщими
пожеланиями
В. С. Летохов.

# CHEMICAL AND BIOCHEMICAL APPLICATIONS OF LASERS

VOLUME III

# Contributors

R. V. Ambartzumian
V. S. Letokhov
C. Bradley Moore

# CHEMICAL AND BIOCHEMICAL APPLICATIONS OF LASERS

### Edited by C. BRADLEY MOORE

*Department of Chemistry*
*and*
*Materials and Molecular Research Division*
*of the Lawrence Berkeley Laboratory*
*University of California*
*Berkeley, California*

## VOLUME III

1977

ACADEMIC PRESS    New York   San Francisco   London
A Subsidiary of Harcourt Brace Jovanovich, Publishers

*ACADEMIC PRESS RAPID MANUSCRIPT REPRODUCTION*

ACADEMIC PRESS, INC.
111 Fifth Avenue, New York, New York 10003

*United Kingdom Edition published by*
ACADEMIC PRESS, INC. (LONDON) LTD.
24/28 Oval Road, London NW1

**Library of Congress Cataloging in Publication Data**

Moore, Charles Bradley, Date
 Chemical and biochemical applications of lasers.


 Includes bibliographies.
  1.    Lasers in chemistry.   2. Lasers in biochemistry.
 I. Title. [DNLM:  1. Lasers. TK7871.3 M821c 1974]
 QD63.L3M66            542            73-18978
 ISBN 0-12-505403-3 (v.3)

# Contents

# List of Contributors

Numbers in parentheses indicate the pages on which the authors' contributions begin.

R. V. Ambartzumian, (167) Institute of Spectroscopy of the Soviet Academy of Sciences, Podolsky Rayon, 142092 Moscow, USSR

V. S. Letokhov, (1, 167) Institute of Spectroscopy of the Soviet Academy of Sciences, Podolsky Rayon, 142092 Moscow, USSR

C. Bradley Moore (1) Department of Chemistry, University of California, Berkeley, California

# Preface

Laser isotope separation and multiple photon infrared photophysics and photochemistry are topics of intense current research activity. Progress in these areas requires that many fascinating fundamental problems in chemistry and physics be solved. This work is driven by practical aims as well, for laser enrichment of uranium promises to be the first major chemical application of lasers in industry. Laser separations have already been demonstrated in the laboratory for many elements using many different methods. The phenomenon of selective multiphoton infrared excitation of molecules was discovered less than three years ago. A satisfactory qualitative understanding of how a single molecule can resonantly absorb tens of photons is just now emerging. The broad significance of this discovery is illustrated by recent publications of organic and inorganic chemists, physical chemists, and optical physicists. The selective photoprocesses now being exploited for laser isotope separation should find wide application throughout the physical and biological sciences.

The fundamental principles and methods of selective photophysical and photochemical processes are described in Chapter 1. Isotopic separations and related research are discussed for each of eight classes of laser methods. Applications of these selective techniques in chemistry, biology, and materials science are discussed. Chapter 2 gives a thorough description of the experimental results on multiphoton infrared processes and their theoretical interpretation. The book is intended to serve both as an introduction to selective photophysical and photochemical processes and as a review of current research.

The entire camera-ready copy for this book was prepared by Luce J. Denney. Her great patience and high standards in this task are very much appreciated. The illustrations were drawn with exacting care by Nancy Monroe. This volume and the preceding volumes of this series have been indexed thoroughly and thoughtfully by Penny Percival. The authors are grateful for all the efforts of these three people in producing this book. We also thank Michael Berman and Mark Johnson for assistance in preparation of the manuscript.

# 1

# Laser Isotope Separation

## V. S. LETOKHOV

INSTITUTE OF SPECTROSCOPY OF THE SOVIET ACADEMY OF SCIENCES

## C. BRADLEY MOORE

UNIVERSITY OF CALIFORNIA

The problem of laser isotope separation (1 - 5) is a part of a more general problem of laser separation of substances at the atomic and molecular level. The great current interest in laser separations rests on firm expectations of new, less costly methods of isotope separation. Large decreases in the costs of separated isotopes would open many new applications in research, in medical diagnosis, in environmental and agricultural studies, and in technology. Even modest cost decreases in materials for nuclear power production would have a substantial economic impact. The fundamental and practical possibilities of laser separations are just beginning to emerge for the preparation of materials, for chemical and biochemical synthesis, for nuclear isomer separation and for highly selective atomic and molecular detection systems. Thus this chapter begins with a general formulation of the problem in section I, is followed by a detailed consideration of the methods applied to isotope separations, II - XI, and concludes with a discussion of some other applications and new possibilities.

I.   INTRODUCTION AND CLASSIFICATION OF METHODS

A.   Laser Separation of Substances at the Atomic and Molecular Level

If we have a gaseous mixture of various particles A,B, C, ... (atoms, molecules), which differ little or not at all in chemical properties, it is difficult and sometimes impossible to separate them by ordinary chemical methods. Let the

1

quantum levels of the particles differ just enough so that we can excite selectively by monochromatic laser radiation certain particles, the A particles for instance, leaving all other particles (B,C, ...) unaffected. The criterion for this selective excitation is not difficult to meet; it is enough to have one absorption line of the selected particle shifted with respect to the spectral lines of the other particles. Now the excitation of the A particles changes their chemical and physical properties and, hence, may be used to separate substances by any method based on differences in the characteristics of excited and unexcited particles. The general concept is illustrated in Fig. 1.

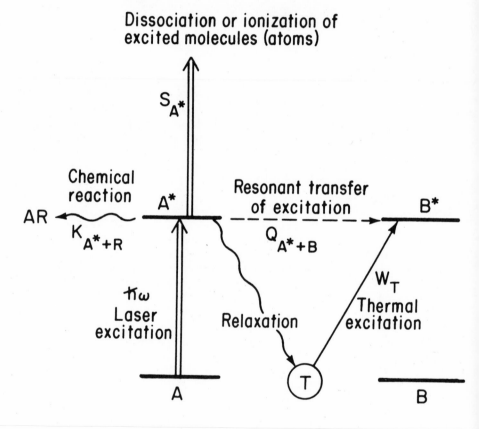

Fig. 1.  General scheme for laser separations. Selective laser excitation of A particles is followed by chemical reaction of A* to give products designated AR, or by absorption of a second photon to ionize or dissociate A*. Nearly resonant transfer of energy to B giving B*, and loss of energy to the thermal reservoir with subsequent thermal population of B* are diagrammed.

The excitation energy is meant here, of course, to be much higher than the thermal energy of the particles $kT_0$ which is responsible for nonselective thermal population of the lower-energy states of all particles in the mixture. This is the general concept of separations at the atomic and molecular level using lasers.

This general concept of separations requires that we address the following two questions:

1. What changes in atomic and molecular structures cause changes in absorption spectra and, thus, can be used for laser separation?

2. What properties of atoms and molecules change as a result of excitation and, thus, provide a means of laser separation?

Successful application of a specific concept of photoseparation requires that the following four conditions be met:

I. There must be at least one absorption line, $\omega_A^{(i)}$, of the "A" particles being separated which does not overlap significantly any absorption lines of the other particles in the mixture.

II. Monochromatic radiation at the chosen frequency of selective absorption, $\omega_A^{(i)}$, must be available with the characteristics of power, duration, divergence, and monochromaticity necessary for the separation method in use.

III. A primary photophysical or photochemical process must be found which transforms the excited particles into species easily separated from the mixture.

IV. The selectivity obtained for the "A" particle must be maintained against all competing photochemical and photophysical processes throughout the entire separation.

The remainder of section I introduces the fundamental physical and chemical phenomena which may be used to meet these four conditions. Methods of separation are classified according to the concepts applied in condition III. General conditions on final separation selectivity are formulated and compared for the classes of methods.

B.   Selective Photoexcitation

Since the spectral linewidths of atomic and molecular absorption in gases are rather small, the spectral criterion of particle difference is very sensitive to the smallest details of atomic and molecular structure.

Spectral differences in atoms may be caused by the following effects:

1. A difference in atomic numbers, or in the number of protons Z, being responsible for a cardinal difference in atomic absorption spectra. This difference is fundamental for

atomic absorption spectral analysis.

2. A difference in the number of neutrons N, the number of protons Z being the same, resulting in a small isotope effect which shows up clearly in atomic spectra.

3. A difference in nuclear excitation level with its consequent change in the nuclear spin causing an isomeric effect in the hyperfine structure of the atomic spectrum. Here the other parameters of the nucleus (Z, N) and the electron shell are unchanged. The value of the isomeric effect is usually even larger than that of the isotope effect.

Spectral differences in molecules may be caused by the following effects:

1. A difference in chemical composition, that is, in the number and the type of atoms in a molecule, resulting in a substantial change of the electronic and vibrational molecular spectra. This difference is basic for molecular absorption spectral analysis. Very small substituent changes which give negligible chemical effects may cause useful spectral shifts.

2. A difference in the isotopic composition of molecules, their other parameters being the same, bringing about an isotope effect in vibrational and, in some cases, in electronic spectra of molecules.

3. A difference in the three-dimensional structure of molecules, their chemical composition being the same, resulting in a difference of vibrational and electronic spectra. Of course, the spectra of many structual isomers are completely different, cyclopropane *vs* propylene. For cis and trans geometrical isomers the differences are still quite apparent. Differences in the secondary and tertiary structure of biomolecules give some spectral shifts. Right and left-handed stereoisomers exhibit spectral differences in polarized light.

4. A difference in the nuclear spin orientation of atoms in a molecule, that is, the difference in the total nuclear spin of a molecule, resulting in a change of the hyperfine structure in the molecular spectrum. This difference appears most vividly in the structure of the spectra of molecules containing identical atoms in symmetrically equivalent positions. The ortho and para forms of homonuclear diatomics are a familiar example. $CH_4$, $NH_3$ and $H_2CO$ are similar.

Thus we may consider the possibility of atomic and molecular separations by lasers of:

1.   Chemical elements
2.   Isotopes
3.   Nuclear isomers
4.   Molecular isomers and
5.   Ortho- and para-molecules.

Detailed consideration of isotopic shifts and isotopically selective excitation is given in section II.

## C.    Laser Radiation

Tunable laser radiation is now available at almost any
wavelength from the vacuum ultraviolet through the infrared.
In principle the fulfillment of requirement II is thus a standard problem in quantum electronics.  In practice it is often
difficult to produce sufficient laser energy for laboratory
separation research.  Of course, years will be needed until
the production of laser radiation at a given frequency with
desired parameters becomes a routine problem solved with
standard devices without involving Quantum Electronics Ph.D.'s.
Lasers sufficiently powerful and efficient for commercial production purposes are now available only for rather limited
wavelength ranges and applications.  In practice the relative
merit of different methods of isotope separation depends very
much on the type of laser required.

Let us enumerate the essential properties of laser radiation which make lasers very valuable and efficient tools in
the field of *selective laser photophysics and photochemistry.*
They are:

1. *Radiation frequency tunability,* which permits the production of radiation at any frequency in the infrared, visible,
ultraviolet and vacuum ultraviolet spectral regions.

2. *High intensity,* sufficient to saturate the absorption
of quantum transitions, i.e. to excite a large portion of the
atoms or molecules.

3. *Controlled radiation duration*, which can be made
shorter than the lifetime of excited atomic or molecular
states.

4. *Spatial coherence of radiation,*  which makes it possible to form directed beams of radiation and to irradiate in
longpath cells.

5. *Monochromaticity and temporal coherence*, which allow
an extremely high selectivity of excitation with a very small
difference in absoprtion frequencies of species being separated.

The combination of all these valuable properties in an
efficient source of optical radiation makes development of
optical methods for materials technology at the atomic and molecular level most promising.


## D.    Laser Separation Methods

The selectivity of laser excitation may be preserved by
using the changes in the properties of atoms and molecules
caused by photoexcitation, Fig. 1.  Laser separation schemes
may fulfill requirement III based on the property that:

1. The *chemical reactivity* of atoms or molecules is often increased by excitation.

2. The *ionization energy* of an excited atom or molecule is smaller than that of unexcited ones.

3. The *dissociation energy* of an excited molecule is smaller than that of an unexcited one.

4. *Predissociation* occurs when an excited molecule passes spontaneously into a dissociative state.

5. The excitation of a molecule may result in *isomerization*. The isomer, because of its different internal structure, has different chemical properties.

6. The *recoil* of an atom or molecule occurs when it absorbs a photon with its momentum $\hbar\omega/c$. This gives a very small but observable particle photodeflection.

7. Excited atoms and molecules may have a higher *polarizability*, a different *symmetry* of wave function, etc. This may cause changes in cross sections for scattering by other particles, in its motion in external fields, etc.
Many possible photochemical and photophysical methods of laser separation follow from this classification.

The field of selective photochemistry has a long history. Changes in the reactivity of atoms and molecules due to photon absorption are well known and have long been used in photochemistry, including *isotopically selective photochemistry*. Isotopically selective excitation of atoms and molecules followed by photochemical reaction was conceived as a method of isotope separation soon after isotopes and the isotope effect in atomic and molecular spectra were discovered (6). The first attempt to bring about photochemical isotope separation was made as far back as in 1922 (7). In this work $^{37}Cl_2$ molecules were exposed to the light of a common source which passed through an absorbing filter containing mainly $^{35}Cl_2$ molecules. The first successful experiment was conducted ten years later by Kuhn and Martin (8) who exposed phosgene ($CO^{35}Cl_2$) molecules to a strong spectral line of an aluminium spark at $\lambda = 2816.2$ Å. At about the same time, Mrozowski proposed (9) a method of selective excitation of mercury isotopes by the 2537 Å mercury line after passage through a resonantly absorbing mercury filter. He suggested that this method should be used in isotopically selective photochemical reactions of excited mercury atoms with oxygen. The method was first realized and investigated by Zuber (10,11) in 1935. After the war the photochemical method of mercury isotope separation was worked out in detail in some countries (12 - 15), and experimental setups were devised, and are currently used, to separate mercury isotopes in small amounts (14,15).

With the advent of laser sources of intense monochromatic radiation it became possible to selectively excite many

atoms and molecules without depending upon accidental coinci-
dences between strong lines of spontaneous radiation and
absorption lines of atoms and molecules.  With lasers, photochemical isotope separation entered a new experimental domain.
The first attempt to realize photochemical isotope separation
with a laser was made by Tiffany, Moos and Schawlow in 1966
(16,17).  High-power IR lasers permitted excitation of molecular vibrational levels and opened the possibility of vibrational photochemistry.  The first attempt at laser isotope
separation by vibrational photochemistry was carried out by
Mayer, Kwok, Gross and Spencer in 1970 (18).  Both of the
experiments relied on the photochemical approach-increased
chemical reactivity of excited atoms and molecules.  Both
failed because the overall chemistry did not preserve the
initial excitation selectivity-Requirement IV.  Laser photochemical separations have since been achieved and photochemical methods are discussed in detail in sections VII and VIII.

Selective photophysical methods based on dissociation and
ionization have met with greater initial success.  The first
proposal for laser isotope separation appears in a patent by
Robieux and Auclair for two-step photoionization of $UF_6$ molecules in 1965 (19).  While this method remains impractical,
two-step atomic ionization and selective two-step molecular
photodissociation were suggested and demonstrated in works
(20 - 22).  Selective photopredissociation gave simple, clean
isotope separations (23,24).  Recently, selective molecular
dissociation by high-power, infrared laser radiation was
discovered (25,26).  All of these methods have yielded good
isotope separations and are actively being developed for
commercial application.  They are discussed in depth in sections III - VI.

There are many methods for laser separation of atoms and
molecules of different isotopic composition.  A general classification of possible methods is given above.  At the same
time they can be classified from another standpoint, according
to the type of particles used, and thus we may subdivide all
the methods into atomic and molecular ones.  Figure 2 shows
schematically possible general approaches of laser isotope
separation based on selective excitation of atoms, i.e. different atomic methods:  1) photochemical reactions;  2) ionization by a second photon and/or by an external electric field
or by collision;  and 3) velocity change of selectively
excited atoms.  Figure 3 gives possible approaches of laser
isotope separation based on selective excitation of molecules.
Of course, the excitation of molecules offers new possibilities:  selective photodissociation and photoisomerization.

An entirely different concept of separation may be employed in processes where the selectivity exists not due to
selective excitation but due to an inherent isotopic selecti-

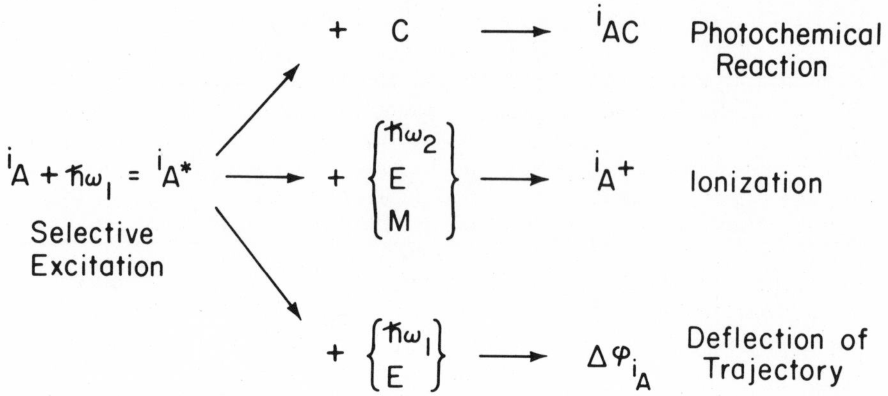

$$^iA + \hbar\omega_1 = {}^iA^*$$

Selective
Excitation

$+ \quad C \quad \longrightarrow \quad {}^iAC$   Photochemical Reaction

$+ \quad \begin{Bmatrix} \hbar\omega_2 \\ E \\ M \end{Bmatrix} \quad \longrightarrow \quad {}^iA^+$   Ionization

$+ \quad \begin{Bmatrix} \hbar\omega_1 \\ E \end{Bmatrix} \quad \longrightarrow \quad \Delta\varphi_{iA}$   Deflection of Trajectory

*Fig. 2. Atomic laser isotope separation schemes.*

$$^iAB + \hbar\omega_1 = {}^iAB^*$$

Selective
Excitation

$+ \quad C \quad \longrightarrow \quad {}^iAC$   Photochemical Reaction

$+ \quad \begin{Bmatrix} \hbar\omega_2 \\ M \\ H,E \end{Bmatrix} \quad \longrightarrow$   $^iA, {}^iA^+$ Photodissociation or Photoionization

$^i\widetilde{AB}$   Photoisomerization

$+ \quad \begin{Bmatrix} \hbar\omega_1 \\ E \end{Bmatrix} \quad \longrightarrow \quad \Delta\varphi_{iAB}$   Deflection of Trajectory

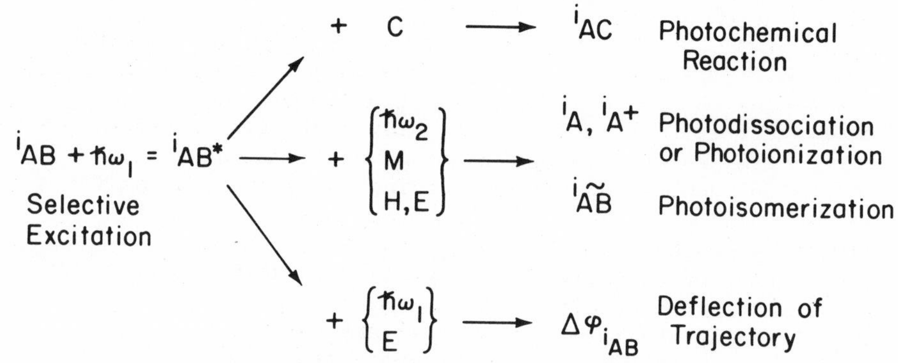

*Fig. 3. Molecular laser isotope separation schemes.*

vity in the kinetics of the process. Kinetic isotope effects
in chemical reactions are a well-known example. For instance,
isotope separation is possible in molecular mixtures with high
vibrational temperatures and low translational temperatures.
Non-selective vibrational heating may be accomplished by laser
or by electric discharge excitation (27 - 30). In these sys-

tems the isotope selected is defined by kinetic rate constants.
The selection of isotope cannot be interchanged, Sec. X.


E.   Selectivity Loss Processes

In every step of the separation process from excitation
through final physical removal of the desired component there
are processes which degrade the initial excitation selectivity.  An understanding of these processes is essential to the
development of a separation method and to comparisons among
alternative methods for any particular separation.

For quantitative comparison of the separation methods
let us define the *coefficient of separation selectivity* of A
and B molecules by the relation:

$$K(A/B) = \frac{[{}^{N}_{AR}]f}{[{}^{N}_{BR}]f} \bigg/ \frac{[{}^{N}_{A}]o}{[{}^{N}_{B}]o} , \qquad (I.1)$$

where $[{}^{N}_{A}]o$ and $[{}^{N}_{B}]o$ are the initial concentrations of A and
B molecules in the mixture, $[{}^{N}_{AR}]f$ and $[{}^{N}_{BR}]f$ are the final
concentrations of new resultant molecules in the mixture which
contains the A and B atoms or molecules being separated.  With
no separation effect present, the selectivity coefficient
$K(A/B) = 1$.  The values $K(A/B) \gg 1$ correspond to a high selectivity of separation.

Figure 1 outlines the processes resulting in conservation
and loss of the selectivity attained in exciting the particles
of type A.  Let the electronic or vibrational level of A
mixed with B molecules be selectively excited by laser radiation.  (All this extends to atoms in equal measure).  The
problem consists in selective binding and separation of the A
molecules from the mixture.  Two processes cause the loss of
selectivity:

1. *Thermal nonselective excitation.*  Excited molecules
may relax to the lowest state before A* is acted upon by the
primary separation process.  For vibrational excitation the
relaxation is radiationless, and hence causes the gas mixture
to heat to some temperature $T > T_0$. As a result, the vibrational levels of both A and B molecules at energy $E^{(i)}_{(A,B)}$ are
thermally populated with the Boltzmann probability
$\exp[ - E^{(i)}_{(A,B)} /kT]$.  Thermal excitation is completely nonselective, and therefore the molecules of both types A and
B are processed at the same rate under irradiation.  Of
course, the laser excitation must be sufficiently strong that
the thermal population at $T_0$ is negligible.

2. *Resonance excitation transfer.*  This process occurs
when an excited A molecule collides with an unexcited B

molecule:

$$A* + B \rightarrow A + B*$$

The cross section of the resonance excitation transfer for molecules with nearly resonant levels is often gas-kinetic, and for atoms it may be hundreds of times larger. Thus it is essential for the primary separation process to occur on a timescale short compared to that for A-B collisions.

3. *Scrambling processes.* Selectivity may be reduced by processes which act after the primary separation step of chemical reaction, dissociation, ionization or particle deflection has occured. Free radical chain reactions initiated by the products of $A* + R$ or by the dissociation fragments of $A*$ may attack B. B may react directly with R on the surfaces of the container. Charge exchange, $A^+ + B \rightarrow A + B^+$, occurs with large cross sections near resonance.

The factors which most severely limit the selectivity of any specific separation process will depend on the details of the individual system. We will compare here some general features of processes based upon the seven different principles for the primary photochemical or photophysical separation step listed in I.D.

1.    *Photochemical Reaction*

Photochemical methods are based on the sequence of processes:

$$\begin{pmatrix} \text{Selective} \\ \text{photoexcita-} \\ \text{tion of } A \\ \text{particles} \end{pmatrix} \rightarrow \begin{pmatrix} \text{Chemical reaction} \\ \text{of excited } A* \\ \text{particles with} \\ \text{acceptor } R \end{pmatrix} \longrightarrow \begin{pmatrix} \text{Physical or} \\ \text{chemical iso-} \\ \text{lation of} \\ \text{products} \end{pmatrix}$$

To produce high selectivity of separation, a chemical reaction between excited $A*$ molecules and suitable acceptors $R$ must be found which occurs at a rate, $K_{A*+R}$ (total time averaged production rate for $AR$), substantially exceeding the rate of selectivity loss by resonance transfer of excitation, $Q_{A*+B'}$ and by thermal excitation, $W_T$:

$$K_{A*+R} \gg Q_{A*+B'}, W_T \tag{I.2}$$

It is also necessary that the reaction rate of the reagent $R$ with unexcited $A$ and $B$ molecules should be much less than that with excited A molecules:

$$K_{A*+R} \gg K_{A+R} \simeq K_{B+R} \tag{I.3}$$

In principle, the fulfillment of conditions (I.2) and (I.3) is quite possible even though in each specific case $R$ must be especially selected. To ensure a high degree of selectivity in separating $A$ and $B$ molecules, it is of importance to fulfill conditions (I.2) and (I.3) with a fairly large margin. In the absence of scrambling processes the approximate coefficient of separation selectivity is determined by a minimum of the values:

$$K(A/B) \simeq min \left\{ \frac{K_{A*+R}}{K_{B+R}}; \ \frac{K_{A*+R}}{Q_{A*+B}}; \ \frac{K_{A*+R}}{W_T} \right\} \qquad (I.4)$$

Such a process is not fully controlled by laser radiation because it depends crucially on the relation between rates of chemical reaction, deexcitation, and excitation transfer in elementary collisions. The efficiency and selectivity of photochemical separations is limited by our ability to find reaction systems which satisfy conditions (I.2) and (I.3) well and do not give scrambling through subsequent undesired reactions.

### 2.    Two-Step Photoionization and Photodissociation

This class of photophysical methods (2 and 3, I.D.) is based on the sequence of processes:

$$\left( \begin{array}{l} \text{Selective} \\ \text{photoexcita-} \\ \text{tion of } A \\ \text{particles} \end{array} \right) \rightarrow \left( \begin{array}{l} \text{Photoionization or} \\ \text{photodissociation} \\ \text{of } A* \text{ particles} \end{array} \right) \rightarrow \left( \begin{array}{l} \text{Physical or che-} \\ \text{mical stabiliza-} \\ \text{tion and isola-} \\ \text{tion of the} \\ \text{products of} \\ \text{ionization or} \\ \text{dissociation} \end{array} \right)$$

A second laser is used to induce the transition of $A*$ particles to another state for which the selectivity loss rate is much smaller. This can be done by photoionization of selectively excited atoms or photodissociation of selectively excited molecules. The second laser radiation does not select between $A$ and $B$, only between excited and unexcited particles. To preserve the initial selectivity the rate of laser-induced loss of $A*$, $S_{A*}$, must exceed the rate of resonance transfer of excitation and of thermal excitation,

$$S_{A*} \gg Q_{A*+B}, \ W_T \qquad (I.5)$$

as well as the rate of photodissociation and photoionization of unexcited $A$ and $B$ particles,

$$S_{A*} \gg S_A \simeq S_B. \qquad (I.6)$$

Unlike the method of excited particle chemical binding, this method is universal as regards the production of high separation selectivity. A high selectivity may always be achieved here by using sufficiently high laser power that photoionization or photodissociation of the excited atoms occurs rapidly

compared to collison processes.  Thus all but the final stage
of separation is fully controlled by laser radiation.  This is
a very important feature of these photophysical methods.  In
the absence of scrambling processes and for perfect excitation
selectivity, the approximate coefficient of selectivity of
separation is determined by a minimum of the values:

$$K(A/B) \simeq min\left\{ \frac{S_{A^*}}{S_B}; \; \frac{S_{A^*}}{Q_{A^*+B}}; \; \frac{S_{A^*}}{W_T} \right\}. \qquad (I.7)$$

When the rate of ionization or dissociation is high enough,
the selectivity of photophysical methods is much higher than
that of photochemical ones,

$$K_{photophys.} \; (A/B) \gg K_{photochem.} \; (A/B). \qquad (I.8)$$

The scrambling processes which compete in the final stage of
separation are often analogous to those which destroy selectivity in photochemical processes but are usually less effective.  The great advantage of two-step excitation is that the
ionized or dissociated $A^*$ particle, compared to $A^*$ itself, has
properties which differ much more strongly from the properties
of unexcited $A$ and $B$.

### 3.    One-Photon Unimolecular Processes

The methods of molecular photopredissociation and photoisomerization (4 and 5, I.D.) seem to combine the advantages
of single-laser, photochemical processes and two-step, collision-free, photophysical processes.  The excitation and
primary separation steps are combined because the dissociation
or rearrangement of the selectively excited molecule occurs
spontanesously without collisions.  Since the spontaneous rate
may be slow and is not controlled by laser radiation, it is
often more effective to induce the process by collisions.  In
practice these methods are intermediate between the photochemical and  two-step  photophysical ones.

### 4.    Particle Deflection

The sixth and seventh (I.D.) effects use quite small
changes in particle properties during excitation which may be
seen in experiments with atomic and molecular beams.  Excitation of particles in a beam leads to a small change in
trajectory.  Even weak, grazing collisions are sufficient to
alter trajectories and destroy selectivity.  Thus these
methods call for low-density atomic or molecular beams in a
very high vacuum.  The mass throughput is much less than for
the first five effects.  It should be emphasized that these

shortcomings are not due to the use of atomic or molecular beams but to the fact that selective excitation brings about only very small changes in particle properties. For instance, with the methods of selective photodissociation or photoionization the dissociation fragments or accelerated ions have drastically altered trajectories. In these cases relatively high density molecular or atomic beams may be used efficiently.

We have noted above the general application of laser radiation to the separation of elements with similar chemical properties, of isotopes, of nuclear isomers and so forth. In the remainder of this chapter we will restrict the discussion to isotope separation. Nevertheless, the much wider application of these methods should be kept in mind.

## II.  SPECTRAL SHIFTS AND SELECTIVE EXCITATION

The first step of any laser isotope separation scheme is absorption of a laser photon by the desired isotopic atom or molecule at a wavelength for which the undesired isotopic species is relatively transparent. Thus, a transition with an isotopic shift greater than its linewidth is needed. The spectrum should not be so dense that the shift only results in a coincidence with another absorption feature of the unwanted isotopic species. The laser should have a spectral resolution and tunability sufficient to excite the desired transition within its linewidth. Following the excitation it is then essential that the separation step occur before nearly-resonant energy transfer to the undesired isotopic species takes place. In this section we discuss the sources and magnitudes of isotopic shifts in atomic and molecular spectra. Linewidths of transitions are compared to isotopic shifts and to spacings characteristic of the spectral structure of various types of transitions. The importance of temperature, pressure and physical state for these comparisons is noted. The line narrowing and spectral simplification achieved in gas dynamic cooling are considered. Non-linear optical line narrowing techniques such as two-photon and saturation-resonance spectroscopy are mentioned. Finally, the spectrum of $UF_6$ is discussed as an illustration of the practical problems of selective excitation.

## A.  Isotopic Shifts

The electronic energy levels of atoms are shifted by

changes in the number of neutrons in the nucleus through the change in total nuclear mass, the change in nuclear volume and therefore charge distribution, and the change in nuclear spin angular momentum (31,32). The effect of nuclear mass on the reduced mass for electronic motion about the nucleus gives an isotope shift of order of magnitude

$$\Delta\omega \sim \omega \frac{m\ \Delta M}{M^2} \qquad\qquad (II.1)$$

where $m$ is the electron mass, $M$ the nuclear mass and $\Delta M$ the isotopic mass difference. It is important only for light elements. Shifts of about 1 cm$^{-1}$ are observed for $^6$Li, $^7$Li. For heavier elements the shifts are dominated by the change in hyperfine structure, interaction of nuclear and electronic angular momentum, and the change in nuclear volume. The change in nuclear volume shifts the energies of s electrons and therefore, gives substantial isotope shifts for transitions involving s electrons. The volume shift is dramatically illustrated by the uranium II transition shown in Fig. 4(a).

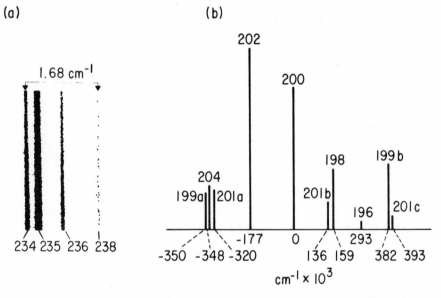

Fig. 4(a). The isotope structure of the 4244.4 Å line of UII (Ref. 32). Fig. 4(b). The isotope and hyperfine structures of the 2537 Å line of Hg.

Figure 4(b) shows the combined effect of volume shifts and hyperfine structure. Hyperfine structure multiplets occur for $^{199}$Hg and $^{201}$Hg and single lines for masses 198, 200, 202 and 204.

As the number of neutrons varies, the nuclear spin variation may change the selection rules.  As a result, atoms of
some isotopes may have lines which are forbidden for other
isotopes.  If the nuclear spin of an atom makes the total
angular momentum $F \neq 0$, then the electron-nucleus interaction
can even induce electric dipole transitions for J=0 → J'=0.
The forbidden lines 2270 Å ($6^3P_2-6^1S_0$) and 2656 Å ($6^3P_0 - 6^1S_0$)
found by Rayleigh in 1927 and Wood in 1928 respectively in the
odd isotopes $^{199}Hg$ (I=1/2) and $^{201}Hg$ (I=3/2) are classic
examples of this effect.  Yu. B. Zeldovich and I.I. Sobelman
have recently noted (33) the possibility of using this effect
in selective excitation and laser separation of even or odd
isotopes of Hg, Sr, Ba, Zn and Cd.  A spectral line shift,
much larger than the isotope shift is normal for this selective excitation with respect to nuclear spin.

In molecular spectra (34 - 36) the structure of vibrational and rotational energy changes is superimposed on the
electronic energy changes.  In addition to these rovibronic
transitions there are also transitions in which only vibrational and rotational energies change.  The isotopic shifts
are dominated by the effect of nuclear mass on the vibrational
energy spacings (inversely proportional to the square root of
the vibrational reduced mass) and the rotational energy level
spacings (inversely proportional to the moments of inertia)
(5, 34 - 36).  The spectra of linear molecules are generally
quite simple;  each vibrational or vibronic band consists
of a regularly spaced set of individual rotational transitions.
The space between lines is much greater than the linewidth for
molecules containing the lighter elements and only begins to
approach the linewidth for molecules with moments of inertia
as large as $I_2$.  The rotational structure in the bands of nonlinear molecules, asymmetric tops, spherical tops, and, to a
lesser extent, symmetric tops, is far more complex.  Fortu-
nately, for many simple molecules, gas phase electronic and
vibrational spectra are still sufficiently well resolved that
the spectra of different isotopic molecules are clearly
resolvable (Fig. 5).  But for many molecules the rotational
structure is so closely spaced that lines overlap within their
Doppler widths. Such a situation is demonstrated by the laser
spectroscopy of the 10μ transition of $SF_6$ (37,38) and $OsO_4$
(39) and $UF_6$ (38,40).  However, the presence of a great number
of absorption lines, does not exclude the possibility of an
accidental occurrence of a section in the absorption spectrum
in which the absorption line of just one isotopic molecule
falls.  Figure 6 shows the vibration-rotation absorption
spectrum of monoisotopic $OsO_4$ obtained by saturation resonance
spectroscopy using several $CO_2$ laser lines (39).  The typical
interval between absorption lines is much smaller than the
20 MHz Doppler width.  Nevertheless, the $R̂(8)$ line of the

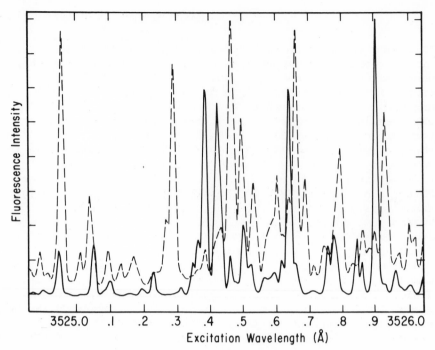

Fig. 5. *Rovibronic structure in the fluorescence excitation spectrum of $H_2{}^{12}CO$ (———) and $H_2{}^{13}CO$ (----) molecules
obtained by scanning the frequency of a pulsed dye laser
(Ref. 3). A single vibronic band covers about 60 Å.*

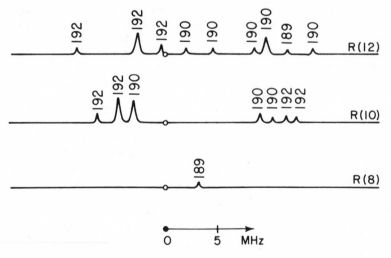

Fig. 6. *The vibration-rotation spectrum $OsO_4$ monoisotopic molecules in the frequency tuning region of the R(8) -
R(12) lines of a $CO_2$ laser. Doppler broadening has been eliminated by saturation-resonance spectroscopy (Ref. 39).*

$CO_2$ laser at 10.6 μm is absorbed mainly by $^{189}OsO_4$ molecules. For heavy molecules the problem with closely spaced rotational lines is compounded by the overlapping absorption spectra of molecules populating the low lying vibrational states (see Sec. IV). The electronic excited states of many simple molecules are often strongly perturbed and extremely complex. For example, it was a *tour de force* to resolve and analyze a single vibronic band from the entire visible spectrum of $NO_2$ (41).

For molecules the nuclear hyperfine structure splittings are usually smaller than Doppler linewidths (42). However, nuclear spin parity has striking effects on molecular spectra through selection rules. For example, ortho-para molecular modifications have different sets of electronic-vibration-rotational lines that allow selective excitation of isotopes with even and odd total nuclear spins. In infrared vibration-rotation spectra, isotopic substitution of one atom in a set of identical atoms can eliminate the forbiddenness of electric dipole transitions. This effect is clearly illustrated in the IR spectrum of HD, HT and DT molecules.

B.    Linewidths

Gas phase absorption lines are broadened by the Doppler effect, by molecular collisions, by radiative decay, by radiationless processes, by passing through the region of interaction with the optical field and by power broadening. At low pressures the Doppler effect usually dominates the linewidth. The full-width at half-maximum is given by (43)

$$\Delta\omega_D = 7.2x10^{-7} \omega (T/M)^{1/2}, \tag{II.2}$$

where $T$ is the temperature in °K and $M$ is the molecular mass in atomic mass units. The other broadenings result from the finite lifetime, $\tau$, of the pair of levels interacting with the radiation field. The full-width at half-maximum is given by

$$\Delta\omega_L = \tau^{-1}. \tag{II.3}$$

The absorption coefficient as a function of frequency is given by (43)

$$k(\omega) = k(\omega_o) exp[-2\sqrt{ln2}(\omega-\omega_o)/\Delta\omega_D] \tag{II.4}$$

for Doppler broadening, and by (43)

$$k(\omega) = k(\omega_o)\left\{1 + \left[\frac{2(\omega-\omega_o)}{\Delta\omega_L}\right]^2\right\}^{-1} \tag{II.5}$$

for lifetime broadening. An upper limit on the operating pressure in bulk gases is often imposed by pressure broadening. Figure 7 shows the line shapes for the 350 nm transition of

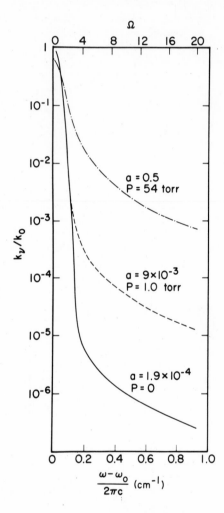

*Fig. 7. The profiles of $H_2CO$ absorption lines near 350 nm showing combined Doppler and lifetime broadening at various pressures of formaldehyde. Here $k_o$ is the maximum absorption coefficient for the case of Doppler broadening, $a = \sqrt{\ln 2}\ \Delta\nu_L/\Delta\nu_D$, $\Delta\nu_D = 0.06\ cm^{-1}$, and $\Omega = (\omega - \omega_o)/\Delta\omega_D$ (2).*

formaldehyde at several pressures. Notice that if particularly high selectivity ratios are required that the tails of the lines become important. They are dominated by pressure broadening and zero pressure lifetime broadening since the Lorentz profile decreases quadratically rather than exponentially with displacement from line center.

The absorption spectra of molecules trapped in matrices of rare gases or other inert molecules at low temperature are often quite sharp (44). Molecular rotation is usually stopped and the spectra exhibit only changes in electronic and vibrational energy. Spectral linewidths of less than 1 $cm^{-1}$ can often but by no means always be obtained. Isotope shifts for Cl and lighter elements are normally well resolved. An observable shift has even been reported for a uranium compound, Fig. 8 (45). Matrix spectra are often complicated by the pre-

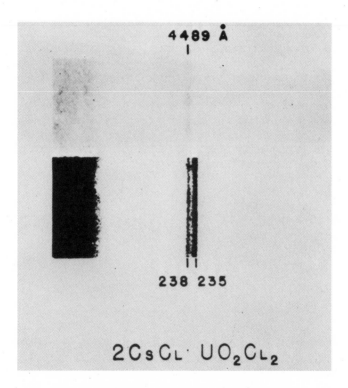

*Fig. 8. The isotopic structure of the absorption line 4489 Å of the uranium compound 2CsCl·$UO_2Cl_2$ at low temperature (45).*

sence of more than one type of trapping site and by the presence of dimeric and polymeric units of the isolated molecule. Resonant and nearly resonant transfers of energy are found to occur on sub-microsecond and microsecond timescales (46). The spectra of liquids are usually so broad that isotope shifts are resolvable only for the first few elements. However, in the case of $SF_6$ dissolved in liquid $O_2$ or liquid Ar, spectral widths of less than 0.3 cm$^{-1}$ have been reported for the $\nu_3$ band at 940 cm$^{-1}$ (47). Since the isotope shift between $^{32}SF_6$ and $^{34}SF_6$ is 17 cm$^{-1}$, highly selective excitation should be quite simple.

## C.    Methods of Excitation Selectivity Enhancement

In some systems Doppler broadening causes the spectra of different isotopic components of a mixture to overlap. Either the isotopic shifts themselves are very small as for alkali atoms or alkaline earths or the spectra are highly congested as for larger polyatomic molecules.

The Doppler width may be largely eliminated by using non-linear optical excitation resonances whose frequencies are independent of molecular velocity. Two-photon excitation in the field of two counter-running light waves with equal frequencies may be applied (48) (Fig. 9a). A narrow absorption peak at the center of a Doppler-broadened absorption line permits, in principle, selective excitation of the atoms and molecules for which the isotope shift is hundreds or thousands of times smaller than the Doppler width. This method necessitates higher intensities as compared to the method of single-quantum excitation. Efficient use of the laser photons is a severe problem. Absorption line shifts in the strong field must be guarded against. These methods are discussed in Refs. 3, 49 and 50, but have not been used for separations.

Another approach to the elimination of Doppler broadening is to eliminate the velocity spread. One can use atomic and molecular beams (51) where the Doppler width of spectral lines is given by

$$\Delta\omega_{beam} \simeq \frac{\phi_b}{\pi} \Delta\omega_D, \qquad (II.6)$$

where $\Delta\omega_D$ is the Doppler width for equilibrium gas from Eq. (II.2), $\phi_b$ is the angular divergence of particle beam, and the excitation is by a collimated laser beam, with the angular divergence less than $\phi_b$, at right angles to the particle beam (Fig. 9b). Very high resolution spectra of $I_2$ have been produced in this way (51). Two-step ionization of Ca atom beams has been reported (52). Bernhardt et al (53) have separated Ba isotopes by photodeflection. For practical separations low beam densities can be a serious disadvantage since long pathlengths are required for good photon utilization and large volumes are needed for high throughput (3). This method of line narrowing is particularly useful for the separation of

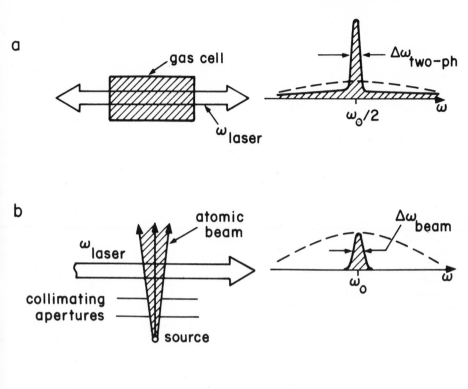

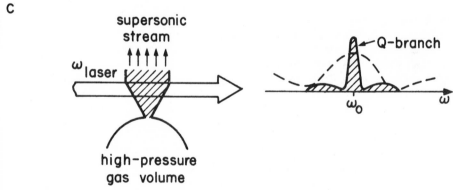

*Fig. 9. Selectivity enhancement methods. On the right the dashed lines indicate the normal spectral profiles. The shaded spectrum is the enhanced one. To the left the arrangement of laser and sample (shaded) is shown for different methods of enhancement: (a) two-photon excitation by oppositely directed laser beams; (b) excitation at right angles to a collimated molecular beam; (c) gas-dynamic cooling by expansion through a shaped nozzle.*

isotopes by selective photoionization when other considerations make the use of an atomic beam most reasonable.

Clearly the Doppler width may be reduced by lowering the
gas temperature. Low temperatures may be achieved at gas
densities above the equilibrium vapor pressure by dynamic
cooling during adiabatic expansion through nozzles such as
have been used for spectroscopy (38, 54 - 57), for molecular
beam sources (58) and for gas dynamic lasers (59 - 61), (Fig.
9c). On expansion at a supersonic rate, a considerable portion of the energy from the internal degrees of freedom is
transferred to bulk translational motion of the gas. Translational and rotational temperatures may drop below 1°K (57).
The vibrational temperature also drops precipitously. This
method has many advantages. It allows selective excitation
in the most complicated case of continuous vibration-rotation
or rovibronic absorption bands of complex molecules in which
the Q-branch width exceeds the value of isotope shift and the
spacing between lines in the spectrum is much smaller than the
Doppler broadening. The low translational temperature narrows
the Doppler width. More importantly, the low rotational
temperature brings all of the molecules into the lowest molecular rotational states. The P, Q and R branches of each
vibrational band are collapsed into a few single lines. And
most importantly, in many cases, vibrational relaxation is
sufficiently rapid that essentially all molecules reach the
lowest vibrational energy level and all hot band absorption
is eliminated. This lowering of the translational, rotational
and vibrational temperatures results in a tremendous simplification of the spectrum and increase in the effective optical
cross section. A large fraction of the molecules is now in
resonance at each of the few remaining absorption frequencies.
The gas density and mass throughput in the expanding flow are
much greater than for an effusive molecular beam. Collisions
occur and cause transitions among the populated quantum states.
It appears that this method may make highly selective excitation of $UF_6$ possible (38).

Selective excitation of molecules in the gas phase is
usually done by lasers whose spectral width is much smaller
than the isotope shift and absorption linewidth. Single-frequency lasers with a narrow radiation line are required to
excite selectively those molecules having a broad electronic-
vibrational-rotational spectrum with overlapping absorption
regions of two isotopic molecules. This sharply reduces the
number of simultaneously excited molecules owing to their
distribution over rotational sublevels. For isotopically
selective excitation one can use filtered, broadband radiation
without the spectral components corresponding to the absorption lines of those isotopic molecules the excitation of which
is undesirable. The radiation of lasers with a broad genera-

tion spectrum and with an absorber in the cavity has a suitable spectral composition (62,63).    If such a laser irradiates
a molecular mixture, the molecules not in the cavity will be
excited for the most part.    Glyoxal has been selectively
excited in this manner using a pulsed dye laser (64).    Datta,
Anderson and Zare (65) have enriched $^{37}Cl$ by selectively exciting $I^{37}Cl$ with a cw dye laser of 3 Å bandwidth in whose
cavity a cell containing $I^{35}Cl$ was placed.    Eerkens (40) has
reported work on uranium isotope separations using a $CO_2$ laser
with an intracavity absorber of $^{238}UF_6$ to excite $^{235}UF_6$.

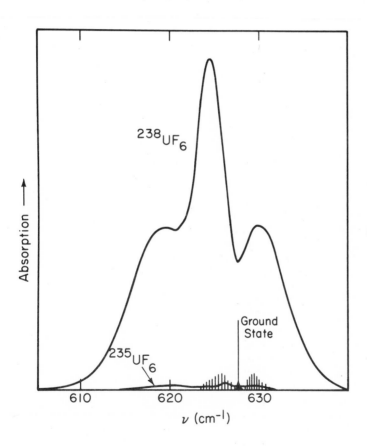

Fig. 10.    Spectrum of $^{238}UF_6$ at room temperature with
ground-state PQR structure of $\nu_3$ band; spectrum of $^{235}UF_6$ is
shown at bottom for comparison (38).

D.    An Example – $UF_6$

The spectrum of $UF_6$ illustrates the difficulties caused by spectral complexity in polyatomic molecules and the great power of dynamic cooling for overcoming these problems. T. Cotter suggested work along these lines in the Los Alamos Scientific Laboratory some time ago.  Some of their elegant results (38) are illustrated here.  Figure 10 shows the quasi-continuous spectrum of overlapping lines of the $\nu_3$ spectrum of $UF_6$.  Since the isotopic shift between $^{235}UF_6$ and $^{238}UF_6$ is about 1 cm$^{-1}$, highly selective excitation is not possible. This broad spectrum is made up of many individual transitions which overlap within the Doppler width to give a nearly smooth contour.  Figure 11 shows the relative populations of vibra-

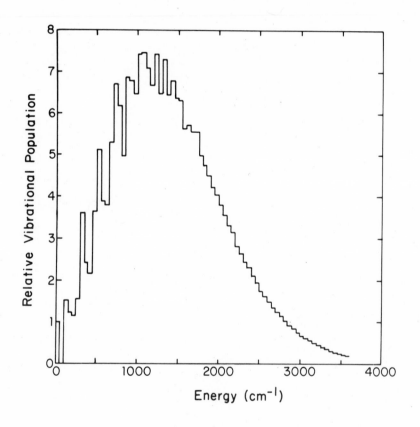

Fig. 11.  Relative population of $UF_6$ vibrational states as a function of vibrational energy (38).

tional energy levels as a function of their total vibrational energy. Since $UF_6$ has 15 vibrational modes with six distinct frequencies, $\nu_1 - \nu_6$, 667, 533, 624, 186, 202 and 142 $cm^{-1}$, the average vibrational energy is large. The spectrum is further complicated by the addition of Coriolis splittings to the usual PQR branch structures.

The temperature of an expanded gas, $T$, relative to the initial temperature, $T_o$, is given by

$$T/T_o = (P/P_o)^{(\gamma-1)/\gamma} \qquad (II.7)$$

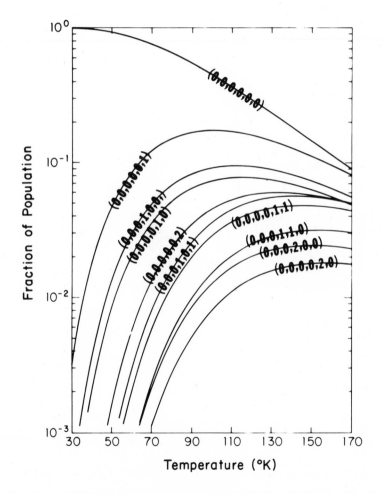

Fig. 12. Fractional population of the ten lowest lying vibrational states of $UF_6$ as a function of temperature (38).

where $\gamma$ is the ratio of constant pressure to constant volume
heat capacities.  The large expansions required to reach temperatures below 1°K are not really practical for isotope separations.  However, Fig. 12 shows that by 95°K half of the $UF_6$
molecules are in the lowest vibrational state and at 40°K only
2.5% of the molecules remain in excited vibrational energy
levels.  The narrowing of the spectrum on cooling to 50°K is
dramatic, Fig. 13.  Figure 14 shows individual resolved transitions in cooled $UF_6$ obtained with a tunable diode laser.
Extremely high excitation selectivity is thus possible.

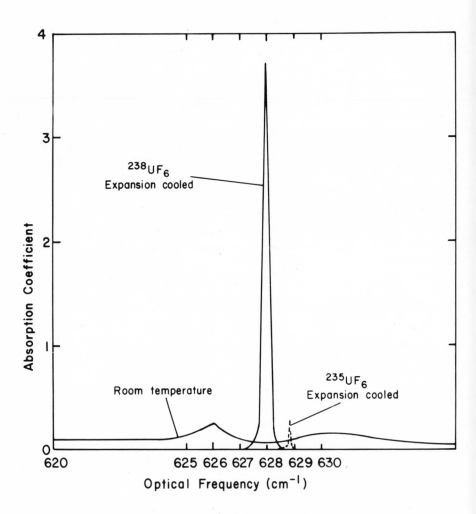

*Fig. 13.  Absorption spectrum of expansion-cooled $UF_6$
compared with room temperature spectrum (38).*

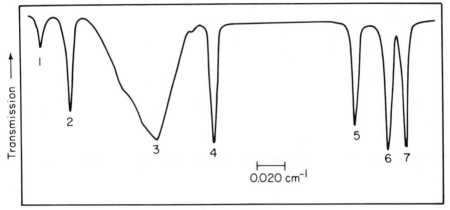

Fig. 14.   Composite spectrum of flow-cooled $UF_6$ taken with tunable diode lasers.   Peaks 2, 4 and 5 are R-branch lines of $^{238}UF_6$, peak 3 is the Q branch of $^{235}UF_6$ and peaks 1, 6 and 7 are from reference gases (38).

Other examples of selective excitation will be discussed along with the resulting isotopic separations in the following sections.

## III. SELECTIVE MULTISTEP PHOTOIONIZATION

Selective photoionization of atoms probably is the most universal photophysical method for selective separation of substances, in particular isotopes, at the atomic level.   The common feature of all schemes for selective ionization is the sequence of the two processes:
   1. Isotopically selective excitation of atoms.
   2. Ionization of the excited atoms.
Figure 15 illustrates some possible schemes of selective atomic ionization of special interest for laser isotope separation. The two-step photoionization scheme is the simplest of them (20). The three-step scheme may be of use, say, for atoms with a high ionization potential (66).   The photoionization cross section can be increased by tuning the frequency of secondary radiation to that of the transition to an autoionization

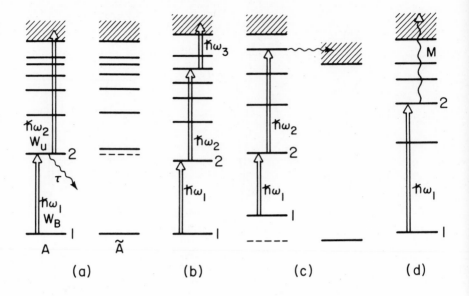

Fig. 15. *Schemes of selective atomic photoionization by laser radiation;  a) two-step photoionization;  b) three-step photoionization;  c) two-step photoionization through auto-ionization state;  d) ionization during collision of an excited atom with another particle.*

(spontaneous (67) or electric-field-induced (68)) state. Finally, excited atoms can be ionized when collided with other particles (electron acceptors (69) or excited atoms (70)).

## A.    Kinetics of Two-Step Process

First let us consider in detail the two-step atomic photo-ionization (Fig. 15a).  In the two-step photoionization scheme the radiation of the first laser transfers atoms from the ground state "1" to an excited state "2."  At the same time the atoms are exposed to the pulse of the second laser.  The quantum energy $\hbar\omega_2$ is inadequate to photoionize the atoms from the ground state but sufficient to photoionize them from an excited level:

$$\hbar\omega_1 + \hbar\omega_2 > E_i \; ; \; \hbar\omega_2 < E_i. \qquad (III.1)$$

The rate of atomic excitation under the action of continuous radiation is given by

$$W_e = \frac{\sigma_e P_e}{\hbar \omega_1} , \quad \sigma_e = \frac{g_2}{g_1} \frac{\lambda_1^2}{4} \frac{A_{21}}{\Delta \omega} , \qquad \text{(III.2)}$$

where $\sigma_e$ is the cross section of the radiative transition 1-2, $A_{21}$ is the Einstein coefficient of transition, $\Delta \omega$ is the line bandwidth of transition, and $P_e$ is the power of exciting radiation ($W/cm^2$) falling within the spectral range of the absorption line. If the relaxation time of an excited level is $\tau$, the population of the intermediate state under excitation will be:

$$N_2 = N_1 \frac{g_2}{g_1} \left[ 1 + \frac{g_2}{g_1} (W_e \tau)^{-1} \right]^{-1} . \qquad \text{(III.3)}$$

To saturate the transition 1-2 the power of exciting radiation should comply with the condition:

$$P_e \gtrsim P_e^S = \frac{\hbar \omega_1}{2 \sigma_e \tau} , \qquad \text{(III.4)}$$

where $P_e^S$ is the saturation power. When the relaxation time of excited level population is determined only by radiative decay to the ground state, $\tau^{-1} = A_{21}$, the saturation power is independent of the lifetime:

$$P_e^S = 2 \frac{g_1}{g_2} \frac{\hbar \omega_1}{\lambda_1^2} \Delta \omega \qquad \text{(III.5)}$$

The photoionization rate of an excited atom under a continuous radiation with its power $P_i$ is given by the expression:

$$W_i = \frac{\sigma_i P_i}{\hbar \omega_2} \qquad \text{(III.6)}$$

where $\sigma_i$ is the photoionization cross section. For each excited atom to be ionized with a probability near unity the power of ionizing radiation should comply with the condition analogous to (III.4):

$$P_i \gtrsim P_i^S = \frac{\hbar \omega_2}{\sigma_i \tau} = \frac{\omega_2}{\omega_1} \frac{\sigma_e}{\sigma_i} 2 P_e^S \qquad \text{(III.7)}$$

To ionize all excited atoms the radiation power required should exceed the power of exciting radiation by approximately $2\sigma_e/\sigma_i$.

The requirement for the power level of the ionizing radiation falls off markedly if the intermediate level is metastable. The saturation power of a resonance transition, provided the relaxation mechanism is radiation, does not depend on the lifetime of the upper level and the power, $P_i^S$, is inversely proportional to the lifetime of the intermediate level. This is true as long as the level lifetime is shorter than the residence time of an atom in the light beam. The residence time for an atom moving with the velocity, $v_o$, perpendicular to a light beam of a diameter, $a$, is $\tau_{res} = a/v_o$. For example, if a hot, heavy atom moves with $v_o = 3 \times 10^5$ cm/sec and the beam diameter $a = 1$ cm the interaction time $\tau_{res} = 3 \times 10^{-6}$ sec. Therefore, it is advisable to use intermediate states with $A_{21} \simeq 1/\tau_{res} = 3 \times 10^5$ sec$^{-1}$; that will reduce the $P_i^S$ power to a value on the order of $10^5$ W/cm$^2$.

It is practicable to bring about two-step photoionization under the action of short radiation pulses with their duration $\tau_1$, $\tau_2$ shorter than the relaxation time $\tau$ of intermediate level population; they are time-synchronized with the accuracy $\Delta\tau$ also shorter than $\tau$. To excite a considerable part of atoms and then to photoionize the greatest portion of excited atoms under such conditions the density of pulse energies should comply with conditions analogous to (III.4) and (III.7):

$$E_e \gtrsim \varepsilon_e = \frac{\hbar\omega_1}{2\sigma_e} , \qquad (III.8)$$

$$E_i \gtrsim \varepsilon_i = \frac{\hbar\omega_1}{\sigma_i} . \qquad (III.9)$$

As an example consider the two-step photoionization of Rb-atoms through the first intermediate level $5p\ ^2P_{1/2}$. For Rb-vapor at 100°C (the Doppler width $\Delta\omega = 4 \times 10^9$ sec$^{-1}$) the radiative transition cross section is $\sigma_e = 10^{-11}$ cm$^2$, and the photoionization cross section of the excited level is $\sigma_i = 10^{-18}$ cm$^2$. The saturation power of the absorption line $\lambda_1 = 7950$ Å is $P_e^S = 0.5$ W/cm$^2$, and the power, required to photoionize atoms with the same probability as that of excitation, $P_i^S = 1.6 \times 10^7$ W/cm$^2$ (the radiative lifetime of the $^2P_{1/2}$ level is $\tau = 2.6 \times 10^{-8}$ sec, $\lambda_2 = \lambda_2^B = 4730$ Å). Thus, to excite each atom and ionize each excited atom, substantially different powers are wanted. This is quite natural since the first process is resonant and the second one is non-resonant. For the case of short laser pulses, the energies of the exciting and ionizing radiations should exceed the values $\varepsilon_{exc} = 1.2 \times 10^{-8}$ J/cm$^2$ and $\varepsilon_i = 0.42$ J/cm$^2$ and durations must be shorter than the time $\tau = 26$ nsec.

Such an experiment on selective two-step atomic photoion-ization was first conducted in work (20). The beam of a tunable dye laser at $\lambda_1 = 7950$ Å excited Rb atoms in a gas cell at $10^{-3}$ torr to the state 5p $^2P_{1/2}$. Excited atoms were then photoionized by the second harmonic of the same ruby la-ser used for pumping the dye laser. The second-harmonic quan-tum energy was sufficient to photoionize excited Rb atoms $(E_{I.P.} - \hbar\omega_1 = 2.62$ eV) but not sufficient to photoionize the atoms from the ground state $(E_{I.P.} = 4.18$ eV). As the wave-length of the dye laser was tuned to the absorption line of Rb a sharp increase of photocurrent signal was observed. In experiment (20) photoionization was done in a gas cell without ion extraction.

Selective ionization of Rb atoms has been studied in more detail in work (71). Here, inequalities (III.8) and (III.9) were satisfied and every atom in resonance with the exciting radiation was ionized during each pulse. The authors investi-gated ion extraction by an electric field ($\sim 10^{13}$ ions/cm$^3$ in one pulse) and the dependence of photoionization cross section on the excess of the energy of the second photon over the ion-ization limit. In these experiments the density of Rb vapor was $N_O = 3.2 \times 10^{13}$ cm$^{-3}$, the electric field strength for ion extraction was about U = 2.5 kV/cm. Figure 16 shows depen-dence of ion yield on the ionizing pulse energy $E_j$ for differ-ent wavelengths of ionizing radiation for full saturation of the first excitation step. When the photon energy, $\hbar(\omega_1 + \omega_2)$, is just above the ionization limit, the ion current saturates. This indicates nearly complete ionization of the excited Rb atoms. This experiment demonstrates two important effects. First, the cross section of the photoionization from the exci-ted state is much larger than that of the photoionization from ground state. Secondly, by selecting the ionizing quantum energy near the ionization limit we may increase the photo-ionization cross section and thus decrease the energy density necessary for the second pulse. Still the cross section of non-resonant photoionization is much smaller than that of resonant excitation and it is a problem to match the intensi-ties of the two radiations for an optimum process.


B.    Under    General Requirements for Practical Processes


Highly efficient isotope separation on a practical scale by selective ionization requires that:

1. All atoms in an unexcited beam should be in the ground state, and ions must not exist. If atoms of a selected iso-tope are distributed over several levels or sublevels, a multifrequency radiation is required to excite atoms from

every sublevel in order to completely remove the selected isotope from the mixture.  Any thermal ions existing in the

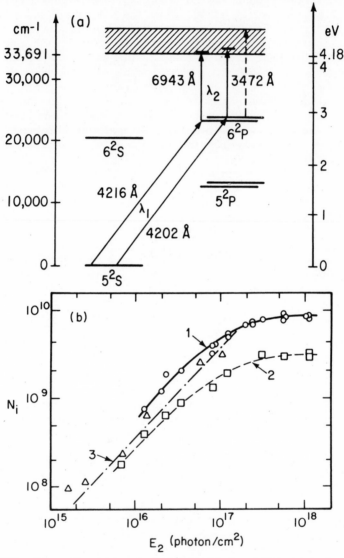

Fig. 16.  Two-step photoionization of Rb atoms in experiments (71):  a) Diagram of the energy levels of Rb atoms and the transitions used;  b) Total number of photoions, $N_i$, as a function of intensity of photoionizing radiation $\nu_2$. Curve 1 - U = 2.4 kV/cm and curve 2 - U = 1 kV/cm for $\nu_1$ = 23799 cm$^{-1}$, and $\nu_2$ = 14403 cm$^{-1}$.  Curve 3 is for photoionization by second harmonic ruby, $\nu_2$ = 28806 cm$^{-1}$.  Here $\sigma_i$ is too small for saturation to occur.

atomic vapor must be removed before laser excitation.

2. The laser radiation should perform selective photo-ionization for each atom of a selected isotope. Powers needed for this purpose are given by relations (III.4) and (III.7).

3. The laser radiation intensity should be used practically in full to excite and ionize atoms of a selected isotope.

4. There should not be transfer of excitation or charge between the isotopes being separated.

Let us consider now the possibility of fulfilling conditions (3) and (4). To do this one must keep in mind the most interesting elongated geometry of installation for isotope separation. In this case the atomic beam is a sheet emitted by a long "line" source parallel to the laser radiation.

The coefficient of linear absorption per unit length for the exciting radiation is $K_e = N_{10} \sigma_e$, where $N_{10}$ is the density of atoms in the ground state. The absorption coefficient for the ionizing radiation is $K_i = N_2 \sigma_i$. Thus, the penetration of exciting radiation with a weak absorption saturation is $\ell_e^o = (N_{10} \sigma_e)^{-1}$. For a strongly saturated transition, that is when $P_e \gg P_e^S$, the penetration rises according to the value $\ell_e \simeq \ell_e^o (P_e/P_e^S)$. As absorption is saturated at the transition $1 - 2$ the intermediate level population $N_2 \simeq 1/2 \, N_{10}$ and, hence, the penetration of ionizing radiation will be:

$$\ell_i \simeq \frac{2}{N_{10}\sigma_i} = \ell_e^o \, \frac{2\sigma_e}{\sigma_i} \qquad \text{(III.10)}$$

The penetration of ionizing radiation is deeper by the factor $2 \, (\sigma_e/\sigma_i)$ than that of exciting radiation. According to (III.7) the same factor appears in the ratios between the powers needed to excite and ionize each atom. Physically this is quite clear. Because of the difference between the cross sections for resonance excitation and for photoionization, materially different powers $P_e^S$ and $P_i^S$ are required to excite and photoionize them. However, the excitation and ionization energies $\hbar\omega_1$ and $\hbar\omega_2$ are of the same order. Because of this the exciting radiation will be absorbed in a path which is $\sigma_e/\sigma_i$ times shorter than that for the ionizing radiation.

To use the exciting radiation in full, the path length should be $L \simeq \ell_i$. For example, for the case of two-step photoionization of Rb-atoms in a beam with the atomic density $N_{10} = 10^{13} \, cm^{-3}$ the value $\ell_i = 200 \, m$ and $\ell_e^o = 10^{-2} \, cm$. Thus it is impractical to ensure similar absorption of exciting and ionizing radiations as they travel along a sheet atomic beam. It is possible to use a transverse propagation of the exciting beam and a longitudinal propagation of the ionizing beam, but this presents a severe problem in introducing the optical radiation into the vacuum chamber of the unit.  Therefore a considerable part of publications on selective atomic ionization deals with the search for methods which increase the cross section of excited atom ionization.  The main ones are listed below (III.D).

There are two collisional processes which can reduce the ionization selectivity of atoms of a particular isotope and hence the coefficient of laser isotope separation:

1. Excitation transfer as excited atoms A* collide with unexcited atoms B of another isotope;

2. Charge transfer between ions $A^+$ of a selected isotope and unexcited atoms B of another isotope.

The cross section of resonance excitation transfer as slow atoms collide will be:

$$\sigma_{tr} = 2.3 \frac{\pi}{\hbar v} d_{12}^{2} , \qquad (III.11)$$

where $v$ is the relative velocity of the colliding atoms, and $d_{12}$ is the dipole moment of the transition 1-2.  For allowed transitions $\sigma_{tr}$ may be as great as $10^{-13}$ to $10^{-14}$ cm$^2$.  Under continuous excitation the atoms of a selected isotope are kept in the excited state one half of the time during which they are in the light beam, $\tau_{res} = a/v_0$.  Let us define the coefficient of ionization selectivity $I$ as the ratio between the number of ionized atoms of a selected isotope and the number of ionized atoms of unwanted isotopes.  The value $Q = <\sigma_{tr} N_0 v>$ determines the rate of resonance excitation transfer to the atoms of unwanted isotopes.  Therefore, to obtain the coefficient of ionization selectivity of atoms of a selected isotope $I$ the following conditions should be met:

$$(Q\tau_{res}) \le I^{-1} \quad or \quad \Lambda_{tr}/a \ge I \qquad (III.12)$$

where $\Lambda_{tr} = (N_0 \sigma_{tr})^{-1}$ is the mean free path of atoms with respect to excitation transfer.  For pulsed excitation or for continuous excitation with $\tau = A_{21}^{-1} < \tau_{res}$ and (III.7) satisfied, the excited atoms live in the isotope mixture over a shorter period of time $\tau < \tau_{res}$.  Instead of (III.12), not nearly so rigid conditions need to be met:

$$(Q\tau) \le I^{-1} \quad or \quad \Lambda_{tr}/(\tau v_0) \ge I \qquad (III.13)$$

To cite an example, when the relaxation time of the intermediate level $\tau \simeq 10^{-7}$ sec and the atomic density in a beam $N_0 = 10^{13}$ cm$^{-3}$ the values $I \simeq 10^3$ can be obtained.

The cross section of resonance ion charge exchange, $\sigma_{ch}$, varies with atomic velocity between $10^{-14}$ and $10^{-15}$ cm$^2$ (72).

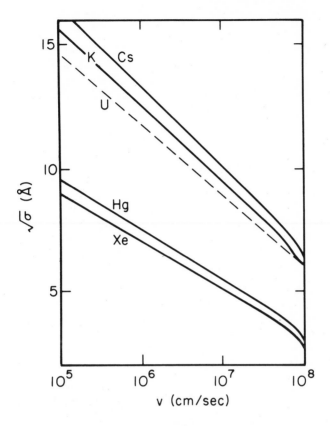

*Fig. 17. The dependence of charge transfer cross section on velocity for different ions.*

Figure 17 shows some typical data. The ion produced by selective ionization must be removed from the atomic beam. To obtain the selectivity coefficient $I$ the mean free path of ions with respect to charge exchange $\Lambda_{ch}$ and the diameter of the atomic beam, $a$, should obey the condition (III.12):

$$\Lambda_{ch}/a \gtrsim I \qquad\qquad (III.14)$$

For example, to obtain $I = 10$ with $\sigma_{ch} = 10^{-14}$ cm$^2$ and $a = 1$ cm this condition imposes a rigid restriction on the overall density of atoms in the beam $N_o \lesssim 10^{13}$ cm$^{-3}$. In charge exchange there is practically no momentum exchange. This may be used for suppressing selectivity losses owing to charge exchange in the following manner (67). If the velocity of the ions is increased sharply by a short pulse of electric field such that the ion displacement during the acceleration is less than $\Lambda_{ch}/I$, then a subsequent charge exchange will not affect escape of selected atoms from the beam.

Spatial separation of ions of a selected isotope at a considerable density can be done in crossed electric and magnetic fields (67). It is necessary to use a magnetic field here because of initiation of a space charge, which prevents ions from being removed under the action of an electric field alone.

## C.    Isotope Separation Experiments

Actually the most attention has been paid to uranium isotope separation. Results have been published from research programs at Avco Everett Research Laboratory and at the Lawrence Livermore Laboratory. In 1974 at the VIII International Conference on Quantum Electronics the results of the first Livermore experiments were presented (73). This experiment is schematically shown in Fig. 18. A collimated beam of $^{235}$U

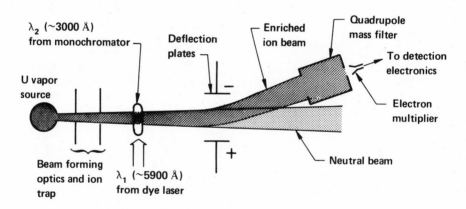

Fig. 18. *The experimental scheme for detection of the separation of U-235 and U-238 isotopes through two-step photoionization of uranium atoms in a beam (73).*

and $^{238}$U atoms mixed was irradiated by a tunable dye laser the radiation of which at $\lambda_1$ = 5915.4 Å excited the atoms of only one U-isotope. At the same time the atoms of the beam were illuminated by UV-radiation of a mercury lamp over the spectral range 2100 Å - 3100 Å. The short-wave boundary of this range is conditioned by the absence of photoionization of unexcited atoms, while the long-wave one is determined by the threshold energy for photoionization of excited atoms. The ions of $^{235}U$ formed were separated from the beam of neutral atoms by the electric field and entered a quadrupole mass spectrometer. Experiments were conducted with natural uranium, 0.7% $^{235}U$, as well as with a mixture enriched in $^{235}U$. Figure 19 shows the results of experiments on selec-

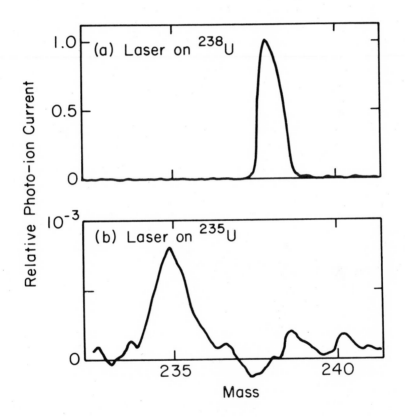

Fig. 19.  The mass spectrum of uranium ions in selective laser excitation and ionization of (a) U-238 and (b) U-235 atoms in the natural isotope mixture (73).

tive photoionization of various uranium isotopes in the natural mixture. When the laser frequency was tuned to the absorption line of a particular uranium isotope , the mass spectrometer detected only that isotope.  The photocurrent signal in selective photoionization of $^{238}U$ was $10^3$ times more than that in photoionization of $^{235}U$. There are two factors: Firstly, the content of $^{235}U$ in natural mixture is 140 times less than that in photoionization of $^{238}U$, and secondly, the absorption line of $^{235}U$, in contrast to the line of $^{238}U$, is split into eight hyperfine structure components only one of which is excited by a given laser frequency.

At the II International Conference on Laser Spectroscopy, 1975, Livermore presented the results of its experiments on two-step ionization of uranium atoms using excitation with the 3781 Å line of a xenon ion laser and ionization with the 3500 Å line of a krypton ion laser (74). The ion yield rate of $^{235}U^+$ in their experiments was $2 \times 10^{-3}$ g/hr, that is $10^7$ times higher than the rate obtained in the early experiments.

An industrial research and development program is being carried out jointly by the Avco Everett Research Laboratory and the Exxon Nuclear Co. (Jersey Nuclear - Avco Isotopes). Some results of this work obtained, as the authors say, in the early stage of investigation of the method in 1971 have been recently reported at the II International Conference on Laser Spectroscopy (75). The intensity of these researches may be revealed by the announcement made in the press by Exxon Nuclear in March of 1976 (76): Exxon Nuclear Company, Inc. announced, that it "has filed a license application with the Nuclear Regulatory Commission for the operation of a facility that would accomodate advanced experimentation and testing on a proprietary uranium enrichment process based on laser technology. The initial phase of the experimental test facility, planned for operational start-up in 1978-79, is expected to cost approximately 15 million dollars." Dr. A.R. Kantrowitz, Chairman of the Board of Avco Everett Research Laboratory, said that "We are quite encouraged by the results of our work. All basic technical go-no go issues have been resolved thus far. As our scale-up program continues to yield favorable results, we believe laser isotope separation eventually could make an important contribution toward accruing adequate supplies of enriched uranium at lower cost and increased efficiency."

Experiments on selective atomic photoionization with the object of working out methods of isotope separation are being conducted now by laboratories in several countries. By way of example, Ca-isotope ionization was demonstrated by scientists from West Germany (52). Excitation was done from the metastable level $4p\ ^3P_2$ to the level $5s\ ^3S_1$ ($\lambda_1 = 6182$ Å) by radiation of c.w. dye laser, while excited atoms were ionized by radiation of argon laser at $\lambda_2 = 4888$ Å. The quantum efficiency for two-step ionization of metastable atoms ranged between $10^{-3}$ and $10^{-2}$. $^{154}Sm$ separation has been reported (77).

D.    Use of Autoionization States

The use of autoionization states is an effective way for increasing the photoionization cross section.  Autoionization is a process of spontaneous transition of an atom from a discrete excited state above the ionization limit into the ionization continuum.  This occurs, for instance, in simultaneous excitation of two valence electrons or in excitation of an inner-shell electron.  Since the autoionizing states are narrow, their optical cross sections are often much larger than for the continuum.  In this case, both laser outputs must have frequencies resonantly tuned to absorption transitions.  Autoionization levels exist in spectra of many atoms (CuI, AgI, SnI, CdI, HgI, etc.).  In work (78) autoionization resonances have been detected in the spectrum of Mg I transitions from the state $3s3p$ $^1P^o_1$ with the cross section $\sigma_j = 5 \times 10^{-16}$ cm$^2$.  To illustrate the existence of autoionization resonance peaks superimposed on the continuous absorption, Fig. 20  shows the spectrum of photoionization from the excited state of the uranium atom obtained in work (73) using a tunable dye laser.  It is seen that by selecting a wave-

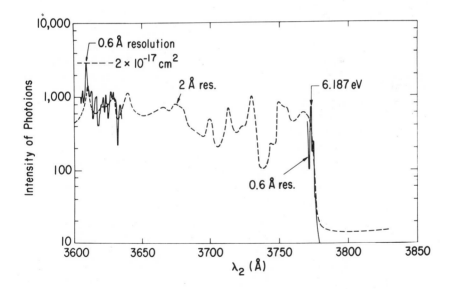

Fig. 20.  The spectrum of photoionization of uranium atoms from a state, excited by laser radiation at $\lambda_1 = 4266.324$ Å, produced by the use of a tunable dye laser (73).

length we can increase the value of ionization cross section $\sigma_i$. Nevertheless, the use of spontaneous autoionization usually cannot solve the problem of increasing the photoionization cross section. The values of $\sigma_i$ are still orders of magnitude less than $\sigma_e$ for allowed transitions and for many atoms autoionization is not observed in the accessible spectral region.

The use of the autoionization levels arising in the atomic spectrum under external electric field (68), we think, may overcome this difficulty. With the use of two or three tuned lasers it is possible to excite in succession an atom to a state with a large value of the principal quantum number $n$, the excitation efficiency being rather high. The electric field distorts the electronic spectrum of the atom so (Fig. 21) that some levels of the discrete spectrum closest to the

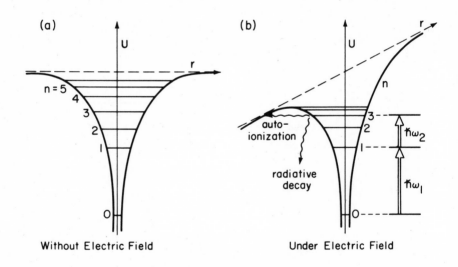

*Fig. 21. Autoionization of highly excited atomic states by an external electric field; a) scheme of normal electronic levels; b) potential distortion in the external electric field and formation of autoionization levels.*

ionization limit fall within the continuum. Slightly lower levels become autoionized and the probability of their ionization decay rises with increase in the principal quantum number. The optimum case is for the laser to excite an autoionized state whose autoionization probability is greater than its radiative decay probability but less than its

Doppler width.   In this case each excited atom will be ionized, and the cross sections of such an ionization are similar to those of resonant transitions between discrete energy levels. The calculations done in work (68) show that this optimum case may be achieved with relatively weak electric fields (below 30 kV/cm) which cause no electric breakdown in the atomic beam.

The first successful experiment on ionization cross section increase by electric field was carried out in work (79) using Na vapors.   Resonant two-step laser excitation produced Na atoms in each of the highly excited states 12d through 18d which were autoionized in an electric field of 10 kV/cm.   For illustration, Fig. 22 shows the variation of photoelectric

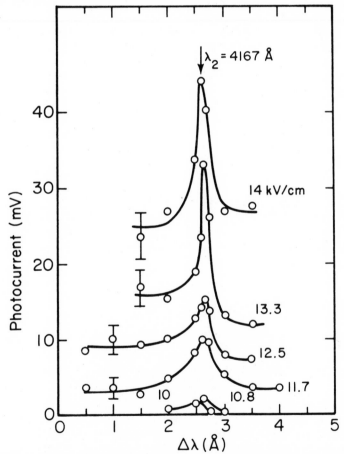

Fig. 22.   The dependence of the signal of Na atom two-step photoionization through the autoionization state 15d, induced by electric field, on the wavelength of the second-step laser for varied values of electric field strength (79).

current for several fixed electric field strengths as the wavelength, $\lambda_2$, of the second laser is tuned through the transition $3^2P_{1/2} - 15^2D_{3/2}$. The state $15^2D_{3/2}$ is 488 $cm^{-1}$ lower than the ionization limit, but the electric field causes autoionization of the atoms. The total cross section of two-step ionization obtained in the experiment (80) is about $10^{-14}$ $cm^{-2}$, that is $10^5$ times higher than the photoionization cross section for Na.

A comprehensive study of the autoionization of excited Na atoms was carried out in works (80,81). Tunneling calculations of the autoionization rate of hydrogen-like atoms in different sublevels of level n = 15 give values which range from $10^6$ to $10^{11}$ $sec^{-1}$ with fields as low as 10 -12 kV/cm. Experimental data are in rather good agreement with the theory for autoionization of individual Stark sublevels of highly excited states where strong mixing of quantum states occurs.

From the viewpoint of the comparative ease of ionization, the highly excited states of atoms are of great interest for isotope separation. Recently a comprehensive study of high-lying levels of uranium atom Rydberg states has appeared (82). Levels within 1000 $cm^{-1}$ of the ionization limit were accessed by time-resolved step-wise excitation using dye laser pulses tuned to resonant transitions. Atoms excited to these states were then photoionized by intense infrared radiation from a pulsed $CO_2$ laser (Fig. 23). The resultant photoion production was monitored. By delaying the infrared ionizing pulse, and thus discriminating against the shorter-lived valence states, Rydberg levels with principle quantum numbers, n, exceeding 60 were preferentially detected. Series convergence yields a value of the ionization limit of 6.1941 ± 0.0005 eV improving upon the value of 6.1912 ± 0.0025 eV obtained in photoionization studies.

E.    Other Ionization Schemes

Ionization of levels near the ionization limit by an infrared laser, Fig. 23a, has been proposed (66) and demonstrated (82). The requirements (III.7) on the ionizing laser are not so difficult since at present IR laser radiation is accessible with much higher power and efficiency by comparison to the power of visible radiation. Thus requirement (III.10) for efficient use of the ionizing radiation may be relaxed without a significant increase in consumption of energy.

Some collisional processes, involving two excited atoms result in ionization of one atom with a fairly large cross section. In work (70) consideration was given to the possibility of ionizing excited atoms A* of a selected isotope on collision with excited atoms C* specially prepared in a gas

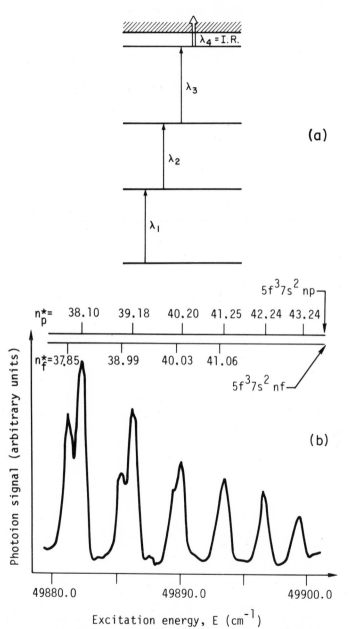

Fig. 23. *Selective laser photoionization through Rydberg states:  a)* $\lambda_3$ *tuned to a high lying state near the ionization potential (49935 ± 20 cm$^{-1}$ in UI) where the subsequent ionization is achieved by an IR photon;  b) The doublet series progression of UI.  The n\* values were computed using 49958.1 cm$^{-1}$ as the limit value (82).*

mixture. In such a process the excited atoms C* induce a high-frequency dipole moment in the excited atoms A*. The amplitude of this moment may be 3 to 4 orders of magnitude in excess of that which can be induced directly by a laser at the corresponding frequency. The estimates (70) show that in this way we can increase the ionization cross section of the excited atoms A* by about 3 to 4 orders of magnitude.

In report (69) attention is drawn to the possibility of effective ionization of excited atoms of a selected isotope as they collide with molecules having a high electron affinity. For example, a molecule of $SF_6$ can attach an electron from an atom lying about 1 eV below the ionization limit with a cross section of $2 \times 10^{-15}$ cm$^2$ (83). In mixtures of excited atoms A* with molecules of high electron affinity, the process of atomic ionization can be carried out simply.

The methods proposed for increasing photoionization cross sections of excited atoms of a particular isotope, basically, up to values of $\sigma_i \simeq \sigma_e \simeq 10^{-12}$ cm$^2$ (with the use of autoionization in the electric field) or at least up to values of $\sigma_i = 10^{-14}$ cm$^2$ will probably make it possible to use the ionizing laser radiation efficiently in isotope separators of reasonable dimensions. For example, if $\sigma_i = 10^{-14}$ cm$^2$ in a beam of atoms with their density $N_O = 10^{13}$ cm$^{-3}$ and with a selected isotope content of 2%, the exciting beam will be absorbed within $\ell_i = 10$ m.

Selective ionization of molecules can also be carried out. In one of the earliest proposals on laser isotope separation (19), it has been suggested for separating uranium isotopes using $UF_6$. One can selectively excite a vibrational level of a particular isotopic species of a molecule. Then due to the red shift in the molecular photoionization limit the excited molecules can be selectively photoionized. However, it is difficult to do this in practice because the vibrational shift of the photoionization band is small compared to its red edge width and because there are no suitable lasers in the VUV region for molecular ionization. A certain improvement might be achieved if molecular electronic states are selectively excited and then photoionized, but this approach is applicable only to a small number of molecules which have isotopic shifts in electronic absorption bands.

Recently in work (84) the first experiment on two-step photoionization of $H_2CO$ molecules has been reported. An $N_2$-laser at $\lambda_1 = 3371$ Å excited the state $^1A_2$ and an $H_2$-laser at $\lambda_2 = 1600$ Å photoionized the excited molecules. The ionization potential of an $H_2CO$ molecule is $E_i = 10.87$ eV, and the total energy of two laser quanta $\hbar\omega_1 + \hbar\omega_2 = 3.7 + 7.7$ eV = 11.4 eV, which is sufficient to photoionize this molecule. The time delay of the $H_2$-laser pulse (its duration is shorter than 1 nsec) was varied with respect to the $N_2$-laser pulse (its duration about 2 nsec), and the dependence of photoion yield on delay time was measured. However, the main aim of this experiment is the selective laser detection of complex molecules, rather than isotope separation.

## IV.  SELECTIVE TWO-STEP MOLECULAR PHOTODISSOCIATION

Selective two-step molecular photodissociation is accomplished by isotope-selective laser excitation of vibrational or electronic molecular states and subsequent laser photodissociation of only the selectively excited molecules. The photodissociation step must be rapid compared to excitation transfer and relaxation, Figs. 1 and 24. This idea was suggested in the first work on selective atomic photoionization (20), then considered and experimentally accomplished in works (21,22). The process of selective two-step molecular photodissociation is possible, provided that during molecular excitation a shift in the continuous photoabsorption band occurs which results in molecular photodissociation. In this case, by selecting the radiation frequency $\omega_2$ in the shift area where the ratio between the absorption coefficients of unexcited and excited molecules is maximum, Fig. 24, the molecules excited selectively by the $\omega_2$ -frequency radiation may be photodissociated.

In principle, both vibrational and electronic excited states may be used as intermediate states. Each case has its own advantages and disadvantages. For excitation of vibrational states the shift in the electronic absorption band is sometimes rather small and low vibrational levels are populated appreciably by nonselective thermal excitation. Therefore it is sometimes difficult to photodissociate primarily the laser excited molecules. On the other hand, isotopic shifts are usually quite distinct in the vibrational spectrum. A rather large shift of the photodissociation band may appear for an electronically excited intermediate state, but the electronic spectra of most molecules have no absorption lines with clearly resolvable isotopic shifts. Where shifts are resolvable the methods of one-step selective photopredissociation (Sec. V) and selective electronic photochemistry (Sec. VII) compete with that of two-step photodissociation. Because of this, the method of selective two-step photodissociation through intermediate vibrational states is of most practical interest.

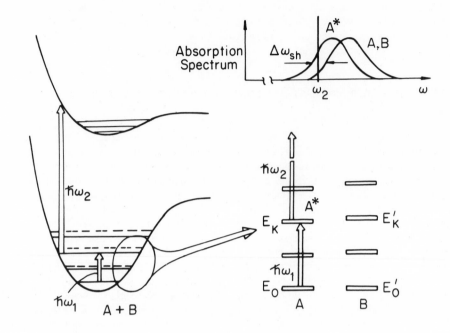

*Fig. 24.  Main features of the process of selective two-step photodissociation of A molecules mixed with B molecules through the intermediate vibrational state v = 2.  The diagram at the top right shows schematically the red shift of the photodissociation boundary for vibrationally  excited A\* molecules and the selection of the radiation frequency $\omega_2$.*

## A.    Important Features of IR-UV Photodissociation

The probabilities of excitation and dissociation of excited molecules are given by relations of type (III.2) and (III.6), in which $\sigma_e$ denotes the excitation cross section of the corresponding intermediate level, and by $\sigma_i$ is meant the photodissociation cross section $\sigma_{pd}$.  At the same time in the case of molecules the whole picture of the selective process is much more complicated due to three effects governing the selectivity and rate of the process (21):  First, the thermal, non-selective excitation of vibrational levels;  second, the diffuseness of the band edge for photodissociation continuous absorptions; and thirdly, the bottleneck due to the rotational structure of vibrational levels.  First we should consider these problems.

1.  *Thermal Nonselective Excitation of Vibrational Levels.*  The thermal equilibrium population of the intermediate

vibrational level must be small; thus

$$\hbar\omega_1 \gg kT. \qquad \text{(IV.1)}$$

In the simplest case of a diatomic molecule when no more than half the molecules $(N_2^{max} \simeq \frac{1}{2} N_{10})$ may be excited from the ground state under the resonance radiation $\omega_1$, a more rigid condition should be met so that the coefficient of excitation selectivity $S = N_2/N_{20} \gg 1$ could be obtained:

$$(\hbar\omega_1/kT) \gg \ln 2S \qquad \text{(IV.2)}$$

For polyatomic molecules the photodissociation selectivity drops in addition owing to the fact that the levels of a single normal vibration are selectively excited by laser radiation, while thermal excitation populates the levels of all normal vibrations. Many of these other excited states can absorb the photodissociating radiation, $\omega_2$, as well. The vibrational energy distributions for $UF_6$ shown in Figs. 11 and 12 illustrate this well (85). As for $UF_6$ dynamic cooling should be useful for many molecules with low vibrational frequencies.

Consideration must be given to the increase in thermal excitation as a result of gas mixture heating by IR radiation. This process is illustrated in Fig. 25, which shows the time history of population of the first vibrational level of an $NH_3$ molecule, when gaseous $NH_3$ is excited by a pulse of $CO_2$-laser radiation. The pulse saturates the vibrational transition $v = 0 \rightarrow v = 1$. Excitation is observed by the IR-UV double resonance method (86 - 89). At low pressures there is no vibrational relaxation during the laser pulse and the population of the level $v = 1$ is determined by laser excitation. As the vibrational relaxation proceeds, the gas is heated and reaches equilibrium at a higher temperature. The consequent thermal vibrational excitation lasts until the gas cools down. The top trace of Fig. 25 shows the V → T relaxation of $NH_3$ at 2 torr pressure. At the higher pressures in the second and third traces 23 torr and 35 torr, the non-thermal excitation is relaxed significantly during the laser pulse. The thermal excitation is shown by the shift from the baseline on the left to the increased steady signal level on the right which lasts for milliseconds (89). Detailed observations of this thermal heating and of shock wave effects have been reported for $CD_4$ (90), $SF_6$ (91) and $C_2H_4$ (92). In order for the selectivity losses of two-step photodissociation due to heating to be eliminated, the $\omega_2$-frequency laser pulse should also be short, no longer than the time of V → T relaxation. (See Sec. VIII).

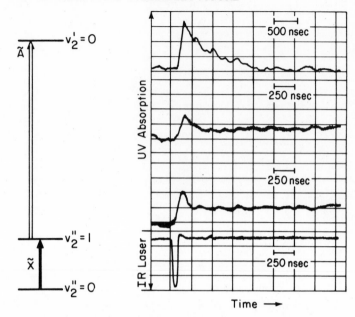

*Fig. 25. The time dependence of the population of the first excited vibrational state of $NH_3$ excited by a $CO_2$ laser pulse. Laser excitation pulse, bottom trace. Top three traces are U.V. absorption by $NH_3 (v_2 = 1)$ vs time at 2 torr (top), 23 torr (second) and 35 torr (third)    (89).*

   *2.  Diffuseness of the Photoabsorption Band Edge.*  Unlike the two-step atomic photoionization where the absorption edge is sharp, for two-step molecular photodissociation we must be concerned with the shift of the electronic absorption band $\Delta\omega_{sh}$ relative to the width of the red wing edge of this band, $\Delta\omega_{wing}$. Highly selective photodissociation of vibrationally excited molecules is accomplished only when $\Delta\omega_{sh} \gg \Delta\omega_{wing}$. Fig. 26 illustrates the typical form of continuous absorption bands for transitions from the ground and vibrationally excited states to a repulsive excited electronic state. The photodissociation band spectrum of a diatomic molecule is given by the formula

$$I_{v\nu} \sim \nu \left| \int_{-\infty}^{\infty} \psi_v(x)\, \psi_\nu(x)\, dx \right|^2, \qquad \text{(IV.3)}$$

where $\psi_v(x)$ is the nuclear wave function of a vibrational level of ground electronic state, $\psi_\nu(x)$ is the nuclear wave function of a level in the continuum of an excited electronic state. The excited state absorption band may be shifted to the red by as much as the vibrational excitation energy, $\hbar\omega_1$. However, the bandwidth is usually greater than the shift. For selective photodissociation it is necessary that there

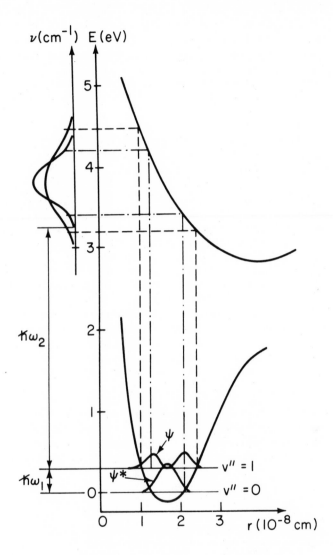

*Fig. 26.   The difference between the dissociative conti-
nua for the ground and excited vibrational states of a diatomic molecule.*

should be a frequency range in which $I_{1\nu} \gg I_{0\nu}$. Estimates
show that this is possible only on the red side of the absorption band where the photodissociation cross section decreases sharply.

Although the photodissociation selectivity may be increased by tuning to the far edge of the band, this greatly

decreases $\sigma_{pd}$ and the photodissociation rate. Therefore it is advisable to apply other methods for increasing the band shift. Overtone vibrational levels may be directly excited by the $\omega_1$-frequency radiation. This method was used in the first experiment on two-step photodissociation of HCl molecules (21). Work (93) has demonstrated isotopically selective excitation of the second overtone of $H^{35}Cl$ and $H^{37}Cl$ molecules with laser radiation at 1.19 μ.

A new possibility for resonant excitation of higher vibrational levels exists for polyatomic molecules. An intense monochromatic IR field can resonantly excite polyatomics by multiple photon absorption (25, 26). Moreover, recently it was discovered that relatively low intensity fields (about $10^4$ W/cm$^2$) can excite higher overtone levels due to multi-quantum rotational-vibrational resonances (94). Thus the difficulty of the broad wing of the UV absorption band can probably be overcome by means of these new effects.

3.    *The Rotational Relaxation Bottleneck.* Due to the Boltzmann distribution of molecules over the rotational levels of the ground vibrational state, the monochromatic laser radiation excites just a small portion of molecules, $q$, from a single rotational sublevel. This decreases the excitation rate. For the excitation rate to be increased, the radiation intensity may be raised. However, if the laser excitation rate $W_{exc}$ becomes much larger than the rotational relaxation rate, $\tau_{rot}^{-1}$, the lower sublevel gets depopulated quickly and the upper one filled. Further excitation is possible only after the lower sublevel is populated again and the upper sublevel depleted by rotational relaxation (Fig. 27). As a result, the vibrational energy, $E_{vib}$, stored in the molecules under IR resonance radiation complies with the condition (95):

$$\frac{dE_{vib}}{dt} < \frac{\hbar\omega_1}{2}\frac{q}{\tau_{rot}}, \qquad (\text{IV.4})$$

where $q << 1$ denotes the portion of molecules interacting with the $\omega_1$ frequency radiation directly. In order that half the molecules be excited by a pulse of duration $\tau_p$, the condition $\tau_p > \tau_{rot}/q$ should be met. Thus, the characteristic time of vibrational excitation will be:

$$\tau_e = \frac{1}{q}\tau_{rot} = \tau_{coll}/P_{rot}q, \qquad (\text{IV.5})$$

where $P_{rot}$ is the probability of rotational relaxation in one gas-kinetic collision; $\tau_{coll}$ is the mean time between gas-kinetic collisions, that is $0.1 < P_{rot} < 1$.

The key factor $q$ may be expressed in terms of the statistical sum of rotational states $Z_{rot}$, the sublevel degeneracy $g$ and the energy $E_{rot}$:

$$q = \frac{g}{Z_{rot}} \, exp \, ( \, - \, \frac{E_{rot}}{kT} \, ) \qquad\qquad (IV.6)$$

For simple molecules the $q$ value can be calculated. Usually it varies between $10^{-1}$ and $10^{-2}$. But for polyatomic molecules, when the levels interacting with the radiation are not identified, calculations are impossible. In some cases one

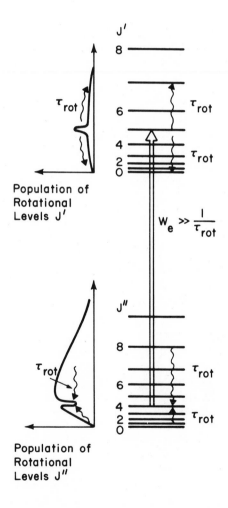

Fig. 27. Diagram of the rotational relaxation bottleneck in the excitation of vibrations by monochromatic laser radiation.

must even deal with a continuous band of unresolved rotational structure which nonetheless exhibits a distinct isotopic shift (for example, the band $\nu_3$ of $^{10}BCl_3$ and $^{11}BCl_3$ molecules (96 - 98). In this case "rotational sublevel" means the fraction of molecules in a vibrational state which can absorb the radiation. In these cases the $q$ value must be measured experimentally. A simple method for measuring the population factor $q$ has been proposed in work (99). It has been found in this work, for example, that for $C_2F_3Cl$ molecules and $CO_2$-laser light at 9.6 $\mu$ the factor $q = 0.036$. The work has shown that the factor $q$ grows with increasing pressure due to the overlapping vibration-rotation lines. The bottleneck effect is exhibited in the amplification of subnanosecond laser pulses in a $CO_2$-laser amplifier. Rotational relaxation restricts the rate of extraction of stored energy from the amplifier medium (100).

The bottleneck effect also has an influence on the thermal excitation of levels. The value of $\tau_{rot}$ is generally less than but can often approach the vibration-translation time $\tau_{VT} = \tau_{coll}/P_{V-T}$. $P_{V-T}$ is the probability of vibrational excitation relaxation during one collision. Thus with $P_{V-T} > q\,P_{rot}$ and $\tau_e > q^{-1}\,\tau_{rot}$, a gas heating during the excitation process is inevitable. The only way to avoid it is by adding an inert buffer gas which accelerates rotational relaxation much more than vibrational relaxation.

Further theoretical work is given in Refs. (101).

B.    Experiments on Isotope Separation

The first experiment on isotope separation by the method of two-step selective photodissociation was carried out by R.V. Ambartsumian et al. in their works (22,87). In their experiments they used $^{14}NH_3$ and $^{15}NH_3$ molecules which, firstly, can be selectively excited by $CO_2$-laser radiation. Secondly, the IR and UV absorption spectra and the photochemical decomposition of these molecules are well studied. The lowest frequency, $\nu_2$, of $^{14}NH_3$ and $^{15}NH_3$ molecules is in the 10 $\mu$ region. The absorption bands of $^{14}NH_3$ and $^{15}NH_3$ overlap but the spectrum has a rich structure consisting of some hundreds of vibration-rotation lines. This structure has been analyzed for $^{14}NH_3$ (102) and for $^{15}NH_3$ (102). In the spectrum of the $\nu_2$ bands there are some non-overlapping rotational-vibrational lines of $^{14}NH_3$ and $^{15}NH_3$, which coincide closely with $CO_2$-laser lines. In the UV region the absorption spectrum of $^{14}NH_3$ is a vibronic progression in which $\nu_2$ appears. The progression starts in the region of 2168 Å and goes to shorter wavelengths (103). The electronically excited $\tilde{A}$ state is unstable due to predissociation (104). Molecular transitions from the

ground state to an excited electronic state and spectral lines
arising from such transitions are represented schematically in
Fig. 28(a).  The vibronic spectrum is somewhat different for
$NH_3$ molecules vibrationally excited to the level $v'' = 1$ of the
$v_2$ band.  The most important feature is a new absorption line
due to the transition $(\tilde{X}, v'' = 1) \rightarrow (\tilde{A}, v' = 0)$.  It has a
"red" shift equal to $\hbar\omega_1$.  When the vibrational transition
$v'' = 1 \leftarrow v'' = 0$ is fully saturated by IR laser radiation the
molecules are equally distributed between the levels $v'' = 0$
and $v'' = 1$.  Fig. 28(b) shows a theoretical distribution of

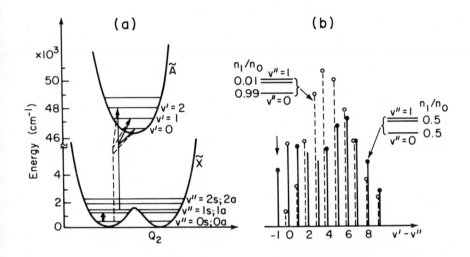

Fig. 28.  (a) Energy level diagram and vibronic transitions for the two-step dissociation of $NH_3$ molecules.
(b)  Spectral lines for normal (---) and laser excited (——)
$NH_3$.

intensity in the vibronic spectrum of ammonia for two cases:
when 99% of the molecules are in the ground state $(\tilde{X}, v'' = 0)$
and when the molecules are equally distributed between the
levels $(\tilde{X}, v'' = 0)$ and $(\tilde{X}, v'' = 1)$.  The new vibronic line at
45250 $cm^{-1}$ has been used to photodissociate excited molecules.
    The experimental scheme, with which the first experiment
on isotope separation by the method of two-step photodissociation was conducted, is shown in Fig. 29.  The $CO_2$-laser pulse
at the P(16) line 10.6 $\mu$ excited $^{15}NH_3$-molecules.  The $CO_2$-la-
ser-pulse-ignited spark whose continuous spectrum matched the
duration of the IR pulse, was used as a source of UV radiation
at the frequency of the new absorption line $(\tilde{X}, v'' = 1)$
$\rightarrow (\tilde{A}, v' = 0)$.  An absorbing cell filled with natural $NH_3$

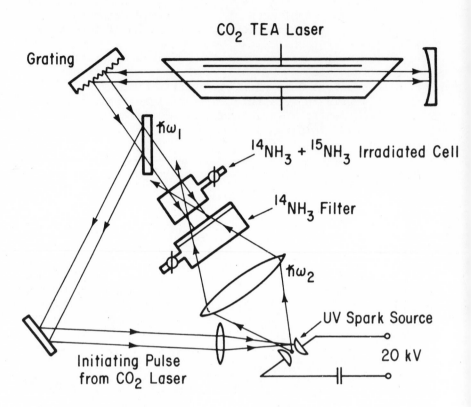

*Fig. 29. The scheme of an experimental setup for nitrogen isotope separation by the method of two-step photodissociation of $NH_3$ molecules (22).*

(99.6% $^{14}NH_3$) was located in front of the cell with the mixture. Since the vibronic lines are broad compared to the $^{14}NH_3$ – $^{15}NH_3$ shift, this removed the UV radiation which may be absorbed by unexcited $^{14}NH_3$ and $^{15}NH_3$ molecules. The pressure in the ammonia filter was chosen so that the spark radiation at the $(v'' = 0, \tilde{X}) \rightarrow (v' = 0, \tilde{A})$ transition and all shorter wavelength transitions could be fully absorbed in the filter. Equimolar mixtures of $^{14}NH_3$ and $^{15}NH_3$ at 10 - 20 torr total pressure as well as mixtures with a buffer gas of Xe or Ne at pressures up to 250 torr were used in the experiment.

The photochemical reactions of ammonia are (104):

$$^{15}NH_3 + \hbar\omega_1 + \hbar\omega_2 \rightarrow {}^{15}NH_2 + H,$$

$$^{15}NH_2 + {}^{15}NH_2 \rightarrow {}^{15}N_2H_4$$

$$^{15}N_2H_4 + H \rightarrow {}^{15}N_2H_3 + H_2 \qquad\qquad (IV.7)$$

$$2\ {}^{15}N_2H_3 \rightarrow 2\ {}^{15}NH_3 + {}^{15}N_2$$

The reactions preserve the isotopic selectivity since none of the radicals reacts with $NH_3$. The mass spectrum of $N_2$ was analyzed in the experiment. The enrichment coefficient, i.e. the $^{15}N$-$^{14}N$ content ratio in molecular nitrogen, ranged from 2.5 to 6 in various experiments. In some experiments with molecular oxygen, used as a buffer gas, much higher enrichment was achieved (up to 50) (105).

The results of the first experiments on nitrogen isotope separation by the method of two-step photodissociation of $NH_3$ according to the described scheme have been recently confirmed by Japanese scientists (106). In their experiment the coefficient of $N_2$ enrichment was also 4 under similar conditions.

Some experiments on an analogous scheme (a $CO_2$ laser and a conventional UV source in the region of 2130 - 2150 Å) in which the $^{10}B$ and $^{11}B$ isotopes were separated through two-step selective photodissociation of $BCl_3$ molecules are described in work (107). The coefficient of enrichment with a light boron isotope in this work equals only 10%, that is, comparable to the typical magnitude of kinetic isotope effect. A molecule of $O_2$ was used as a scavenger of photodissociation products.

## C.   Selectivity Loss and Efficiency

Apart from the effects discussed above of thermal excitation and diffuseness of the electronic absorption band, two collisional processes affect the selectivity of isotope separation: 1) the collisional transfer of vibrational excitation between molecules of different isotopic composition and 2) secondary photochemical reactions of the dissociation products.

The transfer of vibrational excitation is a resonant process the probability of which is usually high. For example, the time for vibrational excitation transfer between $H^{35}Cl$ and $H^{37}Cl$ molecules is 2.1 ± 0.5 nsec·atm (108), while the time of excitation transfer between $^{14}NH_3$ and $^{15}NH_3$ $P\tau_{v-v} = 1.2$ $\pm$ 0.4 nsec $\cdot$ atm. (89). Thus, for the excitation selectivity to be preserved, the pulse duration $\tau_e$ of the photodissociating radiation at the frequency $\omega_2$ should be shorter than the excitation transfer time $\tau_{vv}$. The required value of their ratio is determined by the requisite selectivity value $S$:

$$\tau_{vv}/\tau_{exc} \gtrsim S \qquad\qquad (IV.8)$$

In the $NH_3$ experiment discussed above this condition was realized by selecting sufficiently short IR and UV radiation pulses (below $10^{-6}$ sec) and low pressures of $NH_3$. The enrichment factor in $BCl_3$ experiments (107) is believed to be low because of insufficient UV power to avoid V $\rightarrow$ V transfer between $^{11}BCl_3$ ($\nu_3$) and $^{10}BCl_3$ ($\nu_3$) (isotopic scrambling). The time for V $\rightarrow$ V transfer is less than 0.7 nsec $\cdot$ atm (109).

It is crucial to find a chemical reaction system which binds photodissociation products without selectivity loss. As in the case of $NH_3$, the undissociated molecules must not be drawn into the chemical cycle. Reactions which do not ordinarily occur may well be possible for the highly excited initial dissociation fragments. These excited fragments may transfer energy to undissociated molecules and allow them to be photodissociated at $\omega_2$. Thus it will often be necessary to add a buffer gas to thermalize these fragments. In systems where the dissociation fragments react with the undissociated molecules and initiate chain reactions a scavenger must be carefully chosen. For example, in the $BCl_3$ experiments (107) it was a problem to identify the best scavenger for $BCl_2$ or $BCl$. A scavenger must be: 1) unreactive with $BCl_3$, 2) transparent to 10.6 $\mu$ IR radiation, 3) transparent to 2130 - 2150 Å UV radiation, and 4) have a simple chemistry with no chain reactions or scrambling. The $O_2$ molecule is probably the best scavenger for fragments of $BCl_3$ photodissociation (107). Ideally then one adds a gas which thermalizes the hot dissociation fragments, stabilizes them chemically while preventing chain reactions, and eliminates the rotational relaxation bottleneck.

The scavenger problem may sometimes be simplified by the proper choice of photodissociation wavelength. If there are several channels of photodissociation, it may be possible to select the one with the simplest chemistry, e.g. $H_2CO \rightarrow H_2$ + CO, stable molecules, rather than $H_2CO \rightarrow H + HCO$ free radicals.

To accomplish a high-efficiency process of isotope separation by the method under consideration, it is self-evident that the best use of the laser radiation at $\omega_1$ and $\omega_2$ should be made. This problem is complicated by the difference in the cross sections of excitation $\sigma_{exc}$ and photodissociation $\sigma_{pd}$, just as in the case of two-step atomic photoionization, but this difference is much less than for photoionization. Quantitative analysis of the kinetics of two-step molecular photodissociation in bulk gas shows (110) that given proper parameters it is possible to both photodissociate a considerable part, at least 50%, of the molecules interacting with the laser radiation and make practically full use of the radiation at both $\omega_1$ and $\omega_2$.

The possibility of efficient selective excitation of several vibrational levels by the process of multiple photon absorption from an intense IR pulse looks very promising. The first experiment with $OsO_4$ molecules showed a large red shift of the UV absorption band as a result of excitation by an intense IR pulse (111). This process of selective photodissociation (several IR photons + one UV photon) should be studied experimentally in detail.

In the first experiments on selective two-step (IR + UV) photodissociation the UV source was the conventional UV spark which has an extremely low energy flux in the narrow spectral band for $\omega_2$ and in the short time during which the selectivity is conserved. Recent progress in UV and VUV laser sources (excimer lasers (112, 113), diatomic molecular lasers with intense e-beam pumping) opens many new possibilities for this method of isotope separation.


V.    PHOTOPREDISSOCIATION


Isotope separation by photopredissociation (23, 24, 114) requires a molecular excited state which exhibits a resolvable isotope shift, which decays primarily by dissociation, and whose dissociation products

$$AB + \hbar\omega \rightarrow AB^* \rightarrow A + B$$

are simply removed from the starting material. The lifetime of the state must be long enough so that lifetime broadening does not cause overlap of the spectra of the two isotopes.

$$\Delta\omega_{isotope} \gg \tau^{-1} \tag{V.1}$$

The reciprocal of $\tau$ is the sum of the rates for all decay processes. The quantum yield of dissociation,

$$0 < QY = \tau k_{dissociation} \leq 1 \tag{V.2}$$

must be large enough to give an acceptable overall process efficiency. To satisfy equation (V.1) lifetimes longer than $10^{-10}$ or $10^{-11}$ sec are generally satisfactory. For lifetimes longer than this Doppler broadening will usually limit the selectivity of excitation. Equation (V.2) may be satisfied for lifetimes much longer than $10^{-10}$ sec. Rate constants for fluorescence emission, inverse radiative lifetimes, usually range from $10^5$ to $10^8$ sec$^{-1}$. Often dissociation is the only nonradiative decay channel. Thus acceptable values of $k_{dissociation}$ range over several orders of magnitude for a particular molecular electronic state. There are a number of molecules with one or more excited states known to satisfy the requirements of Eqs. (V.1) and (V.2).

Predissociation may be collision-induced. For example, a molecule may be excited to a vibration-rotation level just below a dissociation limit. The spectrum is perfectly sharp since the state is bound but dissociation can be induced with energy from a collision (17, 115). At sufficiently high pressures collision-induced dissociation will compete with spontaneous predissociation. Some molecular excited states may be induced to dissociate by the application of modest magnetic or electric fields (116). In very strong fields shifts in potential curves may create curve crossings and hence dissociation pathways (117). For most molecules suffi-cient spectroscopic and photochemical information is not available to know whether Eq. (V.2) is satisfied. It is clear though that the method is not as general as the two-step pho-toprocesses described in sections III and IV.

To complete the separation process the dissociation pro-ducts must be stabilized and removed from the reaction mixture. The problems are similar to those encountered in two-step dissociation (see IV.4). In the optimum case the dissociation products are chemically stable and may be separated by simple physical procedures. The molecular dissociation of formalde-hyde, $H_2CO \rightarrow H_2 + CO$, is such a case. More frequently, and always with diatomic molecules, the dissociation fragments are chemically reactive. They initiate free radical chain reac-tions with the undissociated molecules which scramble the selectivity. In this situation a chemical scavenger may be used as discussed for Br separation below. Y.T. Lee and his coworkers (118) are studying predissociation in molecular beams. Here the recoiling dissociation fragments are ejected from the beam on account of their excess translational energy and trapped separately. This scheme appears particularly pro-mising for infrared-induced predissociation of Van der Waals molecules. These molecules with binding energies of a few hundred cm$^{-1}$ are formed in relatively dense, dynamically cooled, supersonic molecular beams (57, 119).


A.    Diatomic Photopredissociation

Isotopically selective photochemistry has been demonstra-

ted by Leone and Moore (4, 120) for bromine.  $Br_2$ is selectively excited to levels of the $^3\Pi_0{}^+{}_u$ state (Fig. 30).  The
excited molecules dissociate and the reactive, isotopically
selected atoms produced are scavenged by HI to yield isotopically enriched HBr.  The enriched HBr was indentified by its
infrared chemiluminescence but not actually separated from the

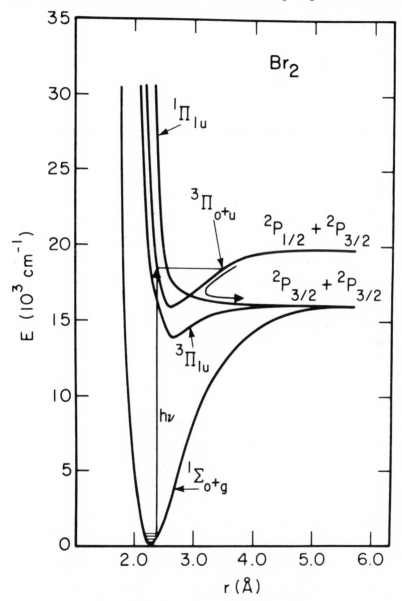

*Fig. 30.  Excitation and predissociation of $Br_2$ (120).*

gas flow.

The $Br_2$ spectrum is sufficiently well-resolved with a
0.7 GHz laser bandwidth at 558 nm that fluorescence of each
of the three isotopic molecules, $^{79,79}Br_2$, $^{81,81}Br_2$ and
$^{79,81}Br_2$, may be selectively excited as the laser is tuned.
The presence of continuum absorption along with the banded
spectrum at 558 nm is shown by the fact that Br atoms are
produced at wavelengths for which no fluorescence is observed.
At 558 nm 30% of the absorption was due to the continuum and
at 532 nm 80%. For a laser bandwidth of 0.3 GHz or less at
the line center of the P(49), 19 ← 1 transition near 558 nm
the continuum absorption is 1/9 of the total absorption (121).
The continuum absorption gives no isotopic selection

$$^{n}Br_2 + \hbar\omega \longrightarrow {}^{i}Br_2^* \quad banded$$

$$\longrightarrow 2\ {}^{n}Br \quad continuum$$

Fluorescence spectra with a 1 GHz resolution dye laser (Mo-
lectron) revealed no improved line-to-continuum ratios at
longer wavelengths (121). The 18 ← 1 band gave selectivities
less than at 558 nm (122). The peak absorption cross sections
for normal gas at room temperature near 558 nm are about
$10^{-19}$ cm$^2$. Excitation of the $^3\Pi_{1u}$ state slightly below its
dissociation limit would completely eliminate the problem of
continuum absorption. The excited molecules could be dissociated by collisions. Unfortunately, this absorption spectrum is very weak.

The excited $Br_2$ molecules may decay by several paths
(4, 120, 121, 123).

$$^{i}Br_2^* \longrightarrow {}^{i}Br_2 + \hbar\omega \quad (\tau_{rad} \approx 20\ \mu sec) \qquad (V.3)$$

$$\longrightarrow {}^{i}Br + {}^{i}Br \quad (k \sim 10^6 - 10^7\ sec^{-1}) \qquad (V.4)$$

$$^{i}Br_2^* + M \longrightarrow {}^{i}Br + {}^{i}Br + M \quad (\sigma \sim 10^{-14}\ cm^2) \qquad (V.5)$$

$$\longrightarrow {}^{i}Br_2 + M \quad (\sigma < 10^{-14}\ cm^2) \qquad (V.6)$$

They may also transfer energy to $Br_2$ collision partners

$$^{i}Br_2^* + {}^{n}Br_2 \longrightarrow {}^{i}Br_2 + {}^{n}Br_2^* \qquad (V.7)$$

with a resulting loss of isotopic selectivity. This is probably not an important problem here because $Br_2^*$ decays much
more rapidly than Br atoms react and because (V.5) is probably
faster than (V.7). It is important to know the rates of all
of these competing processes in order to establish the optimum

pressures of $Br_2$, HI and possible inert gases.

For the Br atoms there are two competing processes (123).

$$^iBr + HI \longrightarrow H^iBr + I$$

$$(k=1.0x10^{-11} \; cm^3 \; molec^{-1} \; sec^{-1}) \qquad \text{(V.8)}$$

$$^iBr + {}^nBr_2 \longrightarrow {}^iBr{}^nBr + {}^nBr$$

$$(k=4x10^{-11} \; cm^3 \; molec^{-1} \; sec^{-1}) \qquad \text{(V.9)}$$

The ratio of these two rate constants and the desired purity of the product determine how large a ratio of HI to $Br_2$ is needed. Measurements of HBr enrichment as a function of the HI:$Br_2$ ratio indicate that Br + $Br_2$ is faster than Br + HI by roughly a factor of 4 (123). Direct measurement of the Br + HI rate gives reaction once in every 16 collisions (123). Were it not for this very fast scavenger reaction rate the isotopic selectivity of the excitation would be lost.

The disadvantage of these highly reactive chemical systems is that undesired side reactions may proceed at rates sufficient to scramble the isotope selection before the final physical isolation of the chemical products is carried out. In this system for example the reactions

$$Br_2 + HI \longrightarrow HBr + BrI \qquad \text{(V.10)}$$

$$BrI + HI \longrightarrow HBr + I_2 \qquad \text{(V.11)}$$

appear to proceed rapidly on the walls of the vessel. The I atoms produced (V.8) react with $Br_2$

$$I + {}^nBr_2 \longrightarrow I^nBr + {}^nBr \qquad \text{(V.12)}$$

less rapidly than once in each $10^3$ collisions. If the separation may be carried out in a fast flow system with walls sufficiently cold that $Br_2$ and I are trapped, it should be possible to collect the $H^1Br$. For $Br_2$ the fast flow of dynamic cooling automatically achieves this and gives all of the advantages described in Sec. II, C. For such a system to work efficiently the product of $Br_2$ absorption cross section ($\sigma \sim 10^{-19} \; cm^2$) times pressure times path length must approach unity for efficient use of photons. The laser power must be high enough to excite a substantial fraction of the molecules so that trace impurities and non-selectively produced HBr are not the main products collected.

By selective predissociation of ortho-$I_2$ molecules with a 514.5 nm argon ion laser the authors of works (124 - 126) were able to convert ortho to para-$I_2$. At an operating pressure of 5 mtorr the following processes may occur (126):

$$o\text{-}I_2 + \hbar\omega_L \xrightarrow{\ \ W\ \ } o\text{-}I_2^* \tag{V.13}$$

$$o\text{-}I_2^* \xrightarrow{\ \ \gamma\tau^{-1}\ \ } I + I \tag{V.14}$$

$$o\text{-}I_2^* \xrightarrow{\ \ (1-\gamma)\tau^{-1}\ \ } I_2 + \hbar\omega \tag{V.15}$$

$$I + wall \xrightarrow{\ \ (7/12)\,\delta\ \ } \tfrac{1}{2}\,o\text{-}I_2 \tag{V.16}$$

$$I + wall \xrightarrow{\ \ (5/12)\,\delta\ \ } \tfrac{1}{2}\,P\text{-}I_2 \tag{V.17}$$

$$I + wall \underset{k^*}{\overset{k}{\rightleftharpoons}} I_{ad} \tag{V.18}$$

$$I_{3\,ad} \underset{}{\overset{\beta}{\rightleftharpoons}} I_2 + I_{ad} \tag{V.19}$$

Only o-$I_2$ is dissociated while the recombination, (V.16) and (V.17), gives an equilibrium mixture of ortho and para molecules (equilibrium ratio o-$I_2$:p-$I_2$ = 7:5). Thus the gas is enriched in p-$I_2$. At 5 mtorr the I atoms formed by dissociation, (V.14), collide immediately with the wall and either adsorb, (V.18), or recombine with atoms already adsorbed to give equilibrium $I_2$. The typical time dependence of the concentration of o-$I_2$ and p-$I_2$ is shown in Fig. 31. In the experiment the total concentration of $I_2$ is falling during the irradiation as iodine atoms formed by predissociation are adsorbed by the walls of the cell and possibly as they bind with $I_2$ to form $I_{3\,ad}$. As the $I_2$ pressure is increased the enrichment decreases. At 80 mtorr enrichment is not found. The effect of chemical scavengers is discussed in Section VII.

The changes in concentration of o-$I_2$ and p-$I_2$ with time at various initial pressures have been modeled quantitatively (126). The overall dissociation rate $\gamma W = 3 \times 10^{-3}$ sec$^{-1}$ and the rate of equilibration of adsorbed iodine species other than $I_2$, are the slow processes which primarily determine the timescale in Fig. 31. Recently Lehmann and coworkers (127) have made detailed studies of lifetimes and predissociation mechanisms for $I_2$ which give useful information on (V.13) - (V.15). In some cases hyperfine-induced predissociation occurs. They suggest the possibility of separations based on hyperfine levels in which the hyperfine states with the larger predissociation rate would be selectively destroyed.

Predissociation spectra have been observed for numerous

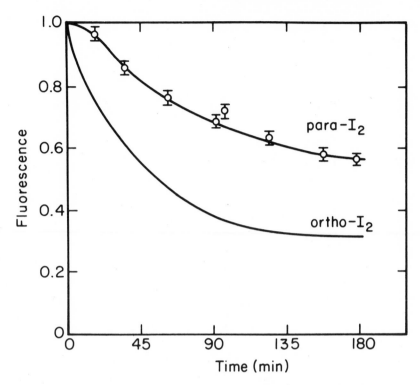

*Fig. 31. Kinetics of selective photopredissociation of ortho-$I_2$ in a natural mixture with para-$I_2$. The mixture is photolyzed with 514.5 nm argon laser radiation. Concentrations of ortho and para-$I_2$ are proportional to fluorescence intensity due to 514.5 and 501.7 nm excitation, respectively (126).*

diatomic molecule transitions.  ICl is another stable molecule for which a low-lying transition exhibits predissociation (128).  Oxygen is predissociated in part of the Schumann Runge bands (129). The A states of OH and OD are predissociated (130).  For $N_2^+$, an accidentally predissociated transition appears for which the predissociation quantum yields vary a factor of ten for $^{14}N_2^+$, $^{14}N^{15}N^+$ and $^{15}N_2^+$ (131).  In a case such as this, in principle, the excitation need not be isotopically selective since the dissociation process itself is inherently selective.  Hyperfine-induced predissociation provides similar opportunities (127).

B.    Polyatomic Predissociation

Photopredissociation has been most extensively studied in

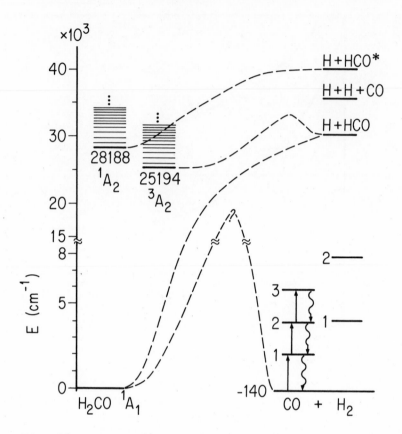

Fig. 32. *Energy diagram for states involved in formalde-hyde predissociation (145).*

formaldehyde (23, 132 - 145). It is known that near the origin of the first excited singlet state formaldehyde (Fig. 32) dissociates with high quantum yield to $H_2$ and CO (133, 143, 144)

$$H_2^{13}CO + \hbar\omega \longrightarrow H_2^{13}CO \ (S_1 \ v_a) \tag{V.20}$$

$$H_2^{13}CO \ (S_1 \ v_a) \longrightarrow H_2 + {}^{13}CO \tag{V.21}$$

$$\longrightarrow H_2^{13}CO + \hbar\omega \tag{V.22}$$

Absorption of a single photon leads to chemically stable dissociation products. For most wavelengths the quantum yield of dissociation is nearly unity (143). Separation of hydrogen from deuterium has been demonstrated using 1:1 mixtures of $H_2CO$ and $D_2CO$ in works (23, 133, 135, 136). Enrichments, limited by the excitation selectivity of the source, were as high as 9:1 (135). Recently, Marling has obtained about 75% $D_2$ from photolysis of mixtures of 2% $D_2CO$ in $H_2CO$ for an enrichment of 180 (141). Enrichments of HD from $HDCO-H_2CO$ mixtures were obtained by careful tuning of selected ion-laser lines. Enrichments of as much as 17 for $^{18}O$, 9 for $^{17}O$, and 34 for $^{13}C$ were obtained with various ion laser lines (141). An 80 fold enrichment was found by Clark et al (137) for $^{12}C$ enrichment using a tunable dye laser. The formaldehyde system shows excellent promise for practical enrichment of $^{13}C$, $^{14}C$, $^{17}O$ and $^{18}O$. Practical application requires a thorough understanding of the photophysics and photochemistry of formaldehyde and the development of efficient tunable UV lasers.

The photophysics and photochemistry of formaldehyde is grossly oversimplified in Eqs. (V.20) - (V.22). In addition to the molecular dissociation channel, Eq. (V.21), there is a free radical channel.

$$H_2^{13}CO\ (S_1\ v_a) \longrightarrow H + H^{13}CO \qquad (V.23)$$

Observations of the kinetics of formation of CO(v) from formaldehyde dissociation at 337 nm show that CO is formed long after $H_2CO(S_1)$ has disappeared and only as the result of collision, Fig. 33. At 0.1 torr the lifetime of $H_2CO$ is 50 nsec

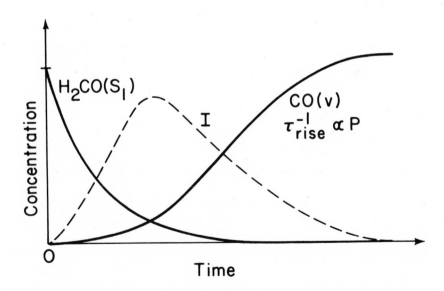

Fig. 33. *Schematic diagram of decay of the excited singlet, $S_1$, and formation of dissociation products, CO. The existence of an intermediate state is inferred (145).*

(146) and CO formation requires 5 μsec. Furthermore only a few percent of the energy available to the $H_2$ and CO appears as CO vibration. Thus $S_1$ formaldehyde decays to an unidentified intermediate state which subsequently yields CO. It is not known whether the H + HCO product is formed by direct dissociation of $S_1$ or through the intermediate.

$$H_2CO \; (S_1 \; v_a) \rightarrow I \rightarrow H_2 + CO(v) \qquad (V.24)$$
$$\rightarrow H + HCO \qquad (V.25)$$

At sufficiently high pressures the excited singlet state is collisionally quenched and the overall dissociation process is

$$H_2{}^{13}CO \; (S_1 \; v_a) + M \longrightarrow H_2 + {}^{13}CO + M \qquad (V.26)$$

$$\longrightarrow H + H{}^{13}CO + M \qquad (V.27)$$

$$\longrightarrow H_2{}^{13}CO \; (S_0 \; v_b) \qquad (V.28)$$

and possibly

$$H_2{}^{13}CO(S_1 \; v_a) + H_2CO \longrightarrow H_2{}^{13}COH + HCO \qquad (V.29)$$

when the collision partner M is $H_2CO$ isotopic selectivity is degraded by:

$$H_2{}^{13}CO \; (S_1 \; v_a) + H_2CO \longrightarrow H_2{}^{13}CO + H_2CO \; (S_1 v_c) \qquad (V.30)$$

$$\longrightarrow H_2{}^{13}CO + H_2 + CO \qquad (V.31)$$

$$\longrightarrow H_2{}^{13}CO + H + HCO \qquad (V.32)$$

To display all possibilities Eqs. (V.26) - (V.32) should be repeated with $H_2{}^{13}CO$ replaced by ${}^{13}I$. From the original work of Yeung and Moore (23) it is clear that these processes contribute less than 20% to the overall photochemistry. Measurements of energy transfer $H_2CO(S_1) \rightarrow D_2CO$ suggest that process (V.30) has a rate less than 3% of that for (V.26) for excitation at the band origin (147). The role of the triplet state in formaldehyde photochemistry must also be considered (148). The free radical dissociation products react chemically and degrade the isotopic selectivity,

$$H + H_2CO \longrightarrow H_2 + HCO \qquad (V.33)$$

$$H{}^{13}CO + HCO \longrightarrow \frac{1}{2} H_2{}^{13}CO + \frac{1}{2} H_2CO + \frac{1}{2} {}^{13}CO + \frac{1}{2} CO \qquad (V.34)$$

unless effective radical scavengers are added (143).

$$H + NO + M \longrightarrow HNO + M \qquad (V.35)$$

$$HCO + NO \longrightarrow HNO + CO \qquad (V.36)$$

$$2HNO \longrightarrow N_2O + H_2O \qquad (V.37)$$

The effectiveness of NO as a scavenger has been demonstrated for carbon enrichment (137). NO scavenging has allowed the absolute quantum yields of the molecular and free radical dissociation channels to be determined as a function of exciting wavelength (143).

The mixing of isotopes by free radical energy transfer processes (V.32) - (V.34) is especially serious when very high purities of an isotope are desired. Table 1 illustrates how S

TABLE 1

## Effect of Mixing on Selectivity

| Initial $^iC, {}^iO$ | Final $^iC, {}^iO$ | | K |
|---|---|---|---|
| | S = 100 m = 0 | S = 100 m = 0.2 | |
| 0.5 | 0.99 | 0.90 | 9 |
| $10^{-1}$ | 0.92 | 0.75 | 27 |
| $10^{-2}$ | 0.50 | 0.40 | 67 |
| $10^{-3}$ | 0.091 | 0.073 | 73 |
| $10^{-4}$ | 0.010 | 0.008 | 79 |

the spectroscopic selectivity is degraded by isotopic mixing. The quantity m is the fraction of the photoproduct which is isotopically mixed and has the isotopic composition of the original mixture. The fraction 1-m has the isotopic composition given by the selectivity of the excitation. In situations where a second enrichment step is required to achieve the desired isotopic purity and where the amount of the desired isotope is comparable to or greater than of the undesired isotope, it will usually be more efficient to selectively photolyze the unwanted isotopic species.

The molecule s-tetrazine photolyzes cleanly in the gas

phase (149), in pure solid (150) and in solid solution (150, 151).

$$H - C \overset{\displaystyle N = N}{\underset{\displaystyle N - N}{\Big\langle}} C - H + \hbar\omega \longrightarrow N_2 + 2HCN \qquad (V.38)$$

The products are chemically stable. The excited state lifetime is several hundreds of picoseconds both in the gas phase (152) and in liquid benzene at 300°K (153). It is presumably similar in solid solutions. In the gas phase this rapid predissociation makes collisions unimportant, even at rather high pressures. The absorption spectrum is covered by visible dye lasers and the absorption cross sections are relatively large. Enrichments have been reported in gas phase at 300°K (149) and in benzene at 1.6°K (150, 151). Isotopic enrichment in the gas phase has been carried out by Karl and Innes (149). The spectrum of this comparatively large molecule is badly congested and the isotopic spectra are overlapped, Fig. 34. The Q-branch bandhead of the origin transition for the lightest isotope, 96% abundant, may be excited with a selectivity of ~ 6 (149). The heavy nitrogen product, $^{15}N^{14}N$, was produced with a selectivity of 2. Karl and Innes achieved overall enrichments as high as 70 for the combined heavy isotopes in the remaining tetrazine by photolyzing away the abundant isotope (149). They photolyzed macroscopic amounts of material with nearly complete absorption of the laser light in a 25 cm cell. Enrichments were measured with a mass spectrometer. Arbitrarily high enrichment of the remaining tetrazine may be achieved at the expense of yield by sufficiently long photolysis (see Sec. VI).

In solid solution with benzene at 1.6°K King and Hochstrasser (150, 151) have found well-resolved spectra of individual isotopic molecules. They have photolyzed particular isotopes with high selectivity, Fig. 35. The absorption spectrum, top curve, shows that the tail of the major peak, $^{12}C_2{}^{14}N_4H_2$, is less intense than the $^{13}C^{12}C^{14}N_4H_2$ absorption at the latter's peak. The spectroscopic selectivity is substantially greater than the natural abundance ratio of 50. The third trace shows an enrichment of the $^{13}C$ tetrazine relative to the abundant isotope of greater than 200. King et al. (154) have gained further insight into the dissociation mechanism by preparing and photolyzing 1,4-s-tetrazine-$^{15}N_2$. Product analysis showed

$$H - C \overset{\displaystyle ^{15}N = N}{\underset{\displaystyle N - ^{15}N}{\Big\langle}} C - H + \hbar\omega \longrightarrow {}^{14}N^{15}N + HC^{15}N + HC^{14}N. \qquad (V.39)$$

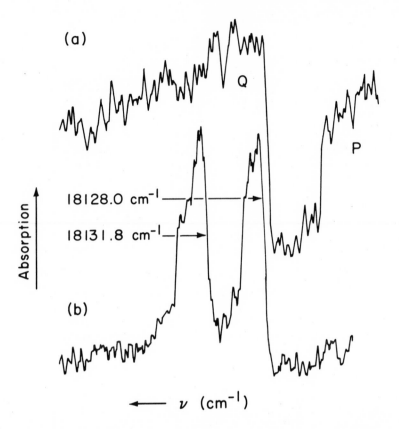

(a)

Q

P

Absorption

18128.0 cm$^{-1}$

18131.8 cm$^{-1}$

(b)

$\longleftarrow \nu \ (cm^{-1})$

*Fig. 34.  Absorption of tetrazine vapor in the region of the Q-branch of the 0-0 band of the naturally abundant isotopic species.  Microphotometer tracings from original photographic plates.  a) Before irradiation; note that the Q-branch is saturated and that, on the low-frequency side, the beginning of the P-branch may be seen.  b) After irradiation; most of the Q-branch intensity of the naturally abundant (and decomposing) species has been lost and a second feature is revealed, namely the nearly superimposed Q-branches of the 0-0 bands of tetrazine-$^{15}N_1$ and $^{13}C_1$ (149).*

In solid solutions one must guard against, or perhaps take advantage of (Sec. X), nearly resonant energy transfer.

The clean photochemistry, visible photolysis wavelength and large absorption cross section make s-tetrazine an attractive candidate for practical enrichment of $^{13}C$ and $^{15}N$. The drawbacks of low spectroscopic selectivity in the gas phase and of high cost for sample preparation and cooling in benzene at 1.6°K, might be avoided by gas dynamic cooling. Smalley, Chandler, Wharton, and Levy (155) have demonstrated

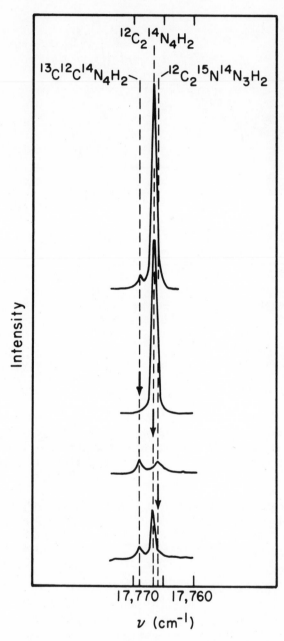

Fig. 35. *Excitation spectra of the isotopically selective photodecomposition of* <u>sym</u>*-tetrazine in benzene at 1.6 K. Four suitably dilute crystals showing the natural abundances of* $^{13}C$ *and* $^{15}N$ *containing tetrazine, in benzene, were placed in an optical helium dewar and cooled to 1.6 K.*

*The $16a^2_0$ transition of the $^1B_{3u}(n\pi^*)$ state of tetrazine was
observed in excitation by monitoring the total fluorescence
intensity as a function of exciting dye laser wavelength.
The excitation and photolysis source was a narrow band ($\sim 0.8$
$cm^{-1}$) tunable dye laser. The transition energies for each
isotopic species of tetrazine were established in the first
sample. Each of the subsequent samples was irradiated at a
specific isotope transition as indicated by the arrows. The
respective excitation spectra of the photolyzed samples demonstrate the high degree of isotopic selectivity achieved in
the decomposition reaction in this system. The situations de-
picted in the figure were achieved by a few minutes irradiation (151).*

more than sufficient spectral simplification of the s-tetrazine spectrum. The cost of preparing s-tetrazine and the pre-
vention of its thermal decomposition could be severe practical
problems.

There are other polyatomic molecules where predissociation might be used for isotope separation. $NH_3$ exhibits a
predissociated spectrum near 210 nm (36). Some promising results have been obtained with $Cl_2CO$ (8, 156). Although spontaneous predissociation of states exhibiting well-resolved
spectra may be relatively rare, it can be expected that collision-induced predissociation is a more general phenomenon.
Collisional quenching of states with energy in excess of the
threshold for dissociation to ground state products is very
likely to proceed by dissociation.

The dissociation of high vibrational levels of ground
electronic state molecules is yet another possibility. After
all, thermally induced dissociation proceeds via collisional
excitation of vibrations until the dissociation threshold is
passed. The molecule, $HN_3$, is reported to exhibit line
broadening in the $v = 3$ and $v = 4$ levels of the hydrogen
stretching vibration (157). If this is so, selective dissociation of $^{15}N$ molecules should be possible with an infrared
dye laser. The very small absorption cross sections for high
overtone transitions present a serious problem in practice.
However, successful experiments on isotopically-selective
excitation of second overtone HCl have been performed in work
(93). Several carefully chosen infrared photons might be used
instead for a stepwise or simultaneous multiphoton absorption.
(See also Sec. VI). As with electronically excited states it
may also be useful with vibrationally excited states to use
collisions to assist in the dissociation process. Molecules
which will be dissociated by absorption of a small number of
vibrational quanta will naturally be ones like $HN_3$ which are
somewhat thermally unstable and therefore dangerous to handle
in large quantities.

If sufficiently short wavelength photons are available, it may be possible to excite autoionizing levels of atoms and molecules.

## C.    Unimolecular Isomerizations

A process closely related to photopredissociation is that of unimolecular isomerization (158, 159). A rearrangement of chemical bonds such as

$$
\begin{array}{c}
H \quad H \\
\backslash C - C \diagdown H \\
\parallel \quad \parallel \\
C - C - H \\
\diagup \quad \backslash \\
H \quad H
\end{array}
\quad\longrightarrow\quad
\begin{array}{c}
H \\
\mid \\
C \\
H_2C \diagup \quad \diagdown C \diagup\!\!\!\!= CH_2 \\
\mid \\
H
\end{array}
\qquad (V.40)
$$

$$
CH_3NC \quad\longrightarrow\quad CH_3CN \qquad (V.41)
$$

renders products which are quite easily separated from starting material. Unfortunately most molecules for which such processes have been studied are too large and complex to be likely to exhibit well-resolved isotopic shifts. By suspending the molecule to be excited in a low temperature matrix the spectrum may be simplified by removing the rotational structure and the possibility of dissociation diminished by confining the fragments in the matrix cage. The photochemical transformation

$$
\begin{array}{c}
H \diagdown \quad \diagup N \\
C \diagup\!\!\!\!\diagdown \parallel \\
H \diagup \quad \diagdown N
\end{array}
\quad\longrightarrow\quad
\begin{array}{c}
H \diagdown \\
\quad C = N = N \\
H \diagup
\end{array}
\qquad (V.42)
$$

was carried out in this way (160). Predissociation might also be carried out in a matrix (114).

Photopredissociation promises to become a practical method of isotope separation. Work on formaldehyde may lead to economically viable methods of enriching $^{13}C$, $^{14}C$, $^{17}O$ and $^{18}O$. While the method is not as generally applicable as the two-photon methods, it may be considerably less expensive in those situations where it can be used.

## VI.    MOLECULAR DISSOCIATION IN AN INTENSE IR FIELD

All the schemes of laser isotope separation discussed above are based on the excitation of electronic states of atoms and molecules by laser radiation in the visible or UV ranges of the spectrum. This method is based on the effect (25, 26) of isotopically selective collisionless and collisional molecular dissociation in the field of a powerful IR radiation pulse the frequency of which coincides with that of the molecular vibrational band. The discovery of this effect was preceded by works (161 - 164) in which the interaction of a powerful IR pulse with molecular gases was investigated.

## A.    Collisionless and Collisional Dissociation

Isenor and Richardson (161) have shown that when focusing a powerful pulse of a $CO_2$ TEA laser into a molecular gas the vibrational absorption band of which coincides with the laser line, there appears visible luminescence even when the focal power density is below the threshold for optical breakdown (usually $10^8 - 10^9$ W/cm$^2$). The intensity of luminescence is 2 - 4 orders of magnitude lower than that of the spark which arises during optical breakdown, and there are spectral lines of free radicals in the visible luminescence. Similar results have been obtained in work (162) on $C_2F_3Cl$ molecules. These works have stimulated more detailed research into the mechanisms of molecular dissociation and visible luminescence in strong IR fields. The first studies of the kinetics of visible molecular luminescence excited by $CO_2$ laser pulses have shown (163) that visible luminescence rises very quickly. The delay time, $\tau_d$, with respect to the leading edge of the $CO_2$ laser pulse is shorter than 30 nsec. This time is much shorter than the time of V-T relaxation and V-V vibrational exchange at pressures less than a torr. Therefore, this effect cannot be associated with either thermal gas heating or collisional excitation of molecules by collisional energy exchange (162, 163).

Figure 36 shows a typical experimental scheme for studying the kinetics of visible luminescence by the action of a short $CO_2$ laser pulse. The beam of a $CO_2$ TEA laser is focused into a molecular gas cell. Luminescence is observed through a side window. Luminescence signals from a photomultiplier and the laser signal from an IR "photon-drag" detector are observed simultaneously. Figure 36 shows also $BCl_3$ luminescence pulses from different regions: focused (A) and partially focused (B). In the region of the laser focus two stages of luminescence kinetics are observable (164), instantaneous and retarded. At low pressures of $BCl_3$ (2 - 3 Torr) these stages are quite distinct. As the pressure is increased, the retarded stage appears to approximate the instantaneous one and they cannot be distinguished any longer. In the partially focused region, that is the region of weaker field, only the retarded luminescence stage is observed. Instantaneous luminescence can be observed down to very low pressures of about 0.03 Torr with its pulses of up to 40 nsec long and has no

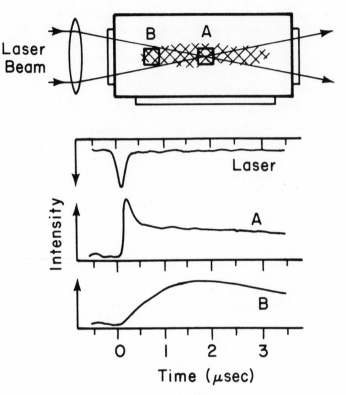

*Fig. 36.  Fluorescence excitation scheme for a molecular gas using a focused $CO_2$ TEA laser pulse:  A - fluorescence from the focal region;  B - fluorescence from the partially focused region.  Data are for 0.8 torr of $BCl_3$ (164).*

observable delay with respect to the laser pulse.  Retarded luminescence is collisional in nature.  This is confirmed by the relation $p\tau$ = const, where p is the gas pressure in the cell and $\tau$ is the delay time of the retarded luminescence maximum with respect to the beginning of the laser pulse.  For $BCl_3$ $p\tau$ = 2.3 µsec torr.  This is somewhat longer than the time for V → V equilibration among the lower vibrational levels, $p\tau_{VV} \leq 0.5$ µsec · torr (109).  Retarded luminescence may be explained by the dissociation of highly excited molecules as they reach the dissociation limit owing to V - V exchange during collisions after the laser pulse is over. Detailed studies on the kinetics of molecular luminescence have been carried out in works on $BCl_3$ (164, 165) and on $SiF_4$ molecules (166).

The essential features of this "instantaneous" luminescence are that it is observed at an intensity of the resonant IR field of about $10^9$ W/cm$^2$ and that it occurs simultaneously

with the laser pulse even though the time between molecular
collisions $\tau_{coll}$ is much longer than the laser pulse. At a
pressure of 0.03 Torr the time between molecular collisions
in $BCl_3$ is $\tau_{coll} \geq 1.3$ μsec. Therefore the probability of one
collision during a laser pulse, $\tau_L = 40$ nsec, is less than
0.03, of two collisions, is less than $(0.03)^2 = 10^{-3}$, etc. So,
the collisional mechanism of molecular excitation up to the
limit of dissociation during a laser pulse is completely
excluded. The "instantaneous" luminescence is caused by dissociation of an isolated molecule acted upon only by a strong IR
field. Dissociation is due to multiphoton absorption.

The effect of collisionless molecular dissociation in a
strong IR field has opened one more possibility for photophysical action on molecules which may be isotopically selective provided that the isotope shift is well-resolved in the
IR spectrum. The first experiments on isotope separation by
this method were run in work (25) on $BCl_3$ molecules. $BCl_3$ has
two nonoverlapping absorption bands of isotopic molecules
$^{10}BCl_3$ and $^{11}BCl_3$. In these experiments the products of $BCl_3$
dissociation reacted with oxygen, and electronically excited
BO radical was formed. The BO radical has an intense system
of lines, known as the α-band, with well-resolved $^{10}BO$ and
$^{11}BO$ features. The isotope shift in the luminescence spectrum
of BO is about 30 Å, which allows simple isotopic analysis of
the reaction products.

The experiment in work (25) was conducted thus: a $CO_2$
laser was tuned to the $\nu_3$ absorption band of either
$^{10}BCl_3$ or $^{11}BCl_3$ molecules, and the line intensity for $^{10}BO$
and $^{11}BO$ was studied. The experiment employed the natural
isotopic mixture ($^{10}B : ^{11}B = 1 : 4.32$). The portion of BO
luminescence spectrum including the (0,2) transition of the
band $^2\Pi \rightarrow {}^2\Sigma$ was studied. The recording system made it
possible to study independently the selectivity of BO formation at times corresponding to both the instantaneous and retarded luminescence stages in pure $BCl_3$ (Fig. 37). On the
basis of the data on BO visible luminescence spectra the enrichment coefficient is evaluated as, R.V. Ambartzumian et al.
(25, 165),

$$K\left({}^{10}B \Big/ {}^{11}B\right) = \frac{\left[{}^{10}BO\right]\left[{}^{11}BCl_3\right]}{\left[{}^{11}BO\right]\left[{}^{10}BCl_3\right]} \simeq 10,$$

when $^{10}BCl_3$ molecules are excited by the R(24) $CO_2$ laser line.
The experiments, in which BO luminescence is observed in the
partially focused region, have also provided support for the
presence of isotopically selective chemical reactions (165).
Note that the dissociation selectivity was found to be the

same both during collisionless and collisional phases of $BCl_3$ dissociation, Fig. 37 (165). The experiments show that even at rather high pressures the dissociation selectivity is high, and that molecular dissociation occurs because of energy storage in the vibrational degree of freedom.

After the experiments on isotopically selective dissociation of $BCl_3$ molecules carried out in 1974, it was clear that the effect was general in nature and could take place in polyatomic molecules with well-separated vibration-rotation absorption bands.

## B.    Isotope Separation Experiments

Important experiments were run in work (167) which have

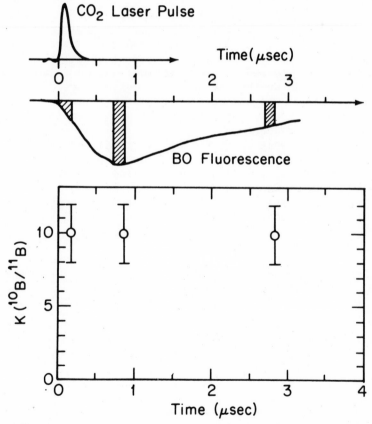

Fig. 37.  Isotopic enrichment of BO radicals for collisionless and collisional dissociation of $BCl_3$ by a $CO_2$ laser pulse.

displayed macroscopic separation of $^{10}$B and $^{11}$B isotopes.  In
these experiments the relative content of remaining $^{10}$BCl$_3$ and
$^{11}$BCl$_3$ molecules was measured using the IR absorption spectrum
of the residual gas mixture.   The results of such experiments
are given in Fig. 38.   Spectrum (a) corresponds to an unirra-
diated natural mixture of $^{10}$BCl$_3$ and $^{11}$BCl$_3$, while the other

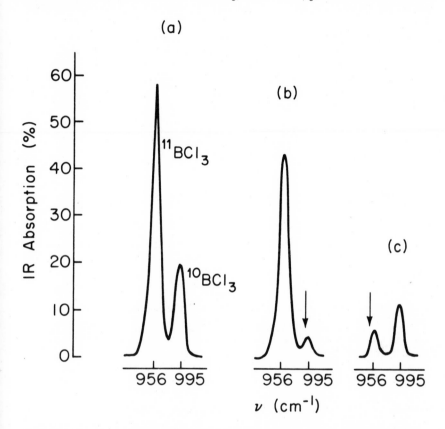

Fig. 38.   IR absorption spectrum for the band $\nu_3$ of a
natural isotopic mixture of BCl$_3$ molecules:   a) before irradiation;   b) after irradiation at the absorption frequency of
$^{10}$BCl$_3$;   c) after irradiation at the absorption frequency of
$^{11}$BCl$_3$.   Data are for 0.44 torr BCl$_3$ with 11 torr air (167).

spectra are given for irradiation of the mixture by a $CO_2$
laser tuned to the absorption band of $^{10}$BCl$_3$ (b) or $^{11}$BCl$_3$
(c).   In these experiments on BCl$_3$ one should be careful in
choosing a scavenging reaction for binding the radicals which
are formed in the dissociation process.   When there is no
scavenger added the radicals recombine.   The identity and
chemistry of the radicals is not known in detail.   Thus in

experiments on boron enrichment it is difficult to take quantitative account of the role of selectivity loss in subsequent chemical reactions. The experiments on boron enrichment in macroscopic amounts by this technique show that the enrichment coefficient in the reaction products varies from 2 to 20 depending on the scavenger added in the reactor (167). Molecules such as $O_2$, $C_2D_2$, HBr and other were used as scavengers. In the case of $O_2$ the resultant BO radicals reacted and formed the stable product $B_2O_3$ which was deposited on the walls as a thin white coating. The typical enrichment coefficient at low pressures was K ($^{10}B/^{11}B$) ≃ 8, but its value decreased sharply with increasing pressure of $BCl_3$.

The experiments on sulphur isotope separation by Ambartzumian et al. (26) based on selective chemically irreversible dissociation of one of the $SF_6$ isotopic molecules in a strong IR field have given more conclusive evidence of the efficiency and simplicity of this isotope separation method. In these experiments the reaction cell was filled with $SF_6$ of natural isotopic composition, and the $CO_2$ laser was tuned to a frequency corresponding to the absorption maximum of either $^{32}SF_6$ (95%) or $^{34}SF_6$ (4.2%). The natural mixture contained also $^{33}SF_6$ (0.75%) and $^{36}SF_6$ (0.017%) molecules. In contrast to $BCl_3$, the process of $SF_6$ dissociation in a strong IR field is irreversible, and there is no need to add radical scavengers. In some cases $H_2$, HBr and NO scavengers were used. The results obtained, with and without scavengers were qualitatively the same. When the $CO_2$ laser frequency was tuned to the maximum of the $^{32}SF_6$ absorption band at 947 $cm^{-1}$ (the P(16) line) and the cell filled with 0.18 torr $SF_6$ and 2 torr $H_2$ was irradiated with 2 x $10^3$ pulses, almost all $^{32}SF_6$ molecules disappeared from the mixture owing to selective dissociation. The mass spectrum of $SF_5^+$ consisted mainly of lines corresponding to $^{32}SF_6$ and $^{34}SF_6$ molecules, Fig. 39. The coefficient of enrichment in    $^{34}S$ with respect to $^{32}S$ was found from the ratio of mass spectral line amplitude before and after irradiation to be about 2800. In analogous experiments the enrichment in $^{36}S$ with respect to $^{32}S$ was K(36/32) ≃ 1200 and in a number of experiments exceeded $10^4$. These values, however, should not be compared directly with analogous parameters in standard methods of isotope separation because enrichment in residual products depends on the extent of dissociation (168). As long as molecules of different isotopic compositions dissociate with different relative rates, the enrichment coefficient in the residual undissociated gas may be as high as one wants at sufficiently complete dissociation of the initial reactant.

Qualitative analysis in the simplest model of dissociation of a two-component isotopic molecular mixture was carried out in work (168). Let the initial concentrations of mole-

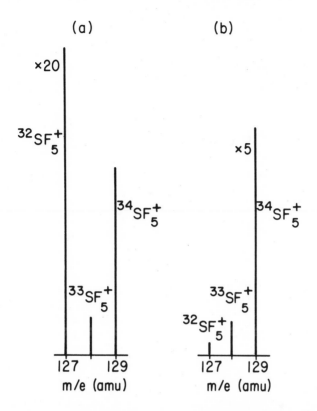

Fig. 39.  A part of the mass spectrum of $SF_6$ (ionic
fragment $SF_5^+$):  a) the natural mixture before irradiation;
b) after irradiation with $2 \cdot 10^3$ pulses of $CO_2$ laser light at
the line P(16) (26).

cules be $N_{ao}$ and $N_{bo}$, and IR radiation pulses be in resonance
with the $N_a$ molecules which contain the isotope "a." Denote
the dissociation rate of "a" molecules by $d_a$ and that of "b"
by $d_b$. The quantity $d_a$ is the product of the volume-averaged
probability of irreversible dissociation of a single "a" molecule under the action of a single pulse and the number of
pulses per unit time. Since "a" molecules are in better resonance with the field, $d_a > d_b$, and the molecular dissociation selectivity is determined by the relation $S = d_a/d_b$.
$S \gg 1$ corresponds to high selectivity. The concentrations of
"a" and "b" molecules decrease exponentially with time or
number of pulses:

$$N_a = N_{ao}\ exp(-d_a t) \ ; \ N_b = N_{bo}\ exp(-d_b t) . \qquad (VI.1)$$

It is assumed that, as the source molecules burn up and the

chemical composition and pressure of the gas mixture change, the dissociation rate constants remain the same. At low gas pressures, which are of most interest for high enrichment, this is quite a reasonable first approximation since the process of dissociation depends mainly on IR field. In this case the coefficient of residual gas enrichment, according to (VI.1), is:

$$K_{res} \left(\frac{b}{a}\right) = \frac{(N_b/N_a)}{(N_{bo}/N_{ao})} = exp(d_a-d_b)t = exp\left[\left(\frac{S-1}{S}\right)d_a t\right]. \quad \text{(VI.2)}$$

The total pressure of residual gas $N = N_a + N_b$ varies thus:

$$N/N_o = \delta_a \, exp(-d_a t) + \delta_b \, exp(-d_b t), \quad \text{(VI.3)}$$

where $N_o = N_{ao} + N_{bo}$ , $\delta_a = N_{ao}/N_o$ , $\delta_b = N_{bo}/N_o$ denote the relative content of isotopic molecules in the source mixture.

The enrichment coefficient of product molecules is determined by the relation:

$$K_{prod} = \left(\frac{N_{ao}-N_a}{N_{bo}-N_b}\right) \bigg/ \left(\frac{N_{ao}}{N_{bo}}\right) = \frac{1-exp(-d_a t)}{1-exp(-d_b t)} . \quad \text{(VI.4)}$$

When the degree of decomposition of the source mixture is small, $d_a t$, $d_b t \ll 1$, we have:

$$K_{prod} \, (a/b) = d_a/d_b = S. \quad \text{(VI.5)}$$

Thus, the enrichment of the product molecules is given by the degree of dissociation selectivity of "a" molecules.

Relations (VI.3) - (VI.5) show that upon exponentially deep "burning" of the original "a" and "b" molecules the small portion of remaining molecules can be enriched with the isotope "b" as much as one wants. Figure 40 illustrates the time or pulse number variation for enrichment coefficients of undissociated molecules and product molecules for three different values of dissociation selectivity factor (S = 1.1 small, S = 2 moderate and S = 10 high selectivity).

In the experiments described in works (26, 168, 169) the coefficients of enrichment in sulphur isotopes K(34/32) and K(36/32) were > $10^3$ for sufficiently deep molecular burning (the pressure of residual $SF_6$ gas was below 1% of the initial one). Under deep molecular burning the enrichment coefficient

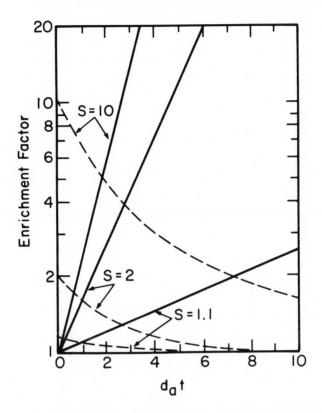

*Fig. 40. Calculated dependence of the enrichment co-
efficient of the residual gas $K_{resid}$ (b/a) (solid curves) and
of the product molecules $K_{prod}$ (a/b) (dashed curves) on
time or the number of pulses during selective dissociation of "a" molecules in a two-component mixture with
different values for the dissociation selectivity coefficient
S (168).*

of residual gas is, of course, not equal to the coefficient
of molecular dissociation selectivity S. The latter must be
evaluated by the use of (VI.3) and (VI.4); that is, the
enrichment coefficient $K_{resid}$ (b/a) and variation of residual
gas pressure must be measured simultaneously.

Figure 41 presents the experimental values of enrichment
coefficients for residual $SF_6$ gas and resulting molecules at
a comparatively high pressure of $SF_6$ (16 Torr) when the enrichment magnitude is far from the maximum possible values.
When approaching the regime of deep molecular burning one can
clearly observe an increase in $K_{resid}$ (34/32) and the beginning of a decrease in $K_{prod}$ (32/34). The value of $K_{prod}$

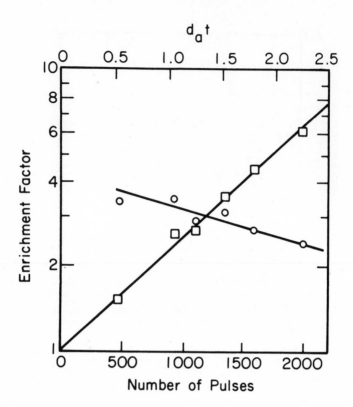

*Fig. 41. Experimental dependence of the enrichment coefficient of residual gas $K_{resid}$ ($^{34}S/^{32}S$), □ , and product molecules $K_{prod}$ ($^{32}S/^{34}S$), ○ , on the number of irradiation pulses during dissociation of $^{32}SF_6$ molecules in the natural mixture with the initial pressure of 1.6 torr (168).*

(32/34) at the beginning of irradiation gives at once the value for the coefficient of $^{32}SF_6$ dissociation selectivity. At an $SF_6$ pressure of 1.6 Torr S = 3.5. As the pressure is decreased down to 0.2 Torr, the dissociation selectivity increases greatly and S can be greater than 10.

In the regime of deep burning marked enrichment may be attained even with a poor dissociation selectivity. This is confirmed by the successful experiments (111, 169) on $^{187}Os$ and $^{192}Os$ isotope separation through dissociation of $OsO_4$ molecules by $CO_2$ laser pulses. The isotope shift is only 1.3 $cm^{-1}$, and thus is much smaller than the envelope of the rotational structure of the vibrational band. The half-width of just the Q-branch of the $\nu_3$ mode of $OsO_4$ is 3 - 4 $cm^{-1}$. The enrichment coefficients obtained are about 12 - 15%.

Successful experiments on isotope separation in macroscopic amounts drew the attention of many laboratories to this method.  In the first experiments (26) the rate of twenty-fold enrichment was $10^{-4}$ g/min. using a simple TEA $CO_2$ laser with its pulse power of 2 J and frequency of 1.5 Hz.  In particular, the experiments on sulphur isotope separation were quickly confirmed by Lyman et al. (170).  Lyman and Rockwood (171) reported boron isotope separation and the first successful experiments on isotope separation of carbon with 62% enrichment through the dissociation of $CF_2Cl_2$ molecules, as well as silicon with 16% enrichment through dissociation of $SiF_4$ molecules.  In 1974 Yogev and Benmair (172) reported a 20% selectivity in decomposition of $CD_2Cl_2$ by a focused $CO_2$ TEA laser in $CH_2Cl_2$- $CD_2Cl_2$ mixtures.  Since the thermal process induced by a cw laser favors decomposition of $CH_2Cl_2$, it now appears probable that selective multiphoton dissociation accounted for a fraction of the observed decomposition.  It is evident that the method can be applied to separation of a rather large number of isotopes using available and inexpensive IR radiation from efficient molecular gas lasers.  Recently many laboratories have done a lot of new experiments on isotope separation by this method:  nitrogen isotopes - $CH_3NO_2$ (173), hydrogen isotopes - $H_2CO$ (174), molybdenum isotopes - $MoF_6$ (175) and others.  However for the development of this method of isotope separation it is most important to understand the main features of multiple IR photon absorption and dissociation processes.  This is the topic of the next chapter by R.V. Ambartzumian and V.S. Letokhov in this volume.  The dissociation process for $BCl_3$ and $SF_6$ molecules was studied most carefully.  These studies enabled us to understand the nature and basic characteristics of both selective dissociation and isotope separation by multiple IR photon excitation. Below we will discuss only the most important characteristic for isotope separation:  the isotopic selectivity of dissociation.

C.    Isotopic Selectivity of Dissociation

Of the most interest are the dependence of dissociation rate W on the IR laser frequency and its correlation with the linear IR absorption band shape.  This frequency dependence for $SF_6$ is shown in Fig. 42 (176).  As can be seen from this figure, the frequency dependence of the dissociation rate (curve 1) is shifted about the maximum of the linear IR absorption of the $\nu_3$ band of $SF_6$ (curve 4) by the anharmonicity value $\Delta\nu_{anh}$ of the $\nu_3$ vibration.  The width of the frequency dependence is about 20 $cm^{-1}$.  Thus the high selectivity of dissociation for two isotopic molecules can be achieved if the

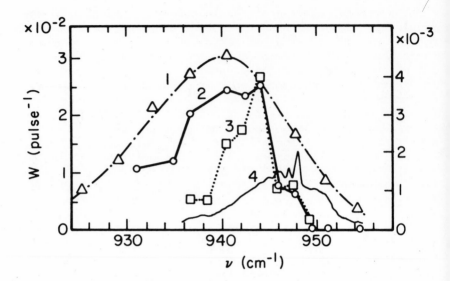

*Fig. 42. Resonance characteristics of the rate of $SF_6$ dissociation in an IR laser field: 1) single-frequency irradiation $P_{av}$ = 31 MW/cm$^2$, T = 300°K; 2) two-frequency irradiation $I(\nu_1)$ = 4 MW/cm$^2$, $P_{av}(\nu_2)$ = 58 MW/cm$^2$, $\nu_2$ = 1084 cm$^{-1}$, T = 300°K; 3) two-frequency irradiation with the same laser parameters but at T = 193°K; 4) linear IR absorption of $SF_6$ at 300°K. The ordinate, curve 1 left and curves 2 and 3 right, gives fractional decomposition per pulse averaged over the entire sample cell (176 and 178).*

isotopic shift is about 20 cm$^{-1}$ or more. The direct measurement of enrichment factor in a product of $SF_6$ dissociation ($SOF_2$) confirmed this conclusion (177) (see Chapter II).

The relatively large width of dissociation rate vs frequency limits the selectivity of dissociation of molecules with small isotopic shifts (as in the case of $OsO_4$ above). A method of dissociation selectivity enhancement was proposed and demonstrated in work (178) which was based on two-frequency IR field dissociation of polyatomic molecules. In this technique a field with the frequency $\nu_1$ is weak and in resonance with the molecular absorption band. This field selectively excites the first few molecular vibrational levels up to the limit of the vibrational "quasi-continuum." The second, intense, nonresonant field with the frequency $\nu_2$ is used for subsequent molecular dissociation. The method is based on the mechanism of polyatomic molecule dissociation in an intense IR laser field suggested in (176). In contrast to single-frequency action by an intense field, in two-frequency dissociation the laser radiation power in the selecting step $\nu_1$ is small and practically does not broaden the vibration-rotation lines of molecular absorption thus affording considerable increase of dissociation selectivity. Results of the measurements of $SF_6$ dissociation rate at 300°K as a function of the low-power (about 1 MW/cm$^2$) frequency $\nu_1$ are shown in Fig. 42 (curve 2). The resonance curve width decreases to 12 cm$^{-1}$ and the high-frequency edge of the resonance curve shifts sharply and becomes more abrupt. The results of the same experiment for $SF_6$ at 193°K are shown in Fig. 42 (curve 3). The resonance curve width becomes about 5 cm$^{-1}$. The considerable drop in the low-frequency edge of the resonance curve with decreasing temperature is associated with the freezing out of hot band absorption. These results show a considerable enhancement of excitation and dissociation selectivity compared to single-frequency fields; this is essential for isotope separation of heavy elements with small isotope shifts.

The existence of distinct isotopic shifts in the linear IR absorption spectra of a molecule is one of the required conditions for isotopic selectivity of dissociation. In work (173) the new effect, occurence of isotopic shift in multiple photon absorption spectra and luminescence excitation spectra, was observed. This effect has been used for nitrogen isotope separation in the mixture of nitromethane molecules $CH_3{}^{14}NO_2$ and $CH_3{}^{15}NO_2$. On Fig. 43 linear absorption spectra over the range 1000 to 1100 cm$^{-1}$ taken with about 0.5 cm$^{-1}$ resolution are shown for $CH_3{}^{14}NO_2$ and $CH_3{}^{15}NO_2$. It is obvious that there is no isotopic shift within the experimental errors. In the intense field the absorption spectra change shape differently in a way equivalent to the creation of an isotopic shift of about 5 cm$^{-1}$ (Fig. 43). The experiments have shown that isotopic shift value does not change noticeably with IR field power density variation between $10^7$ and $10^{10}$ W/cm$^2$. The luminescence excitation spectra exhibit an even greater difference. For dissociation with formation of electronically excited fragments , the isotopic shift reaches 10 cm$^{-1}$ (Fig. 43).

The appearence of the isotopic shift in the intense IR field has permitted successful experiments on nitrogen isotope separation by $CO_2$ laser pulsed excitation of the vibrational band $\nu_{13}$ without an isotopic shift in the linear IR absorption. This effect broadens fundamentally the possibilities of isotope separation by multiple photon dissociation of molecules in an intense IR field. In particular this method may be used in those practically important cases when the vibrations of molecules within the laser tuning range have no isotopic shift in the linear IR spectral band.

The magnitude of the dissociation selectivity decreases with increasing pressure of the dissociated molecules. A typical pressure for which the selectivity is

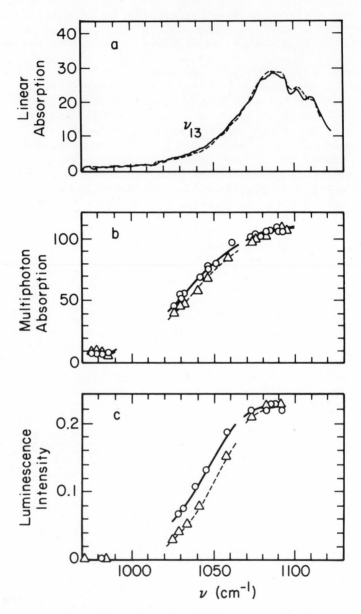

Fig. 43. Isotopic effects for the $\nu_{13}$ band of $CH_3NO_2$ (solid curves - $CH_3{}^{15}NO_2$, dashed curves - $CH_3{}^{14}NO_2$): a) linear absorption of $\nu_{13}$ band, pressure = 20 torr; b) multiple photon absorption spectrum (at power density $10^9$ $W/cm^2$), pressure = 2 torr; c) excitation spectrum of visible luminescence by intense IR field ($10^9$ $W/cm^2$), pressure - 2 torr (173).

decreased by a factor of two is about 0.5 torr in $BCl_3$ and $SF_6$ isotope separation experiments (167, 177). A decrease in the enrichment coefficient with increase in pressure may be caused by increasing contributions of the V-V excitation exchange and nonselective dissociation of molecules due to thermal gas heating by V-T relaxation of those excited molecules which do not reach the dissociation limit. At low pressures these effects are absent, since the time for molecular diffusion to the walls becomes shorter than V-T relaxation and the molecules become deactivated without heating the gas.

## D.  Laser Requirements

The typical power density of IR resonant radiation for dissociation of polyatomics is about $10^7 - 10^9$ W/cm$^2$ for different molecules: 23 MW/cm$^2$ for $SF_6$ (176), several hundreds MW/cm$^2$ for $C_2H_4$ (179). This power density is a very serious requirement for the tunable IR laser even when focusing the laser radiation. However, the potentialities and practical significance of the separation method considered have become greater since the discovery (180) of the effect of selective molecular dissociation when acting with a strong IR field on very weak molecular overtones of $SF_6$ and $CCl_4$. For example, when the combination vibrations $\nu_2 + \nu_6$ and $\nu_2 + \nu_3 - \nu_5$ of the $SF_6$ molecule are acted upon, the molecule can be dissociated. The dissociation rate decreases only as the square root of compound band intensity, a rather slow falloff. Analogous experiments have been conducted on $CCl_4$ molecules. It has been shown (181) that some compound vibrations do not contribute to dissociation. For other molecules isotopically selective dissociation has been observed with the enrichment coefficient of the resulting products for $^{13}C$ being between 7 and 10. These results are of interest in understanding the mechanism of multiphoton absorption by vibrations of polyatomic molecules as well as in widening the scope of the isotope separation method to include many other molecules whose fundamental vibrational absorption bands do not coincide with the radiation frequencies of available high-power IR lasers. Thus, the method of instantaneous selective molecular disssociation in a strong IR field due to multiphoton absorption seems to be very promising and efficient for isotope separation by the use of laser radiation.

The two-frequency technique (178) mentioned above greatly widens the range of molecules suitable for isotope separation, since the necessary intensity of the resonant field is about $10^3$ times smaller ($10^4 - 10^5$ W/cm$^2$) than in the case of a single-frequency field. This makes the use of tunable lasers for the selecting step, $\nu_1$ an interesting possibility.

For the practical application of this method on a large scale the energy losses for dissociation of a single polyatomic molecule are very important. In Fig. 44 the experimental dissociation quantum yields are shown. The number of dissociated $SF_6$ molecules per absorbed $CO_2$-laser photon, or the IR radiation energy (in eV units) expended for dissociation of a single $SF_6$ molecule, is given as a function of power density of the focused $CO_2$ laser radiation. At an intensity of about 200 MW/cm$^2$ dissociation of a single $SF_6$ molecule requires only 12 eV. In a cavity absorption cell most of the laser energy may be absorbed in the gas (182).

Since molecular dissociation in a strong IR field needs a total radiation energy of only 10 - 20 eV, isotope separation by this method appears attractive with regard to energy consumption for large-scale processes of enrichment. If selectivity were sufficiently high, it might even be useful for such rare elements as deuterium. The dissociation of $C_2H_4$ molecules by a $CO_2$-laser (179, 183) makes it possible to use the method for molecules which are available in large amounts (natural gas). This would, however, require an excitation selectivity in excess of $10^3$ (see Sec. XI).

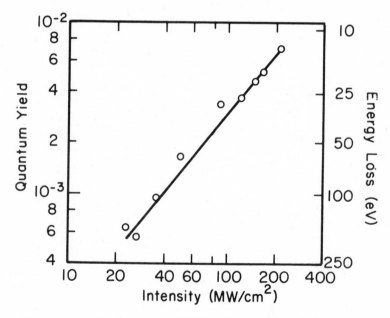

*Fig. 44. The quantum yield (left scale) and energy loss (right scale) for multiple photon IR dissociation of an $SF_6$ molecule as a function of power density of focused $CO_2$ laser pulse.*

## VII. ELECTRONIC PHOTOCHEMISTRY

The chemical reactions of electronically excited atoms and molecules have been an active subject of research for many years. The first photochemical isotope separation was attempted by Hartley et al (7) in 1922. Since the energies of electronic excitation are on the order of chemical bond energies, it is natural to expect that molecular excited states will often be more reactive than ground states. Recently the application of orbital correlation diagrams (159, 184) has considerably improved our understanding of photochemical reactions and our ability to predict the relative reactivity of ground and excited states.

Selective excitation of electronic states is possible for most atoms and diatomic molecules. Herzberg (36) lists spectra of polyatomic molecules with as many as twelve atoms. Although most stable polyatomic molecules do not have transitions with well-resolved rotational structure and isotope shifts in the visible and near ultraviolet there are many which do. There are thus many elements for which isotope separation by electronic photochemistry may be possible.

For successful isotope separation one must find an absorbing atom or molecule and a coreactant which do not react either in the gas phase or on the container walls under the ambient conditions. The excited state must react more rapidly than it loses its excitation by radiation, by predissociation, by energy transfer to molecules containing the undesired isotopes or by quenching. Collisions of the excited molecule with the coreactant should result in chemical reaction more often than in physical quenching. The absolute reaction rate constant should be large enough so that collisions with the coreactant are the primary route of excitation loss at pressures low enough so that pressure broadening does not degrade the excitation selectivity. Finally, the reaction products must be stable and may not undergo further reactions with the unexcited molecules. If free radicals or other unstable molecules are produced in the primary reaction step, they must be scavenged chemically before any isotope scrambling reactions occur. It is unfortunate that most of the photochemistry of small molecules involves the formation and subsequent reactions of free radicals (185, 186).

### A.   Atomic Reactions

The photochemical isotope separation of Hg excited by the 253.7 nm resonance line is a small-scale commercial process (15, 187). Separations have been successfully demonstrated with a variety of reagents. Pertel and Gunning (188) were able to enrich $^{202}$Hg from a 30% natural abundance to 35% in a mixture of Hg and $H_2O$. The addition of butadiene, $C_4H_6$, gave enrichment to 85%. In Hg-$O_2$ mixtures no enrichment was possible. The kinetics of the photochemical reactions of Hg are quite complex. The extensive work of Gunning and his collaborators (188 - 198) does not give a complete mechanism for the Hg-$H_2O$-$C_4H_6$ system. In the Hg-$O_2$-$C_4H_6$ system Nief and his collaborators (199, 200) have enriched $^{202}$Hg to 96.8% purity.

The chemistry of the Hg-$O_2$-$C_4H_6$ system illustrates some of the fundamental photochemical processes and problems with which we must deal in laser photochemical isotope separations. When mercury vapor is excited in the presence of $O_2$ alone no enrichment is found in the solid HgO produced. The dependence of reaction rate on time and light intensity demonstrates that the oxidation occurs on the vessel walls following the photolytic production of an oxidizing species (200). A trace of butadiene removes this species and stops the photolytic oxidation. When substantial amounts of butadiene are added a new reaction occurs. Morand and Nief (200) explain their results quantitatively with the mechanism of Fig. 45. The formation of an oxidized product, HgX, of the selectively excited Hg atoms comes about by the formation of an unstable collision complex HgO$_2$* which then reacts with butadiene to give the oxide and an unidentified product Y. The processes which compete successfully with the desired reaction sequence are quenching of Hg* by $O_2$ and by $C_4H_6$. The $O_2$ quenching limits the quantum yield of the photochemical reaction to $[0.8 \text{ Å}^2/(60 \text{ Å}^2 + 0.8 \text{ Å}^2)] = 1.3 \times 10^{-2}$ at best. The product of the lifetime of the complex and its reaction cross section with butadiene is $5 \times 10^{-6}$ sec $\cdot$ Å$^2$. In this system the pressure of Hg is near $10^{-4}$ torr while the added gases range between 3 and 100 torr. Thus energy transfer in Hg-Hg collisions may be neglected. The added gas pressures are normally large enough to prevent a significant quantum yield of resonance fluorescence at 2537 Å. At their highest pressure, 100 torr, Morand and Nief (200) obtained their best quantum yield of 0.9% mercury oxide product by using 13% butadiene and 87% $O_2$.

Ideally, of course, one hopes to find photochemical reaction sequences which are much less complicated and which give quantum yields much closer to unity. Recently, experiments have been initiated on U atom reactions (106).

B.    Diatomic Molecule Reactions

Isotopic enrichment in diatomics has been carried out by

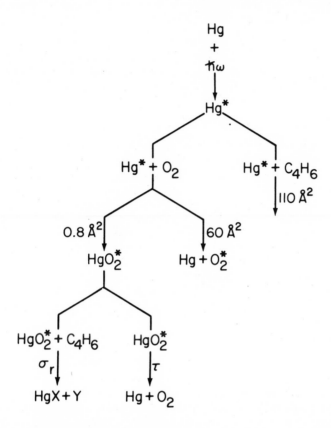

*Fig. 45. Reaction scheme for isotopically selective photo-oxidation (253.7 nm) of Hg (200).*

Harteck and his coworkers using atomic resonance lamps for excitation. Excitation of NO with a Br lamp at 163 nm produced $^{14}N^{15}N$ enriched by as much as a factor of 4 (201). Photolysis of CO with 206 nm light from an I lamp preferentially excites $^{13}CO$ and $C^{18}O$ about a factor of 30 more strongly than the abundant isotope (202). The spectroscopic selectivity was apparrently degraded by about a factor of six due to energy transfer between CO molecules. Both the products $CO_2$ and $C_3O_2$ were enriched in $^{13}C$. Only the $CO_2$ was $^{18}O$ enriched. These results are interpreted in terms of the reaction sequence (202).

$$^{13}CO^* + CO \longrightarrow CO_2 + {}^{13}C \qquad \text{(VII.1)}$$

$$\longrightarrow {}^{13}CO_2 + C \qquad \text{(VII.2a)}$$

$$^{13}C + CO + M \longrightarrow {}^{13}C = C = O + M \qquad \text{(VII.2b)}$$

$$^{13}C = C = O + CO + M \longrightarrow O = C = {}^{13}C = C = O \qquad \text{(VII.2c)}$$

for $^{13}CO$ excitation.  For $C^{18}O$ excitation enriched product is formed only by

$$C^{18}O^{*} + CO \longrightarrow {}^{18}OCO + C. \qquad \text{(VII.3)}$$

Laser sources for photochemical isotope separation have several advantages over incoherent sources.  The number of favorable chance overlaps between isotopic molecular absorption lines and atomic resonance  lines is severely limited. The broad tunability and high ultimate resolution of dye lasers gives a free choice of absorptions in the visible and near ultraviolet and allows the highest possible selectivity. For molecular transitions, especially of relatively rare isotopes only a small fraction of the uncollimated light from an incoherent source will be absorbed in a photochemical reactor.  Laser radiation may be multiply passed in long cells for enormous increases in absorption efficiency.

A successful laser photochemical enrichment has been reported for para-$I_2$ (125, 126, 203).  Ortho-$I_2$ was excited with the 514.5 nm line of an $Ar^{+}$ laser.  In mixtures with 2-hexene the ortho-$I_2$ was reduced to about 5% as a result of a 1 hr irradiation with a 0.2 W laser.  The kinetics were studied (126) by using the time dependence of fluorescence excited by an argon laser at 514.5 nm.  This fluorescence was proportional to the density of ortho-$I_2$ molecules.  The character of the time variation of fluorescence is shown in Fig. 46.  The fluorescence signal decreases by e times in 5 min  and then drops slowly over 60 min  and becomes constant. The fluorescence signal excited by a weak test laser beam at the wavelength of 501.7 nm is proportional to the density of para-$I_2$ molecules.  The concentration of para-$I_2$ changes much more slowly than does that of the ortho-$I_2$.  Thus as the reaction cell is irradiated with 514.5 nm light, the excited ortho-$I_2$ molecules react with 2-hexene and the para-$I_2$ remains largely unreacted.

Halogenation of olefins such as 2-hexene in the gas phase is usually a free radical chain reaction (185, 204).  However, when a cell with a mixture of $I_2$ and 2-hexene is irradiated by the argon laser line at 488 nm, with a quantum energy greater than the dissociation limit, no distinct change in the concentration of $I_2$ is observed (126).  Thus the free I atoms are seen to be actually less reactive than the excited $I_2$ molecule.  One more fact confirms that the contribution of the

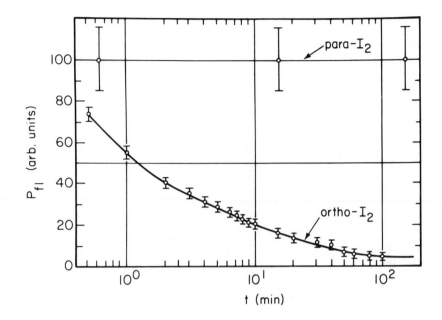

*Fig. 46. Time dependence of ortho-$I_2$ fluorescence excited by the intense line 514.5 nm, 2.2 W, and of para-$I_2$ excited by the weak probe line 501.7 nm of an argon laser for 0.2 torr $I_2$ in 2-torr of 2-hexene (126).*

radical-chain mechanism is very small. Experimentally, little or no para-$I_2$ is consumed. Thus the number of I atoms produced must be very small compared to the number of $I_2$ molecules excited, otherwise the rapid reaction

$$I + p\text{-}I_2 \rightleftarrows o\text{-}I_2 + I \qquad \text{(VII.4)}$$

would mix ortho and para molecules. The reaction is thus a direct reaction between $I_2^*$ and 2-hexene.

This reaction of excited molecular iodine and the ortho-para enrichment was first carried out in 1930 by Badger and Urmston using 546 nm mercury lamp excitation (205). Molecular iodine excited by the 185 nm mercury line is even more reactive and attacks saturated hydrocarbons and hydrogen (206, 207)

$$RH + I_2(185 \text{ nm}) \longrightarrow RI + HI. \qquad \text{(VII.5)}$$

These reactions do not occur with longer wavelength excitation and are not atomic radical reactions. Their quantum yields are a few percent. A reaction with a quantum yield of $4 \times 10^{-4}$ for visible radiation both above and below the dissociation

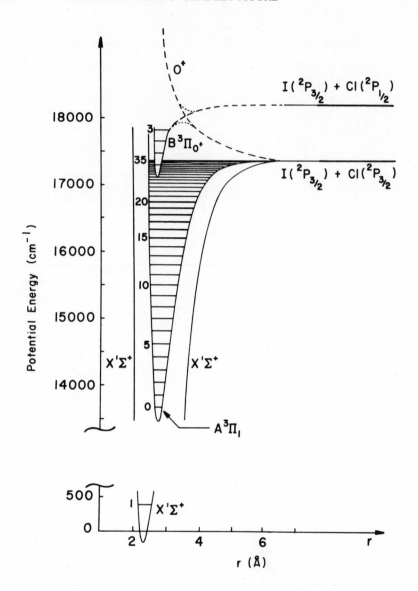

*Fig. 47. Potential energy curves for the lower energy states of ICl (209).*

limit has been reported for $I_2$- propane mixtures (208). This reaction is attributed to the formation of a propane-I complex followed by reaction with I to give the products propyl-iodide and HI. The double bond of hexene is much more reactive than the saturated hydrocarbons and it is plausible that the B-state of $I_2$ would react with it. The kinetics of the reaction

does not appear to be entirely straightforward.  The initial
reaction rate is reported as (126)

$$- \frac{d}{dt} [o\text{-}I_2] \Big|_{t=o} = k \, I_{las} \, [X]^{\frac{1}{2}} \, [I_2]^o \, . \qquad \text{(VII.6)}$$

The overall quantum yield is about $10^{-4}$.  A great deal remains
to be learned about the qualitative reaction chemistry of the
excited states of $I_2$.  A quantitative understanding of the
competition of reaction with quenching processes, of the intermediate states of reactants or complexes if any, of the relevant potential surfaces and of the detailed dynamics of these
reactions requires further study.

Photochemical separation of Cl isotopes has been achieved
by Zare and his coworkers using reactions of excited ICl with
halogenated olefins (65, 209, 210).  Figure 47 shows the potential curves for the lowest states of ICl.  The ICl spectrum
has been extensively studied (211).  The curve crossing near
18,000 $cm^{-1}$ makes this an excellent molecule for the selective
predissociation method described in Sec. V for $Br_2$.  In experiments (65, 209, 210) the $v' = 18$ level of the A state was
excited.  This level lies 784 $cm^{-1}$ below the limit for dissociation to ground state atoms and is thus a stable molecular
excited state.  $I^{37}Cl$ molecules were excited in the $v' = 18$
bandhead using a cw dye laser at 605.4 nm.  Insertion of an
absorber cell containing $I^{35}Cl$ into the laser cavity resulted
in a considerable enhancement of the excitation selectivity
by removing the $I^{35}Cl$ spectrum from the laser output (65).
The laser linewidth of 0.3 nm was much greater than the spacing between rotational lines in the ICl spectrum.  Comparable
selectivity was obtained without the absorber cell when the
laser was narrowed to 0.01 nm.

The overall photochemical reaction of ICl with BrHC=CHBr
yielded mainly ClHC=CHBr and a small amount of ClHC=CHCl (65,
209).

$$ICl(A,v'=18) + BrHC\text{=}CHBr \longrightarrow$$

$$ClHC\text{=}CHBr + IBr + ClHC\text{=}CHCl \qquad \text{(VII.7)}$$
$$(major) \qquad\qquad (minor)$$

The trans-ClHC=CHCl product was 85% $^{37}Cl$ corresponding to an
enrichment factor of 17 (65).  Milligram quantities of total
product were formed, then separated by gas chromatography and
analyzed mass spectroscopically.  Enrichment of the major product was not measured.  The photochemical reaction of ICl with
bromobenzene

$$ICl(A,v'=18) + Br\text{-}\underset{\substack{\parallel \\ \text{C-C}}}{\overset{\substack{H \quad H \\ \text{C=C}}}{C}}\text{C-H} \longrightarrow Cl\text{---}C_6H_5 + BrI \qquad \text{(VII.8)}$$

has been reported on in detail by Brenner, Datta and Zare (210). Figure 48 shows enrichment as a function of inverse ICl pressure for a constant 3.3 torr of bromobenzene. At high

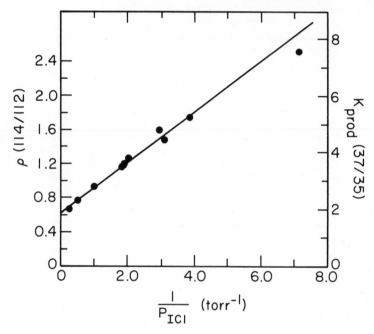

*Fig. 48.  Enrichment of Cl in ICl-bromobenzene mixtures. The left ordinate is the ratio of mass peaks for $^{37}ClC_6H_5$ to $^{35}ClC_6H_5$ in products.  The right ordinate is enrichment factor. The bromobenzene pressure is constant at 3.3 torr  (210).*

ICl pressures the enrichment approaches a limiting value of nearly 2.  It is not clear how much greater an enrichment than 8 would be possible at lower ICl or higher bromobenzene pressures.  The quantum yield of the reaction with bromobenzene at 0.5 torr ICl is about 0.10.

The nature of these photochemical reactions of ICl is not clear.  The simplest possibility and probably the most attractive for isotope separation schemes is the direct reaction

$$I^{37}Cl(A,v') + BrR \longrightarrow {}^{37}ClR + BrI \qquad \text{(VII.9a)}$$

$$\longrightarrow {}^{37}ClR + I + Br. \qquad \text{(VII.9b)}$$

It is also possible that electronic-to-vibration energy transfer

$$I^{37}Cl(A,v') + M \longrightarrow I^{37}Cl(X,v'') + M, \quad v''>>v' \qquad \text{(VII 10)}$$

is induced by collisions before reaction

$$I^{37}Cl(X,v'') + BrR \longrightarrow {}^{37}ClR + BrI \qquad \text{(VII.11a)}$$

$$\longrightarrow {}^{37}ClR + Br + I \qquad \text{(VII.11b)}$$

takes place.  Collision-induced dissociation must also be considered since one collision in fifty is sufficiently energetic
to cause dissociation followed by radical reaction.

$$I^{37}Cl(A,v'=18) + M \longrightarrow I + {}^{37}Cl + M \qquad \text{(VII.12)}$$

$${}^{37}Cl + BrR \longrightarrow {}^{37}ClR + Br \qquad \text{(VII.13)}$$

$${}^{37}Cl + ICl \longrightarrow I^{37}Cl + Cl \qquad \text{(VII.14)}$$

Brenner et al. (210) conclude from their detailed chemical
studies that the free radical and one or both of the molecular
mechanisms is important.
Recently the reaction

$$ICl(A,v') + H_2 \longrightarrow HCl + HI \qquad \text{(VII.15)}$$

has been found by Harris (213) to have a quantum yield of a
few tenths of a percent.  Quantum yield measurements as a
function of wavelength show that electronically excited ICl
reacts directly;  free Cl atoms and vibrationally excited
ground state molecules are not involved.  Also Stuke and
Schäfer (212) have shown that $I^{37}Cl(A,v')$ adds to acetylene
to give $IHC=CH^{37}Cl$ with an enrichment of 48.  This process
has substantial commercial promise.

## C.   Polyatomic Molecules

Cl enrichment has been achieved by selective excitation
of $Cl_2CS$ in mixtures with diethoxyethylene (214).  Mass
spectral analysis of the remaining $Cl_2CS$ after irradiation with
either 465.78 nm $Ar^+$ or dye laser light showed the concentration of $^{35}Cl$ altered from its natural abundance of 75% to 64%
or 80% depending on the isotopic species initially excited.
The photochemical quantum yield in this system was shown to be
as large as 0.5.  The additional chemical complexity of this
system compared to the ICl-bromobenzene system, makes it less

promising for practical Cl separations. The early experiments (8) on Cl enrichment in phosgene suggest that phosgene may be an interesting molecule both for photochemistry and predissociation.

Photochemical isotope separation should be possible for a wide variety of elements. Many of the problems and constraints of the method as well as its virtues are similar to those discussed for predissociation (Sec. V). It is too early to judge the real promise of photochemical methods for isotope separation. Photochemistry already gives us the best practical separations of Hg and Cl. It is clear from the research described in this section that our fundamental knowledge of the photochemistry of simple molecules is inadequate. The search for a scavenger which reacts in such a way as to preserve the selectivity of excitation and to prevent all isotopic scrambling processes is generally a project in fundamental photochemical research. For the photochemist, of course, this is an inviting challenge. Isotopically selective excitation is a powerful new tool for studying the mechanisms of photochemical reactions. Work in this field will surely produce a great deal of new and valuable basic photochemistry as well as some practical isotope separation processes.


VIII. VIBRATIONAL PHOTOCHEMISTRY


The rate of a chemical reaction may be substantially enhanced by vibrational excitation of the reactant molecules. Gibert (215) suggested this as a method for isotope separation in 1963. Gürs (216) later hypothesized that by using an infrared laser source uranium separation might be achieved. In a chemical reaction a bond in one of the reactant molecules is usually broken and a new bond formed to make products. It is thus natural to expect that vibrational excitation of the bond to be broken will cause the reaction to proceed more rapidly. If a large fraction of the energy of an exothermic reaction appears as vibrational excitation of the products, then the rate of the reverse, endothermic reaction, will be greatly increased by vibrational excitation (217 - 220). The dependence of reaction rates on vibrational excitation for atom-diatom reactions has been observed and analyzed in detail for several systems during the last few years (217 - 230). For polyatomic molecules many very interesting laser-driven chemical reactions have been reported (18, 172, 182, 231 - 255). In some, the reactions have been clearly proven to result from non-equilibrium excitation of vibrations. In all cases it has been either very difficult, or impossible, to unravel the complicated competition between laser excitation, vibration-

vibration energy transfer, vibrational relaxation, and chemical reaction. Thus very few rate constants have been determined for known, single quantum states of vibrationally excited polyatomics. Many more references may be found in the papers cited above and in reviews (256, 257).

In this section we will discuss the important general features of the method for a very simple reaction scheme. Several possible methods of excitation are described. The kinetic scheme and practical limitations on isotope separation for the reaction $HCl(v = 2) + Br$ are discussed in detail. Other examples of atom-diatom reactions as well as several more complicated systems are discussed.

## A.    Reaction Kinetic Scheme

A simple reaction scheme for isotopically selective photochemistry is shown.

$$AB + \hbar\omega \longrightarrow A^iB^\dagger \tag{VIII.1}$$

$$A^iB^\dagger + C \xrightarrow{k_L} A + {}^iBC \tag{VIII.2}$$

$$A^iB^\dagger + M \xrightarrow{k_Q} A^iB + M \tag{VIII.3}$$

$$A^iB^\dagger + AB \xrightarrow{k_{VV}} A^iB + AB^\dagger \tag{VIII.4}$$

$$AB + C \xrightarrow{k_T} A + BC \tag{VIII.5}$$

The laser photons, $\hbar\omega$, selectively excite a single isotope of AB to a particular vibrational level. These molecules react, (VIII.2), with reagent C. The laser-enhanced rate constant, $k_L$, applies to reaction of a particular vibrational level. Usually the rate constant will be the same for the $AB^\dagger$ produced in (VIII.4). Two relaxation processes, (VIII.3) and (VIII.4), compete with reaction (VIII.2). In (VIII.3), $A^iB^\dagger$ loses some or all of its vibrational excitation and no longer reacts with C at a rate equal or comparable to $k_L$. This decreases the quantum efficiency of the overall process. The collision partners, M, include the reagents AB and C as well as the products A and BC and any other species present in the system. When $A^iB^\dagger$ collides with AB, the AB molecule may be excited to the reactive state $AB^\dagger$, (VIII.4), and subsequently react by (VIII.2). The result is isotopic scrambling. This may be minimized by increases in the concentration of C relative to AB.

The thermal reaction between AB and C, (VIII.5), will

produce products BC as long as the reagents are mixed.  The
thermal population of $AB^\dagger$, with vibrational energy E, is given
by a Boltzmann factor and thus

$$k_L/k_T \leq exp(E/kT) \qquad \qquad (VIII.6)$$

To the extent that other thermally populated states of AB react $k_L/k_T$ will be smaller.  Or put another way, in the very
probable case that the vibrational excitation is not completely effective in reducing the activation energy, $k_L/k_T$ will be
less than exp(E/kT).  For the reaction

$$NO + O_3 \longrightarrow NO_2(^2B_1) + O_2 \qquad \qquad (VIII.7)$$

$k_L/k_T \simeq 6$ for NO(v = 1) where exp(E/kT) = 8600 (236).  When
$O_3$ is excited by a $CO_2$ laser $k_L/k_T \simeq 5$ while exp(E/kT) = 150
(233, 234).  In both cases the reaction is accelerated by
much less than the upper limit of Eq. (VIII.6).  The nonselective product, BC, reaches a concentration $k_T$ [AB] [C]$t_{process}$,
where $t_{process}$ is the total time during which the
reagents are mixed.  In many situations (258) the value of
$k_L/k_T$ required is proportional to the separation factor needed
and to $t_{process}$.  This is a serious limitation on infrared
photochemical schemes.  If the excitation is by a single
photon of $CO_2$ laser light near 1000 cm$^{-1}$, modest separations
would be possible only if $t_{process}$ were in the microsecond
range.  If the reagents are to remain mixed for many seconds,
several thousand cm$^{-1}$ excitation energy is required.  For many
systems it will be necessary to reach overtone and combination
levels.  Theoretical analysis of the simple kinetic scheme
above and of more complete schemes may be found in references
(255 - 267).


B.    Excitation Processes

A wide variety of excitation processes may be used.  The
infrared active fundamental vibrations of a molecule may be
excited by absorption of a single photon.  (See Sec. IV for a
discussion of thermal effects and the rotational energy transfer bottleneck).  Absorption coefficients are usually between
$10^{-2}$ and 10 cm$^{-1}$ torr$^{-1}$.  Excitation of combination and overtone bands gives two or more quanta of vibrational excitation
on absorption of a single higher energy photon.  This may give
larger values of $k_L/k_T$ and may greatly increase the ratio
$k_L/k_{VV}$ (see below).  The disadvantage is that the optical absorption cross sections decrease by nearly two orders of
magnitude for each additional quantum of vibrational excitation.  By placing the sample within the cavity of a laser

tuned to an absorption line center this disadvantage is greatly reduced. Berry and coworkers were able to pump HCl to v = 6 using a cw dye laser (254).

Higher vibrational levels may also be reached by stepwise excitation through one or more intermediate levels (228, 268). This is particularly convenient when the molecule to be excited may be made to lase. Arnoldi, Kaufmann and Wolfrum (228) used an HCl laser to excite $H^{35}Cl$ from v = 0 to v = 1 and then to v = 2. If rotational relaxation is fast compared to the pulse length and if the laser energy approaches saturation, then a substantial. fraction of the molecules which reach v = 1 will be pumped to v = 2. The scheme is clearly more practical than use of an optical parametric oscillator to pump v = 0 → v = 2 directly. A large-scale, fastflow separator might utilize high-power cw lasers.

The selective multiphoton excitation of vibrational levels (Sec. VI) is demonstrated by direct observation of the excitation of high vibrational levels at comparatively moderate intensities ($10^6$ - $10^7$ W/cm$^2$) (111, 168, 169). This is probably, the only effective method for direct excitation of levels with several eV of vibrational energy. Though the method is used now mainly for direct molecular dissociation, it can be applied, no doubt, with success to selective vibrational photochemistry of highly excited molecules (265). The three- or four-photon resonances suggested by theories of multiphoton excitation and by two-frequency experiments (Sec. VI and Chapter II, Secs. III and IV) suggest the possibility of reaching specific high levels of excitation with intensities in the $10^4$ - $10^6$ W/cm$^2$ intensity range.

Raman excitation may also be used (232, 244, 245, 269, 270). It is the only method for excitation of vibrations with zero transition dipole (e.g. homonuclear diatomics). Two lasers tuned such that their frequency difference is equal to the vibration-rotation frequency can strongly excite gases when powers approach $10^8$ W/cm$^2$ (269). Unfortunately, the overall conversion of laser energy to molecular excitation is not efficient. Raman excitation of $N_2$ has led to isotope enrichment (244, 245).

Vibration-vibration energy transfer processes may occasionally be used to advantage. The fundamental level of a molecule may be excited, and subsequent V → V transfers will populate higher levels. Heavy isotopic species are preferentially excited as described in Sec. X.A, on non-interchangeable methods of isotope separation. Under most circumstances the effect of V → V transfers Eq. (VIII.4) will be to destroy the isotopic selectivity provided by the laser rather than to create a selectivity not provided by the laser.

Energy transfer from vibration into translation and rotation destroys all of the selectivity of the laser excitation, both with regard to isotope and chemical species. Even though this excitation is completely thermal, reaction products may be quite different from those obtained by heating the vessel walls. This is particularly true when a cw laser beam of a few mm diameter passes through a cell of a few cm diameter containing pressures of a few torr or greater. Many experiments have been performed under such conditions (18, 241 - 243, 255, 261, 271 - 274). In such a case the gas in contact with the wall remains cool while the center may be heated above 1000°K (255, 271 - 274). The rates of homogeneous gas phase reactions are accelerated while wall-catalyzed reactions are unaffected. When the pressure is lowered sufficiently non-thermal reactions have been observed by Bauer, Manuccia and Chien and modest isotope selectivity has been demonstrated (238, 239, 251). Significantly non-thermal reactions have been demonstrated by Dever and Grunwald (253) using submicrosecond pulses at pressures of 60 torr. V → V transfer is undoubtedly important here (266, 267) and isotope selectivity except for H vs. D (172) is highly unlikely.

C.  <u>Isotope Enrichment - Br + HCl(v = 2)</u>

The process

$$H^i Cl(v = 2) + Br \rightarrow HBr(v = 0) + {}^i Cl \qquad \text{(VIII.8)}$$

$$^i Cl + Br_2 \rightarrow {}^i ClBr + Br \qquad \text{(VIII.9)}$$

provides an excellent example of isotopic separation kinetics (4, 227 - 229). Competing processes of interest include:

$$H^i Cl(v = 2) + Br \rightarrow H^i Cl(v = 1) + Br \qquad \text{(VIII.10)}$$

$$\rightarrow H^i Cl(v = 0) + Br \qquad \text{(VIII.11)}$$

$$H^i Cl(v = 2) + HCl(v = 0) \rightarrow H^i Cl(v = 1)$$
$$+ HCl(v = 1) - 102 \, cm^{-1} \qquad \text{(VIII.12)}$$

$$H^i Cl(v = 2) + M \rightarrow H^i Cl(v = 1, 0) + M \qquad \text{(VIII.13)}$$

$$HCl(v = 1) + M \rightarrow HCl(v = 0) + M \qquad \text{(VIII.14)}$$

$$H^i Cl(v = 2) + HCl(v = 0) \rightarrow H^i Cl(v = 0) + HCl(v = 2) \qquad \text{(VIII.15)}$$

$$H^i Cl\,(v = 1)\, +\, HCl(v = 0)\, \rightarrow\, H^i Cl\,(v = 0)\, +\, HCl(v = 1) \quad \text{(VIII.16)}$$

$$HCl(thermal)\, +\, Br\, \rightarrow\, HBr\, +\, Cl \qquad\qquad \text{(VIII.17)}$$

$$HCl\,(v = 1)\, +\, Br\, \rightarrow\, HBr\, +\, Cl \qquad\qquad \text{(VIII.18)}$$

Rates have been measured for many of these processes (227, 229). The rate constant for removal of HCl(v = 2) by Br is given by

$$k_8\, +\, k_{10}\, +\, k_{11}\, =\, 1.8 \times 10^{-12}\ cm^3\ molecule^{-1}\ sec^{-1}$$

Of this total approximately 70% is (VIII.10). The remaining 30% is most probably the desired reaction (VIII.8). Thus 30% is an absolute upper limit on the quantum yield. The quantum yield is further decreased by processes (VIII.12) and (VIII.13). The rate of (VIII.12) places an upper limit on useful values of the ratio of HCl to Br concentration. Since

$$k_{12}\, =\, 3.3 \times 10^{-12}\ cm^3\ molecule^{-1}\ sec^{-1},$$

Br pressures should be at least double the HCl pressure. For (VIII.13) and (VIII.14) the most important collision partner is $Br_2$;

$$k_{14}^{Br_2}\, =\, 3.3 \times 10^{-14}\ cm^3\ molecule^{-1}\ sec^{-1}.$$

and $k_{13}^{Br_2}$ is probably about twice this value. Br concentrations should therefore be greater than 2% of $Br_2$. The rate of the isotope scrambling reaction (VIII.15) has not been measured. Since it involves the exchange of two vibrational quanta in one collision, it will probably be at least one order of magnitude slower than (VIII.16) for which

$$k_{16}\, =\, 1.9 \times 10^{-11}\ cm^3\ molecule^{-1}\ sec^{-1}$$

Separation schemes using overtone or combination levels usually have the advantage illustrated by HCl(v = 2) + Br that the single-quantum V → V process (VIII.12) does not cause isotopic scrambling. Reaction (VIII.18) is highly endothermic; thus the reaction of HCl(v = 1) may be neglected. A separation scheme using HCl(v = 1) would require a coreactant with a rate constant comparable to $k_{16}$ or about one tenth gas-kinetic. However, for the HCl(v = 2) + Br system when the concentration of v = 0 and v = 1 are comparable, energy transfer by (VIII.16) followed by the reverse of (VIII.12) gives natural HCl(v = 2) which then reacts by (VIII.8) to scramble the separation.

The thermal rate $k_{17}$ may be calculated from the measured rate of Cl + HBr (275) and the equilibrium constant:

$$k_{17} = 3 \times 10^{-23} \; cm^3 \; molecule^{-1} \; sec^{-1}.$$

Thus $k_L/k_T = k_8/k_{17}$ is greater than $10^{10}$.

Arnoldi, Kaufmann and Wolfrum (228) report separation factors greater than two for this reaction system, Fig. 49.

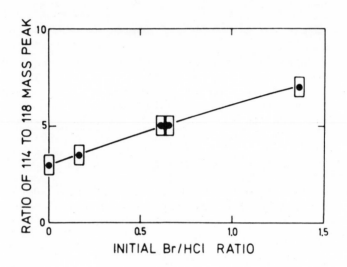

*Fig. 49. Ratio of mass peaks for product $^{79}Br^{35}Cl$ to $^{81}Br^{37}Cl$ for the reaction sequence (VIII.8) and (VIII.9) with $H^{35}Cl$ selectively excited. The natural abundance ratio is 3 (228).*

They used an HCl pulsed chemical laser to sequentially excite from v = 0 to v = 1 to v = 2. For equal pressures of Br and HCl a separation factor of 2 was observed. This sets a limit for $k_{15}$

$$k_{15} \leq k_8 + k_{10} + k_{11} + k_{12}$$

$$\leq 5 \times 10^{-12} \; cm^3 \; molecule^{-1} \; sec^{-1}$$

The efficiency of the processes with respect to the infrared laser is limited by the ratio $k_8/(k_{10} + k_{11})$, by the ratio $k_8(Br)/(k_{12} + k_{15})(HCl)$ and by the molecules excited only to v = 1 and not on to v = 2. For good separation the Br atom pressure should be higher than the HCl pressure. Effective use of this reagent requires excitation of a substantial

fraction of the desired isotope.  An upper limit on total
reagent pressures will be set by

$$Br + Br + M \rightarrow Br_2 + M \qquad \text{(VIII.19)}$$

In many cases it will be useful if the laser pulse duration is
long compared to the chemical lifetime of HCl(v = 2), $[(k_8 + k_{10} + k_{11})(Br)]^{-1}$.

To complete a useful separation the reagents and products
must be physically separated at a low cost before the system
returns to thermodynamic equilibrium.  The scavenging reaction
(VIII.9),

$$k_9 = 1.2 \times 10^{-10} \ cm^3 \ molecule^{-1} \ sec^{-1} \qquad \text{(VIII.20)}$$

is so fast that there is no problem with competing processes
such as:

$$^{i}Cl + HCl \rightarrow H^{i}Cl + Cl \qquad \text{(VIII.21)}$$

$$^{i}Cl + HBr \rightarrow H^{i}Cl + Br \qquad \text{(VIII.22)}$$

On the longer timescales, $10^{-2}$ to $10^2$ sec, of distillations or
selective condensation from the reaction mixture, care must be
taken to prevent thermal reaction (VIII.17) and both homogeneous and wall-catalyzed reversal of the separation by
processes such as

$$^{i}ClBr + HBr \rightarrow HCl + Br_2. \qquad \text{(VIII.23)}$$

It is likely that the attempt (229) to isolate enriched $^{37}Cl$
using this system failed because of (VIII.23).  It should be
noted that surface catalysis often gives changes in activation
energy much larger than can be induced by single infrared
photons.

A modest enrichment of Br has recently been reported by
Badcock, Hwang and Kalsch (276) using the reaction

$$H^{81}Br^{\dagger} + I \rightarrow HI + {}^{81}Br. \qquad \text{(VIII.24)}$$


D.   Isotope Enrichment - Polyatomic Reactions

Experiments on the non-thermal chemistry resulting from
pulsed laser excitation of polyatomics to high vibrational
levels have produced many exciting results and much controversy.  The work on multiphoton dissociation in the absence
of collisions (Sec. VI and Chapt. II) is beginning to be
understood.  The collisions of excited molecules, both during

and after the laser pulse, have produced isotopic enrichments and unexpected chemistry.

In 1972 Lyman and Jensen (231) dissociated $N_2F_4$ (VIII.25) with a $CO_2$ TEA laser at a rate substantially exceeding the rate for thermalized excitation.

$$N_2F_4 \rightarrow 2\ NF_2 \qquad\qquad \text{(VIII.25)}$$

Laser-induced explosions of $N_2F_4$ or $SF_6$ with $H_2$ exhibit shorter induction times than expected for thermal heating (246). The pressures used in these systems were one or two orders of magnitude greater than those for which isotope selectivity via multiphoton dissociation (Sec. VI) was later found (25, 26, 170). The fascinating inorganic syntheses and separations (182, 250), the multitude of vibrationally enhanced reactions and laser driven thermal reactions and the controversy concerning their interpretation is beyond the scope of this chapter.

Isotopic enrichments of boron have been reported by Freund and Ritter (265) using a TEA $CO_2$ laser to excite a mixture of 2 torr $BCl_3$ with varying amounts of $H_2S$ and $D_2S$, Fig. 50. Several hours of irradiation at an $^{11}BCl_3$ frequency enriched $^{10}B$ in the residual material by a factor of 1.7.

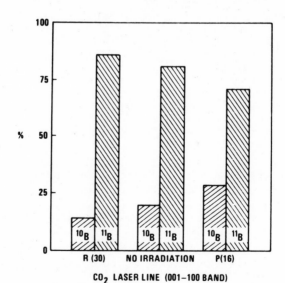

Fig. 50. Optimum isotopic abundances in residual $BCl_3$ for $BCl_3$ + $H_2S$ reaction. Data are for excitation of $^{10}BCl_3$ by the R(30) and of $^{11}BCl_3$ by the P(16) lines of the 10 μ band of a TEA $CO_2$ laser (265).

Irradiation at a $^{10}BCl_3$ frequency produced an $^{11}B$ enrichment of 1.4. The reaction mechanism is rather complicated (265, 277) and involves the absorption of more than one photon. Pulses longer than 1 µsec and cw irradiation are not effective. Under the conditions of the experiment it seems most likely that reactions of vibrationally excited molecules are involved. However, collision-induced dissociation cannot be ruled out.

At sufficiently low densities, a few mtorr, isotopic enrichment has been observed using cw $CO_2$ lasers (238, 239, 251). Highly non-thermal vibrational distributions are achieved by causing vibrational relaxation to occur on the container walls or by V → T transfer to a large excess of buffer gas. Scrambling due to V-V transfer, Eq. (VIII.4), is avoided by making relaxation, Eq. (VIII.3), much faster. Thus isotopic selectivity is preserved. The intensity of the laser must be large in order to keep $A^iB^\dagger$ large. In this way a significant amount of selected product from (VIII.2) can be produced in competition with processes (VIII.4) and (VIII.5). The quantum efficiency of the method is poor. However, it works in the most commonly found situation where the laser-enhanced reaction rate constant is much smaller than the V → V transfer rate constant. Relatively small enrichments of D and B have been reported by Bauer and Chien (239) in the thermal decomposition of $H_3BPF_3$ and $D_3BPF_3$. The quantum efficiency was about $2 \times 10^{-5}$ and the product yield approximately cubic in laser power up to 100 W. Mannucia, Clark and Lory (251) have found enrichments of 3% for $^{79}Br$ and 2% for $^{81}Br$ using the reaction

$$CH_3{}^iBr(\nu_6) + Cl \rightarrow HCl + CH_2{}^iBr \qquad \text{(VIII.26)}$$

$$CH_2{}^iBr + Cl_2 \rightarrow CH_2{}^iBrCl + Cl. \qquad \text{(VIII.27)}$$

In their experiments a few mTorr of $CH_3Br$ flowing with a few torr of Ar and $Cl_2$ was irradiated at 100 W/cm². Cl was produced by an electrical discharge. The purely thermal reaction which occurs with cw irradiation at pressures of tens of torr has been analyzed by Eletskii, Klimov and Legasov (255) for

$$PF_5 + SO_2 \rightarrow POF_3 + SOF_2. \qquad \text{(VIII.28)}$$

The reaction of methanol with bromine pumped by an HF laser has been observed under similar high pressure conditions (18). The experimental conditions appear most unfavorable for the deuterium enrichment which was reported. There have been unsuccessful attempts to reproduce this work (278). Recently, the laser-driven reaction of $UF_6$ with HCl has been reported by Eerkens (279).

The attractive feature of vibrational photochemistry for isotope separation is the promise of using low energy IR photons from an efficient molecular laser to get a good yield of product. Since 1 mole of photons at 3,000 cm$^{-1}$ is 10$^{-2}$ kWh and some IR lasers are 10% efficient, processing of bulk chemicals might even be economic. There are many problems to overcome: thermal reaction rates, vibrational energy transfer, surface catalyzed reactions, fixed frequency sources. Perhaps the greatest problem is our meager knowledge of the effect of vibrational excitation on the rates of chemical reactions.

## IX.  SELECTIVE DEFLECTION OF ATOMS AND MOLECULES

Changes in the magnitude and direction of the velocity of atoms (molecules) of a particular isotopic composition can be used in isotope separation. Isotopically selective changes in the velocity of atoms (molecules) may occur either under resonance light pressure alone or in combination with the action of external fields on the selectively excited atoms (molecules). Both approaches are possible for isotope separation under laboratory conditions. Further, it is possible to act on atoms (molecules) both in a beam geometry and in a bulk gas. In the first case we deal with isotopically selective deflection of particles and in the second one with isotopically selective pressure on one gas component. In any case the change in direction of velocity can be the most efficient elementary act for the separation of particles as far as energy is concerned. Therefore all the methods considered below have the common title:  "selective deflection of atoms and molecules."

## A.    Separation by Light Pressure

It is known that an atom or a molecule (for short, we say "an atom" below) in a gas in a light field is acted upon by light pressure (280). The resulting force causes the atom to recoil as it scatters or absorbs photons (281). In a resonance light field the atom is acted upon by forces of two types:  the force of spontaneous light pressure and that of induced light pressure (see, for example, (282)). The force of spontaneous light pressure manifests itself when the atom absorbs photons from a collimated light beam of a running light wave and then spontaneously emits photons isotropically as a spherical light wave.
    Assume that an atom absorbs the resonant radiation from a collimated light beam and then reemits it isotropically in

any direction.  Then in the light beam direction it is acted
upon by the force:

$$F_{sp} = \frac{I\sigma}{c} \left( |a_1|^2 - |a_2|^2 \right)$$    (IX.1)

where I is the beam intensity, $\sigma$ is the resonant absorption
cross section, $|a_i|^2$ is the probability of atomic occupation
for the i-th level of a transition resonant with the field.
The first term corresponds, evidently, to photon absorption
from the running wave and the second one to stimulated photon
emission into the same running wave.  In a strongly saturating
field, when the time averaged $|a_i|^2 \simeq 1/2$, stimulated emission,
because of the opposite sign for its recoil momentum, almost
precisely compensates for the force due to absorption.  The
difference

$$\left( |a_1|^2 - |a_2|^2 \right) \longrightarrow \left( |a_1|^2_0 - |a_2|^2_0 \right) \left( 1 + \frac{I}{I_{sat}} \right)^{-1}$$

where $I_{sat} = \hbar\omega\gamma/2\sigma$ is the saturation intensity of the transition which depends on $\gamma$ the rate of particle relaxation from
excited state 2 to ground state 1.  So, with $I \gg I_{sat}$ and
$|a_1|^2 = 1/2$ the force acting on an atom tends to the constant:

$$F_{sp}^{max} = \hbar\omega\gamma/2c$$    (IX.2)

The physical meaning of expression (IX.2) is very simple.
During each photon reemission the atom acquires a momentum
with $\hbar\omega/c$.  The number of such reemissions is determined by
the $\gamma$ rate of relaxation to the ground state, and the multi-
plier 1/2 appears due to the fact that the atom spends one
half its time in ground state 1.

If an atom interacts with a light beam of diameter  a,
during $\tau_0 = a/v_0$, it acquires the velocity $\Delta v$ in the light
beam direction and is deflected by the angle $\Delta\phi$, Fig. 51(b):

$$\Delta\phi = \frac{\Delta v}{v_0} = \frac{\gamma a}{2c} \frac{\hbar\omega}{Mv_0^2} \simeq \frac{\gamma a}{2c} \frac{\hbar\omega}{kT_0}$$    (IX.3)

where $v_0$ is the average atomic velocity in a beam determined
by the temperature $T_0$ of the atomic beam source.  For allowed
transitions with $\gamma = 10^8$ sec$^{-1}$, a = 1 cm, $T_0 = 10^3$ °K and
$\hbar\omega_0 = 2$ eV, an atom in a light beam can execute $10^3 - 10^4$
photon reemissions and be deflected by the angle $\Delta\phi \simeq 0.04$ rad;
that is, quite observable.  It is evident that, for observation to be possible, the divergence angle of the atomic beam
$\psi_0$ should not exceed the angle of deflection $\Delta\phi$.

The force of spontaneous light pressure was experimentally studied (281, 283) for photodeflection of Na atoms by a

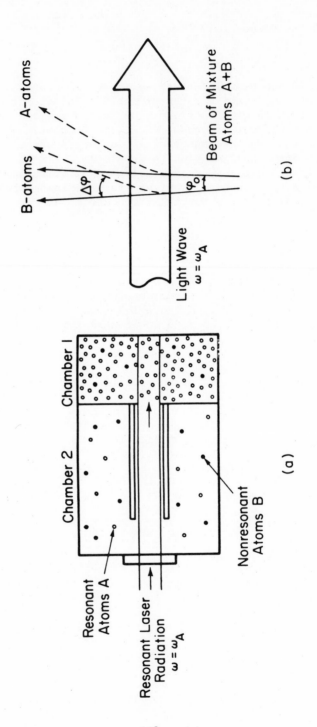

Fig. 51. Isotope separation by selective light pressure: a particle mixture is acted upon in a gas (a) and in a beam (b).

very small angle caused by absorption of single photons from an Na resonance lamp by a well-collimated atomic beam. With the advent of tunable lasers this effect became not only more easily observable but also the subject of discussion for its application to isotope separation on a practical scale. It is probable that the first suggestion for the use of light pressure in selective pumping of gases, including those of a particular isotopic composition, was put forward in work (284). The separation scheme suggested in this work employs not only atomic beam deflection but also the effect of selective light pressure in low-pressure gases in a two-chamber vessel, Fig.51 (a). Laser light acts as an optical pump pushing forward resonant atoms (circles) into chamber 1 without acting on non-resonant atoms (points). As a result, the concentration of resonant atoms appears to be somewhat higher in chamber 1 than in chamber 2. Later isotope separation by the use of light pressure was suggested both for beam geometry (285, 286) and gas mixtures (285).

The first successful experiment on isotope separation by means of atomic photodeflection was done with a beam of barium atoms (287). The resonant transition $6s^2 \, ^1S_0$ - $6s6p \, ^1P_1$ at 553.5 nm was used; its isotopic hyperfine structure is extremely small (288). The addition of neutrons to the filled neutron shell (in the sequence of isotopes from $^{134}Ba$ to $^{138}Ba$) with the magic number $N = 82$ changes the nuclear charge distribution very little and so the isotope shift is extremely small. The hyperfine structure components for the line of BaI are divided by an interval of no more than 30 MHz, which is much smaller than the Doppler width of $10^3$ MHz. Therefore, isotopically selective excitation requires an atomic beam collimated to about 1/200. The lifetime of the excited state $^1P_1$ is 8.4 nsec; thus during the time required to pass through the light beam, $\tau_0 \simeq 10^{-4}$ sec, an atom might reradiate photons $10^4$ times. Actually this does not occur because Ba atoms have a high probability for radiative transition from the excited state $^1P_1$ to the lower metastable state $^1D_2$. The ratio between probabilities of return to the ground vs metastable states equals 60, and so the maximum increase of transverse velocity due to light pressure is equal to $60(\hbar\omega/Mc)$ = 45 cm/sec. Such selective deflection of Ba atoms during selective excitation of a particular isotope has been observed through mass spectral analysis of deflected atoms. Figure 52 presents some experimental results (287). $^{138}Ba$ (71.66%) is the main isotope of Ba in the natural mixture. During selective excitation of $^{137}Ba$ one can observe an increase of its mass peak. Excitation of a component of the hyperfine structure of $^{136}Ba$ gave an increase in the mass peak of $^{135}Ba$ as well which indicates an overlapping of two spectral components.

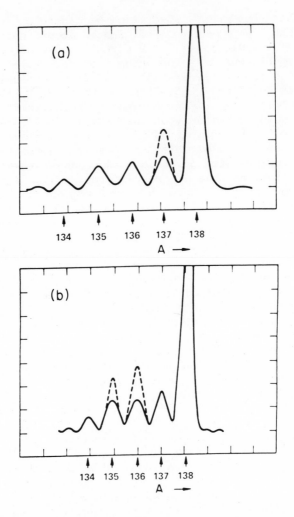

Fig. 52.  The mass spectrum of the natural mixture of Ba
isotopes (solid lines) and photodeflected Ba atoms (dashed
lines) during resonant action of laser radiation on the hyperfine structure components of the $^{137}$Ba isotope (a) and $^{135}$Ba
and $^{136}$Ba atoms (b) (287).

An experiment (289) has been carried out recently in
which the force of spontaneous light pressure acting on an Na
atom in the gas from the running wave of a tunable laser induces directed motion of the gas along the cell.  This can be
considered as a first step towards experiments on laser isotope separation in the non-beam geometry drawn in Fig. 51(a).
As far as the consumption of light energy and selectivity are

concerned, this method is less efficient than the beam
approach since the atoms in a gas move in a random way and
when colliding involve other, unexcited, atoms.

An apparent disadvantage of the photodeflection method
is that an energy $\hbar\omega$ should be consumed to deflect an atom by
a small angle. The atom's energy is increased by an almost
negligible amount equal to the recoil energy. There is a way
of increasing the force of light pressure with no expenditure
of additional light energy and of getting rid of the limitation connected with a fixed rate   of atomic relaxation to the
initial state. For this purpose it is necessary that the atom
should be brought back to the initial state by stimulated
emission into the counter-running light wave. In this case
the force of <u>induced light pressure</u> appears which, similarly
to (IX.1), is expressed by:

$$F_{ind} = \frac{I_0}{c} \left( |a_1|^2 + |a_2|^2 \right) \qquad (IX.4)$$

The maximum of this force is determined by the rate of induced atomic transitions between levels $W_{ind}$:

$$F_{ind} = \frac{\hbar\omega}{2c} W_{ind} \qquad (IX.5)$$

Since the condition $W_{ind} \gg \gamma$ can always be met in a strong
field, the force of induced light pressure may be much greater
than that of spontaneous light pressure. This fact is of
importance for transitions with a low rate of radiative relaxation.

Work (290) offers a particular case to realize the force
of induced light pressure for isotope separation. In this
work atomic excitation and its return to the ground state are
done by the method of fast adiabatic passage under the action
of two counter-running light pulses with variable frequency.
The first pulse whose frequency scanning (chirping) rate
through the atomic absorption line complies with the condition
(291, 292):

$$\left| \frac{d\omega}{dt} \right| \ll \left( \frac{\mu E}{\hbar} \right)^2 \qquad (IX.6)$$

transfers the atom into the excited state with transfer of
momentum $\hbar\omega/c$, and the counter-running pulse carries the atom
back to the ground state in the same stimulated (not spontaneous) way with transfer of the same momentum. It is assumed
that a considerable savings in energy can be obtained provided
the conditions for using a considerable portion of the pulse
energy when propagating along a sheet atomic beam are met
(290). Some other methods employing the force of induced
light pressure in a standing wave for isotope separation are

considered in works (293).

The authors of many works (287, 290, 293) are optimistic about energy consumption in the method of photodeflection for isotope separation as well as about the development of the method for isotope production on a commercial scale. This opinion, however, should be considered with care since the method of light photodeflection demands that matter should be prepared in a very specific state, that is, in the form of a well-collimated atomic beam in high vacuum; in addition high-quality coherent light with controlled parameters should be used. The main thing in this case, of course, will not be expenditure of energy but the price of the equipment for producing atomic beams and highly coherent laser light.

B.    Deflection of Excited Atoms and Molecules by External Fields

Excited atoms and molecules, their initial and final quantum states properly chosen, may have different energies of interaction with external nonuniform electric or magnetic fields owing to the difference in polarizability, magnetic moments, wave function symmetry, etc. It is obvious that, in combination with isotopically selective excitation of such quantum states, these effects may also be applied to isotope separation. Such an approach to laser isotope separation is similar to the classical electromagnetic technique where the effect of selectivity is realized by virtue of the difference in interaction of ions with a field. Although some advantage is gained by not ionizing the material, the limitations of the method of electromagnetic isotope separation (low output, high energy consumption, high chamber vacuum, etc.) will be typical of these laser techniques as well. On the other hand, the methods discussed below have a number of advantages over some other laser methods (short time of separation cycle, directed motion of separated isotopes, the absence of electric charge, etc.) which can make these methods useful in laboratory experiments on radioactive and fast-decaying isotopes, for example.

The possibility of deflection of vibrationally excited molecules in an external electric field has been discussed (294), but the choice of the $SF_6$ molecule was poor (295). The $NH_3$ molecule has been proposed (295). This possibility has not been experimentally studied yet. This method will not be effective for molecules, especially for polyatomic ones, unless their thermal distribution over vibrational and rotational levels is first collapsed by expansion cooling to a very low temperature (56, 57). The use of atomic beams is more promising from this point of view.

To selectively deflect excited atoms we may use an electric field. This is very convenient for highly excited atoms (296). It is known that highly excited atoms may have polarizabilities as much as $10^6$ - $10^7$ times higher than those of atoms in the ground or low-lying states (297). Their polarizability increases as $n^7$ where n is the principal quantum number, and hence reaches huge values for states with n > 10. Therefore highly excited atoms, especially those with a large angular momentum, have a very high energy of interaction with an electric field too weak to cause autoionization (Sec. III). The radiative lifetime of highly excited states is rather long, $10^{-6}$ sec for 10d Na (298), which makes it possible to deflect them by a considerable angle, say, greater than 10° with a comparatively short distance of interaction with a nonuniform electric field. Thus, doing isotopically selective multistep excitation of highly excited atomic states, one can collect a specified isotope by the use of an external electric field. This method is of interest for those cases when the isotopes must be produced as neutral atoms. Short flight paths of atoms for separation (shorter than 10 cm) make the method promising for short-lived isotopes, $\tau \leq 10^{-4}$ sec.

X.    NON-INTERCHANGEABLE METHODS OF LASER ISOTOPE SEPARATION

So far we have considered the methods of isotope separation for which selective excitation of a chosen isotope in the mixture is the key feature. In all these methods it is possible to separate any isotope provided we can excite it selectively. It is this approach that we always mean when speaking of laser isotope separations which employ the unique properties of laser light. As interest in laser isotope separation increased, laser modifications of various methods were also considered. Known methods based on the isotope effect in chemical reaction rates were considered. There are well-known methods of isotope separation under thermodynamically non-equilibrium (kinetic effect) and equilibrium (chemical exchange) conditions based on the difference in chemical reaction rates of molecules with different isotopes (299, 300). The magnitude of isotope effects increases drastically with decreasing temperature, but very low temperatures cannot be used in practice because of the sharp decrease in the rates of chemical reactions at low temperatures. We may use laser radiation to create highly non-equilibrium energy distributions by strong excitation of the vibrational or electronic degrees of freedom. A comparatively low translational temperature may be maintained. It is hoped then that the kinetic isotope effect may be substantially increased

while a high rate of chemical reaction is maintained. Such an
increase in the kinetic isotope effect by use of a laser does
not enable us to separate an arbitrary isotope. The separation relies on the change in rate constants or physical properties with isotope rather than on an isotopically selective
excitation and a change in properties of the selectively
excited species. Therefore this class of isotope separation
methods, with or without the use of laser radiation, may be
called non-interchangeable. That is the choice of isotope
may not be interchanged by changing laser wavelength. The
disadvantages typical of the non-interchangeable methods are
the same with or without a laser. For instance, it is particularly difficult to separate intermediate mass isotopes.
Nevertheless, this approach is worthy of consideration for in
some cases it may prove to be competitive with the methods
discussed above.


A.    Mixtures of Vibrationally Excited Molecules

    The vibration-vibration energy transfer which disrupts
the isotopic selectivity in other methods can bring about an
excitation of high vibrational levels which favors heavy
isotopic species. Separation could then be effected by
reactions as described in Sec. VIII.
    Let the first vibrational level of the AB molecule be
excited by laser radiation:

$$AB(v=0) + \hbar\omega \longrightarrow AB(v=1) \qquad (X.1)$$

The following collisional processes then provide excitation
of the higher vibrational levels of this molecule:

$$2AB(v=1) \underset{\longleftarrow}{\overset{\longrightarrow}{}} AB(v=2) + AB(v=0) + \Delta E_{anh}, \qquad (X.2)$$

$$AB(v=2) + AB(v=1) \rightleftharpoons AB(v=3) + AB(v=0) + 2\Delta E_{anh}, \qquad (X.3)$$

where $\Delta E_{anh}$ is the difference in energy between adjacent
vibrational transitions due to anharmonicity. Processes (X.2),
(X.3) and the subsequent ones are exothermic because of anharmonicity, and the equilibrium constant may be much greater
than 1. For instance, for reaction (X.3) the ratio between
the rates of the forward and reverse processes is
$\exp(2\Delta E_{anh}/kT)$, where T is the translational temperature of
the molecular gas. Thus, molecules become excited to high
vibrational levels. The number in higher levels is much
greater than for an equilibrium harmonic oscillator distribution. This mechanism of non-equilibrium excitation of high
vibrational levels because of anharmonicity was described by
Treanor et al. (301) and by Teare et al. (302). The population of

high levels is limited by V-T relaxation. Simple estimation for competition of the relaxation process at the rate $1/\tau_0$ and V-V processes  (X.2), (X.3) at the rate $1/\tau_{vv}$ gives the following rough estimate of the number of vibrationally excited levels (303):

$$v_{max} \simeq (\tau_o/\tau_{vv})^{\frac{1}{2}}. \qquad (X.4)$$

We consider now a mixture of two isotopic molecules AB and $^iAB$. In such a mixture the collisional processes of type (X.2) and (X.3) favor the production of highly excited heavy molecules, since for them $\Delta E$'s are larger:

$$AB(v=1) + {}^iAB(v=0)$$

$$\qquad (X.5)$$

$$\rightleftarrows AB(v=0) + {}^iAB(v=1) + \Delta E_{is},$$

$$AB(v=1) + {}^iAB(v=1)$$

$$\qquad (X.6)$$

$$\rightleftarrows AB(v=0) + {}^iAB(v=2) + \Delta E_{anh} + \Delta E_{is},$$

$$\cdots\cdots$$

$$AB(v=1) + {}^iAB(v=n)$$

$$\qquad (X.7)$$

$$\rightleftarrows AB(v=0) + {}^iAB(v=n+1) + n\Delta E_{anh} + \Delta E_{is},$$

$$\cdots\cdots$$

$$AB(v=n) + {}^iAB(v=n)$$

$$\qquad (X.8)$$

$$\rightleftarrows AB(v=n-1) + {}^iAB(v=n+1) + \Delta E_{anh} + \Delta E_{is},$$

where $\Delta E_{is} = \hbar\omega - \hbar\omega_i$ is the difference in vibrational energy of the AB and $^iAB$ molecules. Because of the isotopic difference in energies the rates of V-V excitation transfer between the isotopic molecules, i.e. the rates of processes (X.3) and (X.7), differ by the factor $\exp(-\Delta E_{is}/kT)$. As a result, the vibrational temperatures of the two isotopic molecules, $T_v$ for $T_v^i$, become different, $T_v^i > T_v$ for the heavier molecule $^iAB$. Experimentally this isotopic effect was discovered by Dubost et al. (46). They excited the molecule of CO in an argon matrix at about 10°K. After the level $v = 1$ of the $^{12}CO$ molecule was excited by the second harmonic of a $CO_2$ laser, V-V energy transfer processes took place. In times less than 1 msec the energy was transferred to higher vibrational levels and heavier isotopes. In CO samples of natural isotopic composition and concentrations of CO in argon between 0.003 and 0.01, the number of excited molecules $^{13}CO$ and $C^{18}O$ was greater than that of $^{12}C^{16}O$. Thus, the excitation selectivity of the heavier isotopic molecules was over 100. The levels up to v = 7 were excited. This fact suggests that selective reactions with such highly excited molecules are possible.

The difference in the vibrational temperatures $T_v$ and $T_v^i$ was calculated in (28, 29). The authors suggested also that this effect could be applied to isotope separation. In the simplest approximation, when the mixture of two isotopic molecules is considered as a mixture of harmonic oscillators with slightly different frequencies, the following expression is true:

$$\frac{1}{T_v^i} - \frac{1}{T_v} \simeq \frac{\Delta E_{is}}{\hbar\omega}\left(\frac{1}{T} - \frac{1}{T_v^i}\right) \qquad (X.9)$$

In work (29) a more precise expression with due regard for the isotopic composition of the mixture and the kinetics of excitation and relaxation has been deduced. The difference in vibrational temperatures of isotopic molecules must bring about a difference in chemical reaction rates with a proper reagent R:

$$AB^* + R \xrightarrow{\ k_{AB}\ } AC \qquad (X.10)$$

$$^{i}AB^* + R \xrightarrow{\ k_{i_{AB}}\ } {}^{i}AC. \qquad (X.11)$$

The simplest model sets an upper limit on the difference in rates of chemical reactions (X.10) and (X.11) when the rates are described by the Arrhenius-like expression:

$$k_{AB} = k_o\ exp\left(-\frac{E_{act}}{kT_v}\right),\ k_{i_{AB}} = k_o\ exp\left(-\frac{E_{act}}{kT_v^i}\right). \qquad (X.12)$$

with the activation energy $E_{act}$, different vibrational temperatures and the same preexponential factor $k_o$. Since the reaction rate depends exponentially on vibrational temperature, the reaction proceeds effectively at a particular vibrational temperature $T^*$ such that reaction competes with V $\rightarrow$ T transfer. This means that the reaction selectivity is approximately equal to the ratio between the rate constants of reactions (X.10) and (X.11) with $T_v$, $T_v^i \sim T^*$. In this optimal case the following expression holds true for isotopic selectivity in the products of the reactions (AC and $^1AC$) (29):

$$K(^iA/A) \simeq exp\left[\frac{E_{act}}{\hbar\omega}\frac{\Delta E_{is}}{k}\left(\frac{1}{T}-\frac{1}{T^*}\right)\right] \qquad (X.13)$$

It is assumed here that $\Delta E_{is}/kT \gg \hbar\omega/E_{act}$. Relaxation of vibrational excitation limits $E_{act}$ to values not much larger than $v_{max}\hbar\omega$ expressed by (X.4). As a result, in the most optimal case the coefficient of molecular mixture enrichment in a heavier isotope is approximately given by the expression (29):

$$K(^iA/A) \simeq exp\left[\left(\frac{\tau_o}{\tau_{vv}}\right)^{\frac{1}{2}}\frac{\Delta E_{is}}{kT}\right] \qquad (X.14)$$

which should be considered as an upper limit derived assuming the maximum effect of vibrational energy on reaction rate. For reactions whose rates are unaffected by the vibrations excited $K(^iA/A) = 1$. Section VIII discusses the range of possibilities.

The results of several experiments on isotope separation during vibrational excitation of molecules in a mixture have been published. The data are rather contradictory. Basov et al. have published (304, 305) the results of their experiments on $^{15}N$ enrichment of the NO molecule thought to be caused by the chemical reaction:

$$N_2^* + O_2 \longrightarrow NO + other\ products. \qquad (X.15)$$

The reaction proceeded following laser Raman excitation of the vibrational levels of nitrogen molecules in air at liquid nitrogen temperature. The enrichment coefficient reported was of the order of $10^2$. Since it does not matter for the method being discussed which of the isotopic molecules is initially excited, similar experiments have been carried out by Basov et al. with the molecular vibrations excited in an electric glow discharge (306). Again the experiment was done in air at a reduced pressure and at liquid nitrogen temperature. The coefficient of enrichment in $^{15}N$ of the NO formed by reaction (X.15) was stated to be between 10 and 20.

These experiments were repeated by Manuccia and Clark (307) who found much lower enrichment coefficients. In the reaction between $N_2$ and $O_2$ molecules excited by a glow discharge at 77°K NO, $N_2O$ and $NO_2$ were produced. These products were enriched with $^{15}N$ by only 21%; the enrichment coefficient was 1.21. The fact that the results of this work differ from those of (306) by two orders of magnitude is attributed (307) to the presence of water molecules in the experimental mixture (306) which causes protonation of reaction products and hence an error in measuring the enrichment coefficient of $^{14}N$ and $^{15}N$ by mass spectroscopy. In addition, the estimates of the authors of work (307) exclude the possibility of such a high enrichment coefficient reported in (306). These experiments were also carried out at the Institute of Stable Isotopes in Tbilisi (308) and resulted in much smaller enrichment coefficients (about 2 - 3) both in pulsed and continuous glow electric discharges.

Preferential vibrational excitation of heavy isotopic molecules in a mixture can be used, of course, for isotope separation not only in chemical reactions but also in other processes which are strongly affected by vibrational excitation. A marked isotope effect was reported in the condensation of vibrationally excited $CO_2$ molecules (309). The variation of sticking coefficient with vibrational level may provide a useful method of isotope separation in "interchangeable" laser excitation as well.

The experimental results (306 - 309) are very interesting and must be investigated in detail. It is clearly difficult to obtain the ultimate enrichment coefficient given by Eq. (X.14). The poorly controlled conditions of an electric discharge, with the concommitant production of reactive species which accelerate V → T transfer, are less than ideal. In fact, Manuccia and Clark point out that their enrichments may be entirely due to equilibrium isotope effects and not depend at all on a non-equilibrium distribution of vibrational excitation (307). The energy consumption of 0.22 MeV per product N atom (307) is rather large. The general method may yet become attractive if several-fold enrichments can be obtained in non-equilibrium mixtures of very low cost chemical reagents such as $N_2$ and $O_2$.


B.    Mixtures of Electronically Excited Molecules

Isotopic effects upon energy transfer of electronic excitation in crystals at low temperatures are well known (310, 311). These effects were found while studying luminescence in isotopically doped crystals. Due to the difference in the electronic excitation energy of molecules with different isotopic compositions some isotopic molecules become energy traps at low temperatures; this appears as preferential luminescence of such molecules.

Work (312) proposed to use this effect for photochemical isotope separation. Two conditions are necessary here: 1) the temperature must comply with kT < ΔE, where ΔE is the energy trap depth; and 2) transfer of excitation energy between isotopic molecules occurs faster than photochemical transformation. In this work the method was tested experimentally for the example of hydrogen isotope separation in mixtures of $CH_3CHO$ and $CD_3CDO$. Both molecules absorb the uv light of a mercury lamp, but because of fast excitation transfer the $CH_3CHO$ molecules become preferentially excited. Therefore preferential photodecomposition of $CH_3CHO$ molecules can be expected. Photodecomposition of $CH_3CHO$ in the region of $\pi - \pi^*$ transitions occurs in two ways:

$$CH_3CHO + \hbar\omega \longrightarrow CH_3 + HCO \qquad (X.16)$$

$$\underline{\qquad} CH_4 + CO \qquad (X.17)$$

The methyl radicals formed by the reaction (X.16) from $CH_3CHO$ in mixtures may produce both $CH_4$ and monodeuteromethane $CH_3D$. Thus, the degree of photodecomposition of $CH_3CHO$ is determined by the total quantity of $CH_4$ and $CH_3D$, while that of photodecomposition of $CD_3CDO$ is determined by the quantity of $CD_4$ and $CD_3H$. At normal temperature the isotopic enrichment value was 1.9, at 77°K it was 10.4 and at 4.2°K it reached 88. An isotope effect of about 10 to 20 was observed in the electron paramagnetic resonance spectra of methyl radicals $CH_3$ and $CD_3$ as the same mixture was photolyzed at 77°K.

The observed isotope effect (312) is in agreement with the idea of isotopic energy traps. In this case $CH_3CHO$ appears as an energy trap, the trap depth being $\Delta E \approx 120$ cm$^{-1}$. The distribution of excitation energy between isotopic molecules following rapid energy transfer is determined by a Boltzmann factor exp $(-\Delta E/kT)$. An isotope effect of about 10 corresponds to the value $\Delta E \approx 120$ cm$^{-1}$ at 77°K. Thus the experiment (312) demonstrated the principle of using the isotope effect in electronic excitation transfer in a photochemical reaction.

XI.  ECONOMICS AND ENGINEERING OF LASER ISOTOPE SEPARATION

The isotopic enrichment of $^{235}U$ for electrical power generation in nuclear reactors is by far the most economically important objective of laser isotope separation. World fuel requirements for nuclear reactors between now and the year 2000 are estimated to be approximately 6 million tons of $U_3O_8$. Enrichment cost estimates for processing this ore amount to approximately 150 billion $USA (313). An efficient laser enrichment scheme could cut these costs to a small fraction of this amount. Large quantities of $D_2O$ are needed for heavy water nuclear reactors. The current world production capacity of somewhat greater than $10^6$kg/yr is probably not sufficient. The Canadians estimate that they will need in excess of $4 \times 10^6$kg/yr for the reactors which they will build (314). Since the present production costs of D are $10^3$ less than $^{235}U$

on a per atom basis, it will be much more difficult to develop
an economically viable process for D.  The estimates of  enrichment needs given above presume a smooth and relatively
rapid growth of the nuclear power industry.  The safety and
ecological desirability of nuclear power are currently the
subject of vigorous debate.  It seems inevitable that political
forces will significantly slow the development of nuclear
power.

The economic importance of separating isotopes of other
elements with laser methods is neither so obvious nor so
pressing as that for $^{235}U$.  The total annual world sales of
separated stable isotopes other than $^{235}U$ or D is less than
$5 million.  The importance of enriched isotopes of other
elements lies not in present uses but in the possibility of
new uses in scientific research, in medical diagnosis, in
agricultural and environmental testing  and in technology.
These uses are likely to develop only if high volume, low
cost sources of isotopes become available.

A.    $^{235}U$ Enrichment

The economic and industrial scale of the $^{235}U$ enrichment
problem is beyond the scope of the normal consideration of
laboratory scientists and therefore deserves some additional
perspective here.  In considering the urgency of this problem
it is well to bear in mind that the cost of enrichment today
is approximately 6% of the total cost of nuclear power production.  $150 billion over 25 years is not so large when
compared to the recent $50 billion increase in the price of
oil sold in one year.  Nevertheless the magnitude of the
challenge to research scientists, to engineers and to industry
to develop some of the scientific ideas described above into
practical methods, from there into pilot plants and on into
a multibillion dollar industry is enormous.  Success is by no
means certain.  The magnitude of the industry can be illustrated with a few figures from the U.S. nuclear industry.  A
typical light water reactor produces $10^9$ watts.  It requires
processed fuel from 600 tons of $U_3O_8$ ore for an initial charge
and an annual use of 200 tons.  The fuel is enriched to about
3% $^{235}U$ from a natural abundance of 0.75%.  The depleted ore
is discarded at an abundance of 0.2 to 0.3% $^{235}U$.  The enrichment is carried out in large gaseous diffusion plants.
One such plant handles about 20,000 tons of $UF_6$ annually.
Such a plant consumes $2.4 \times 10^9$ watts of electricity.  The
construction cost of the plant is about 3 billion and the
power plants to go with it about $1.5 billion.  The U.S. has
projected the need to put new plants of this magnitude into
operation each 18 months beginning in 1984.  Such a plant is

now being built in France.  There are several countries perfecting gas centrifuge technology.  This may cut enrichment costs by 10 - 30%.  The Becker nozzle process has entered the pilot plant stage and in time might lead to significantly lower costs, perhaps a factor of two.

The potential of laser processes for cost reduction is illustrated by the fact that gaseous diffusion uses 5 MeV of electrical energy for separation of each $^{235}U$ atom.  The overall enrichment cost is about $5. per gram $^{235}U$.  The energy requirement for a laser process can be less than 10 eV of photons per atom of $^{235}U$.  This requires a few keV of electricity for standard visible laser efficiencies, and less for the most efficient IR and uv lasers.  Chemical processing of the ore or vaporization of solid samples requires a similar amount of energy, 5 - 10 eV  divided by the natural abundance of 0.0075.  The power costs of an ideal laser system would be negligible.  However, to produce the 90 tons of $^{235}U$ per year, contained in the enriched product from a diffusion plant, a perfectly efficient use of photons at 500 nm requires an average power of 3kW.  Realistically tens of kW will be needed. Existing tunable laser systems in the visible and ultraviolet operate with only a few watts of power.  The recent development of rare gas halide lasers could be very important.  A great deal of laser engineering and perhaps some fundamentally new laser systems will be crucial to the ultimate success of laser separation of uranium.  If such lasers can be built for orders of magnitude less than the cost of diffusion plants, not only will laser enrichment be very cheap but we may have some very useful laser systems for other areas of science and technology.  The very high selectivity inherent in most laser separation schemes makes it likely that the $^{235}U$ discarded with the $^{238}U$ would be greatly reduced.  This would reduce the requirement for new ore by as much as 40%.  Furthermore, the stockpile of depleted uranium tails contains about as much $^{235}U$ as has been produced for power generation and weapons production since World War II.

At the present time it is not at all clear which method of laser separation of uranium will be the least expensive. An infrared photochemical method requiring mainly the use of one of the efficient lasers already developed or a new and similarly efficient one could be very cheap indeed.  The two-photon ionization method can clearly be used to enrich uranium. But the cost of such a system for commercial production could be very high.  It is clear that a great deal of research on possible methods is needed in order to develop and select the best one.  It is likely that the first laser method to go into commercial production would not be the final one.

Recently there has been some concern about the potential for nuclear weapons proliferation through laser separation of

$235_U$. Much of this has resulted from early *naïveté* concerning the difficulty of practical $235_U$ separations on the multikilogram scale. A moratorium on laser isotope separation development has been suggested (315). Krass (316) has analyzed the proliferation potential realistically in terms of the fundamental scientific and technological constraints on the processes which are being developed. He concludes that without a substantial and unexpected breakthrough laser isotope separation will remain among the more difficult ways of obtaining materials for nuclear weapons. Certainly the enormous size of the research programs at Avco-Exxon, Livermore, Los Alamos and elsewhere suggests that $235_U$ production by lasers will demand very large investments of time and money as well as a high level of technological sophistication.


B.    Deuterium Enrichment

A reduction in the price of heavy water could have a substantial impact on nulcear power generation costs with the CANDU heavy water moderated reactors. These reactors can use unenriched uranium fuel. The 800 tons of $D_2O$ required for a 1 GWatt power reactor costs about $100 million and accounts for 15 - 20% of the total capital cost.

At Ontario Hydro research is under way on laser enrichment of deuterium (314). The main effort is devoted to formaldehyde predissociation (see Sec. V). A clean vibrational photochemistry scheme (Secs. VI and VIII) using an efficient IR laser is also being sought. The challenges are formidable indeed. A heavy water plant which produces 1,000 tons/year of $D_2O$ must process at least $10^7$ tons/year of feed stock (for water). Only water, methane and perhaps hydrogen are available in these quantities at a single location. If the laser acts on a secondary feed such as formaldehyde, it must exchange hydrogen with the primary feed material. If every laser photon yields a D atom in the product stream, then a 1 MW laser is required. Since the abundance of D is 1.4 x $10^{-4}$ or 1 HDCO per 3500 $H_2CO$, at an excitation selectivity of 3500 at least 2 MW of laser power is required. For the selectivities of order 100 which have been demonstrated 35 MW is required. Figure 53 shows cost estimates for heavy water production as a function of optical selectivity and laser efficiency. The curves are lines of constant $D_2O$ cost for two different laser capital costs. The major cost in this and in most laser isotope separation schemes is the capital cost of the laser. In well-engineered systems this cost usually approaches one or a few dollars per watt of input electrical power. Thus the price of laser photons is inversely proportional to the laser efficiency. In order for laser separation

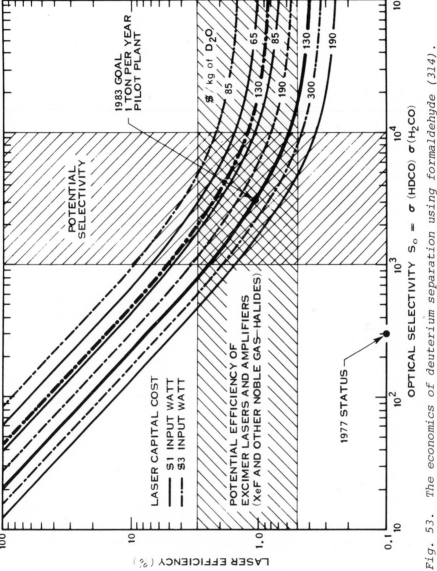

*Fig. 53.  The economics of deuterium separation using formaldehyde (314).*

125

of deuterium to become economically viable lasers must be developed which are 100 times larger and 1000 times more selective than those required for uranium enrichment.

## C.    Elements 2 Through 91

The economics of production of a variety of stable isotopes is illustrated in Table 2.  For the relatively small

TABLE 2

*Current Stable Isotope Prices (approximate)*

| Isotope | Natural Abundance % | Enriched Abundance | Price (US $/mole) | Method |
|---|---|---|---|---|
| D | 0.015 | 99+ | 1 | Chemical Exchange |
| $^{13}C$ | 1.11 | 96 | $10^3$ | Distillation |
| $^{17}O$ | 0.037 | 40 | $10^5$ | Distillation |
| $^{34}S$ | 0.042 | 90 | $6 \times 10^3$ | Distillation |
| $^{37}Cl$ | 25 | 90 | $10^4$ | Calutron[a] |
| $^{48}Ca$ | .185 | 95 | $10^6$ | Calutron |
| $^{57}Fe$ | 2.19 | 90 | $2 \times 10^5$ | Calutron |
| $^{79}Br$ | 50 | 90 | $4 \times 10^4$ | Calutron |
| $^{196}Hg$ | .15 | 48 | $10^8$ | Calutron |
| $^{235}U$ | .75 | 3 | $10^3$ | Gas Diffusion |

[a]*Mass Spectrometer at Oak Ridge, Tenn., U.S.A. (317).*

amounts of material produced other than uranium and deuterium the most efficient methods are fractional distillation or counter current chemical exchange for the lighter elements and magnetic deflection (mass spectroscopy) for the heavier ones (317, 318).  For laser separation a lower limit on cost is set by the photon energy used.  This is only 0.1 kWh for one

mole of photons at 3,000 $\overset{\circ}{A}$. If the overall efficiency of
conversion from electrical energy to laser photons and finally
to separated product is $10^{-3}$, then the electrical energy cost
per mole is about $1. In the near future and for small scale
production the capital cost of lasers and the cost of labor
will certainly be far larger than energy costs. Nevertheless,
the possibility of large reductions in isotope cost is clear.
     Work on formaldehyde suggests two situations where laser
enrichment may most easily be applied to advantage. The rare
isotope $^{17}O$ is intermediate in mass between $^{16}O$ and $^{18}O$. As
a result distillation techniques do not produce cleanly separated material. Formaldehyde predissociation could provide
an effective method of removing $^{17}O$ from mixtures enriched by
distillation. Another application could be to increase the
sensitivity of $^{14}C$ dating. The carbon in the sample to be
dated would be converted to formaldehyde and selectively laser
photolyzed to convert nearly all of the $^{14}CH_2O$ to $^{14}CO$ and
leave behind a maximum amount of $^{12}CH_2O$. Both of these applications would become economically feasible at process costs
which would be too high for mass production of $^{13}C$, $^{18}O$ or D.
The appropriate lasers are well within the reach of current
technology.
     In view of the extensive research required for the de-
velopment of methods of separation for any one element and the
likelyhood that the optimum process will be different for each
element, it is important to identify now those isotopes which
are most likely to be of value in new applications. Some
applications for C, N and O are already being explored since
their prices are relatively low. C and O could be used as
monitors of body functions such as metabolism. Nitrogen depleted of $^{15}N$ is being used in agricultural and environmental
fertilizer tracer studies (319 - 322). Labeled natural
products may be prepared by feeding simple $^{13}C$, $^{15}N$ or $^{17,18}O$
containing food to plants or animals. Such work traces biochemical pathways and can produce labeled compounds almost
impossible to make in the laboratory. The isotopes $^{13}C$ and
$^{18}O$ may be used extensively in $CO_2$ lasers for two purposes.
$^{13}C$ may be used in simple laser devices for measurements of
the isotope ratio $^{13}C/^{12}C$ ($^{12}CO_2$ and $^{13}CO_2$ lasers) (323) which
find wide application in geology for oil prospecting. Carbon
and oxygen isotopes may be used in electric discharge $CO_2$
lasers (324) and optically pumped $CO_2$ lasers (325) to widen
the region of frequency tuning. The isotope $^{40}K$, which is a
natural radioactive isotope, is useful for agricultural investigations. We suspect that the production of some important materials such as $^{34}S$, $^{57}Fe$, $^{48}Ca$, etc. will be developed
provided their cost is reduced by several orders of magnitude.
Isotopes with particularly large or small neutron cross
sections may be important as reactors. Gross (326) has

suggested a separation scheme for $^{50}$Ti.

The development of production facilities for laser isotope separation is generally limited by laser technology. Figure 54 shows laser powers required to match the output of currently used production facilities.

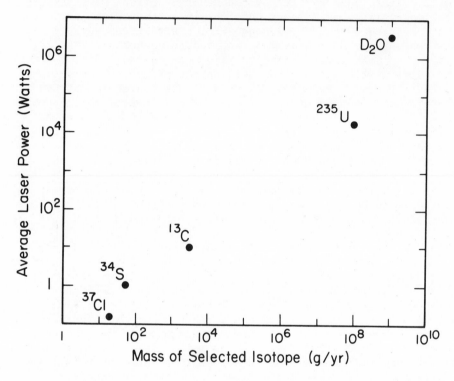

*Fig. 54. Average laser power required to reproduce output of largest present-day production facilities for several isotopes. Output is measured in grams of desired isotope contained in product.*


## XII. CONCLUSION: OTHER APPLICATIONS


In the present review we have made an attempt to consider systematically the main methods of laser isotope separation already demonstrated in research laboratories. They form the basis for further progress not only in science but also in industry. We hope that in the near future other methods will be elaborated. Since nowadays, when the energy crisis has begun to affect the standard of living of people and the progress of technology, some scientists in many countries consider studies into laser isotope separation to be an urgent and important problem. In fact, isotopically selective photophysical and photochemical processes are being investigated very actively. Laser isotope separation has become one of the most urgent topics at international conferences on laser applications (327, 328), at national conferences (329 - 330) and in schools (331, 332).

The greatest number of laser isotope separation methods has been proposed and demonstrated on the basis of the processes in atomic vapors (or beams) and molecular gases. There is some promise for progress with methods based on laser selective processes in condensed phases and heterogeneous systems (at the gas-condensed phase boundaries). It is quite probable that the development of the latter processes would be important not only for isotope separation but also for other separations and transformations by selective laser photophysics and photochemistry.

Selective processes in *condensed media* are as yet poorly understood and seem to be less promising because of line broadening at normal temperatures and because of the high rate of relaxation of vibrational excitation. This dictates certain stringent experimental conditions for selective action: 1) excitation of electronic states with the relaxation rates smaller than for vibrational states; 2) use of low temperatures at which spectral line broadening and V-T relaxation rates are drastically smaller; or 3) use of ultrashort pulses. For the time being we have only one example of such an isotopically selective photoprocess (150, 151). Selective photobiochemical processes (333) of course can only be achieved in condensed media.

In the case of *heterogeneous media* it is possible to act selectively both on the particles in the gas phase and on the atoms or molecules on the surface of the condensed medium. In the first case such processes as photochemical reactions, photoadsorption and condensation of excited atoms or molecules can be used (309, 334). As the atoms or molecules on the surface are excited, their selective photodesorption and evaporation are possible. Selective removal of particular atoms or molecules from the surface or selective chemical reactions on the surface (335) are important though very difficult objectives. Selective ejection of electrons or protons from certain parts of macromolecules may prove to be rather important for a direct identification of the secondary and tertiary spatial structure of biological molecules (336).

In finishing this review we would like to stress that the methods of selective photophysics and photochemistry which are being developed for laser isotope separation will go far beyond the limits of this problem both in fundamental principles and practical applications. We finish here with a brief discussion of the possibility of laser production of highly pure substances, selective laser photochemical processing, selective laser biochemistry, laser separation of nuclear isomers and selective detection of single atoms, molecules and molecular bonds as examples of wider applications.

A.    Materials Technology at the Atomic-Molecular Level

The methods of selective photophysics and photochemistry being developed to separate isotopes make it possible to elaborate a new approach to materials technology at the atomic-molecular level. With laser radiation one can directly manipulate atoms and molecules of a particular sort; thus one can collect macroscopic amounts of a substance "atom-by-atom" or "molecule-by-molecule." The most important process in a universal, atomic-molecular, laser technology of materials is undoubtedly the production of highly pure substances in atomic state, alloys, and molecular compounds. The processes of atomic photoionization and selective molecular photodissociation may be used to produce highly pure substances or to remove selected impurities from a substance. There are different possibilities and fields of applications for both approaches.

*Selective Atomic Photoionization*
This approach to materials technology is the most universal and flexible. An optimal scheme for selective atomic photoionization under the action of two or more laser beams with properly tuned frequencies and properly chosen intensities enables every desired atom to be ionized within $10^{-5}$ to $10^{-8}$ sec. When 20% of the radiation at an average power of $10^3$ W is consumed to photoionize atoms with an ionization potential $E_i \simeq 10$ eV, it is possible to ionize about one mole of material per hour. Thus a comparatively moderate scale setup may ensure production of several tons of a pure substance per year. The method of selective atomic ionization in combination with tunable lasers, with average outputs from 100 to 1000 W, can therefore be considered as a reasonably efficient method of separation at atomic level.

Laser purification of materials by selective ionization (337) will have a number of significant advantages over the existing methods of purification which are based on differences in the bulk chemical or physical properties of substances and impurities:

1. High selectivity or high degree of purification in a single-step process. The degree of purification of a desired element from any admixtures may be higher than $10^3$. This value depends on the process of charge exchange during collisions between the ion of the given element and the neutral

atom of the other elements.  Basically, by decreasing the
atomic density in the beam we may achieve a separation selectivity much higher than $10^3$, with a correspondingly reduced
efficiency.  For example, if we take a mass-production material with a purity of $10^{-7}$%, it is possible to purify it to
$10^{-10}$% by the method of selective atomic ionization.

2.  Universality.  Selective ionization of any element
may be achieved by proper selection of laser frequencies independent of their physical and chemical properties (melting
and boiling temperatures, reactivity, etc.).  When one or more
specified element is to be removed from a substance, it is
possible to ionize only the impurities and to remove them
from the atomic beam of the substance.  In this case a maximum
efficiency is achieved and a minimum of coherent light energy
is needed.

3.  Flexibility which makes it possible to use directly
ion beams to produce pure films and to implant ions into a
homogeneous substance (ion implantation).  Ion beams can be
directed onto a substrate surface to produce a pure film of a
specified element, as shown in Fig. 55.  We think it is possible to simultaneously ionize and deposit on a single surface
two or three elements from different atomic beams.  So, it
will be possible, probably, to produce films of complex atomic
compounds with their stoichiometric composition controlled by
the intensity of photoion beams.  The whole process of selective atomic ionization, extraction of ions from an atomic beam
and their deposition on a substrate can be brought about in an
ultrahigh vacuum.  It is not necessary for the process that
the substance to be purified should make contact with any
reagents or material aside from the substrate for which we may
always  use a material without undesirable impurities.

We should concentrate our attention on the use of selectively formed photoions of boron, arsenic, phosphorus and
other elements in setups for ionic implantation in semicon-
ductors (338).  Electrodeless laser production of certain ions
eliminates, first of all, the necessity of using an electromagnetic mass separator and, secondly, makes it possible to
isolate the high-temperature source of atoms from the ionizer.
The latter is of no small importance since this enables the
atoms to be photoionized near the high-voltage electrode and
hence the construction of electrostatic ion accelerators with
an MeV energy or more may be substantially simplified.

The basis for successful development of the photoionization method is elaboration of optimal schemes for multistep
selective ionization of various elements and also creation of
rather efficient tunable lasers in the uv and visible ranges
with high average powers and long lifetimes.  The multistep
resonant excitation of the states near the ionization limit
and subsequent autoionization of highly excited atoms by a

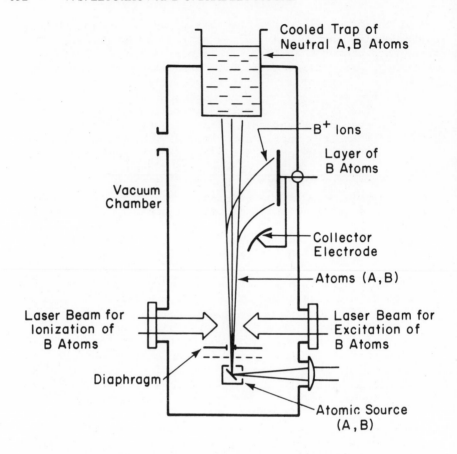

*Fig. 55. Possible scheme for preparation of films of highly pure composition during deposition of $B^+$ ions produced by the method of two-step selective ionization of B atoms from a beam mixed with other atoms.*

pulsed electric field is a universal ionization scheme providing a high excitation cross section and high ionization yield.

*Selective Molecular Dissociation*
The process may be used to purify a substance in the gas phase of molecular impurities the removal of which by standard techniques is not efficient. Purification by the dissociation method is based on differences in the physical-chemical properties of the original substance and the dissociation products. This enables us to transform the impurities into ones which may be separated by standard techniques of purification after the mixture is irradiated.

The possibility of substance purification in a gas phase through dissociation of admixed molecules by intense IR radiation has been recently demonstrated experimentally in work (339) where arsenic trichloride ($AsCl_3$) was purified of 1, 2-dichloroethane ($C_2H_4Cl_2$) and carbon tetrachloride ($CCl_4$) (see also Sec. XII.B and Chapter II).

The method of selective molecular dissociation should be applicable not only to the technology of pure materials but also to the removal of toxic and carcinogenic substances from gas mixtures. If dissociation of such impurities converts them to inactive forms, the method becomes rather simple and direct.

B.    Laser Photochemical Processing

The potential of lasers for industrial photochemical processing has been the subject of much wild speculation, some interesting proposals, a few promising experiments and recently, even serious industrial interest. Some of the possibilities have already been mentioned in Secs. V - VIII. Multiphoton dissociation and multiphoton-induced chemistry appear particularly promising. Recent research in organic photochemistry has shown interesting changes in chemical products with excitation wavelength. Some laser experiments which do not involve the selectivity necessary for isotope separation have yielded interesting chemical products.

Bachmann, Nöth, Rinck and Kompa (250) have irradiated $B_2H_6$ with a continuous $CO_2$ laser and triggered a chain reaction which produces $B_{20}H_{16}$ as the major product. The thermal and uv photochemical reactions give very different products. This laser-induced reaction is much the simplest way to prepare $B_{20}H_{16}$. When mixtures of $BCl_3$ and phosgene, $Cl_2CO$, are excited by a cw $CO_2$ laser, the phosgene decomposes (340). The $BCl_3$ is excited by the laser but the phosgene decomposes. The reaction may provide a practical scheme for industrial removal of the phosgene impurity from crude $BCl_3$.

Photopolymerization initiated by two-photon absorption, simultaneous or sequential, has been proposed as a means of producing three-dimensional images and figures (341). A sequence of points would be defined in space by the intersection of two light beams or by the focal position of a single beam. Three dimensional fluorescence displays have been produced in this way in ICl gas (342).

The possibility of using laser photochemical reactions in the liquid state for nuclear waste reprocessing has been suggested (343). Such a process would be limited by the available selectivity in photochemically active liquid spectra

and by the non-selective reactions induced by the high levels
of radioactivity. Photochemical selectivity has been demonstrated for some rare earth separations (344).


C.    Selective Laser Biochemistry

Selective action of laser radiation on complex molecules
in a condensed medium is an enticing prospect for molecular
biology. The possibilities are largely unexplored and remain
quite uncertain. In a condensed medium at normal temperatures
the requirements for excitation selectivity and for conservation of selectivity are at best very difficult to meet. The
electronic transitions of biomolecules are primarily in the uv
region. They are quite broad and it is quite difficult to
expect a good excitation selectivity of particular molecules
in a mixture. However, it is for these states that selectivity is most easily preserved because lifetimes on the order of
nsec are possible and because reaction often occurs with an
appreciable quantum yield. Vibrational excitation of molecules on the other hand, is more selective but relaxation
occurs in a very short time (at 300°K the relaxation time
$T_1^{vib} \leq 10^{-11}$ sec) (345). Since the energy of one vibrational
quantum is only a few times greater than thermal energy kT,
the contribution to the biochemical reaction rate by short-
lived vibrationally selective excitation may not be as large
as caused by the subsequent thermal nonselective excitation.
Successful experiments on isotopically selective photoprocesses in gas media show that there are at least two ways
for overcoming these problems: 1) combination of selective
vibrational excitation with subsequent electronic excitation
from the vibrationally excited state, that is two-step, IR-uv
excitation; 2) multiple photon vibrational excitation in an
intense IR field. In both cases the process must be carried
out using picosecond laser pulses so that a molecule can absorb a substantial energy, several eV, prior to thermal
relaxation of vibrational excitation. The process of two-step,
IR-uv excitation which takes advantage of the favorable
aspects of both vibrational and electronic excitations may
be quite general. Rapid intramolecular transfer of vibrational energy will be a particularly severe problem for macromolecules at the high levels of vibrational excitation important
in multiphoton IR excitation.
Vibrational excitation selectivity is particularly good
in resonance Raman spectroscopy (346, 347). Strong Raman
scattering occurs only for vibrations affecting the electronic
absorption chromophore with which the exciting laser is in
near resonance. Thus coherent Raman pumping with two lasers
at $\omega_1 > \omega_2$, such that $\omega_1 - \omega_2$ is equal to the vibration fre-

quency and $\omega_1$ is slightly less than the electronic absorption
frequency $\omega_e$, will very selectively pump a particular vibration near a particular chromophore. If $2\omega_1 - \omega_2$ is greater
than $\omega_e$, then the Raman excited molecules will be pumped to
the excited electronic state. For $(2\omega_1 - \omega_2) < \omega_e$ a third
laser frequency $\omega_e > \omega_3 > (\omega_e + \omega_1 - \omega_2)$ is required.

For each particular system, of course, we must first
prove the feasibility of a selective photoprocess for a particular molecule in solution or for a particular portion of a
macromolecule. The scanty spectral information available for
the excited states of biomolecules makes the answer to this
question far from being trivial. We must also deal with other
"tricky" problems, such as: 1) heating of the medium during
selective photoexcitation; 2) absorption of IR radiation by
solvent molecules. These problems are particularly important
for experiments "in vivo."

The estimates in work (333) show that it is possible to
eliminate heating when the concentration of molecules in a
solution is below $10^{-3}$ M and the absorption of IR radiation
by solvent molecules is rather weak. The estimates show that
the required intensities for IR and uv pulses of $10^{-11}$ to
$10^{-12}$ sec duration are between $10^8$ and $10^9$ W/cm$^2$. No doubt,
cooling of molecules would decrease the V-T relaxation rate
and simplify these problems.

Some particular possibilities of selective laser biochemistry (selective excitation of DNA molecule bases, selective excitation of C = C double bonds of lipid molecules,
selective excitation and breaking of hydrogen bonds in DNA
molecules) have been considered in the paper (333) and the
report (348). Below, as an example, the possibility of selective excitation and breaking of hydrogen bonds in DNA molecules will be discussed.

The double helix of DNA is formed by hydrogen bonds
between the bases (guanine-cytosine and adenine-thymine). The
breaking of the hydrogen bonds splits the double helix into
two single helices. Opening the helix permits replication to
begin. Selective excitation of hydrogen bonds and their
selective breaking could be of interest for laser control over
the process of DNA replication. The rates of biological processes might thus be increased by laser stimulation and the
laser may provide the basically new possibility of an external
controlled "start" of the DNA replication process.

Two pairs of bases in DNA have slightly different hydrogen bonds. The pair A - T is linked by two hydrogen bonds
N - H...O, and an energy of about 7.0 kcal/mole is required to
separate this pair of bases. The pair G - C is linked by two
hydrogen bonds N - H...O and one bond N - H...N, and about
9.0 kcal/mole must be consumed to break them (349).

An IR absorption band at about 1720 cm$^{-1}$ corresponds to

the G - C bond in native DNA and a band at about 1700 cm$^{-1}$ to
A - T.  In DNA denaturation when the hydrogen bonds are broken
both bands disappear (350).  Therefore in work (333) it was
proposed to excite these frequencies (5.814 μ and 5.888 μ)
with powerful picosecond IR pulses to stimulate hydrogen bond
breaking.  Of course, there are many more possibilities for
experiments "in vivo," especially as far as the choice of more
convenient wavelengths not absorbed by the solvent.

The potential function for a hydrogen bond has two characteristic minima corresponding to two spatial positions of the
proton (Fig. 56).  Though not experimentally observed, the

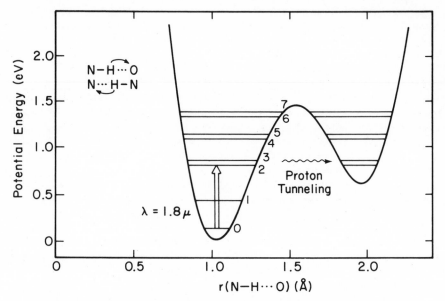

*Fig. 56.  Potential function of hydrogen bond in DNA
(349).*

energy levels of a proton in the hydrogen bonds N - H...N have
been calculated.  The transition of a proton (or rather of two
protons in a pair of adjacent hydrogen bonds) to a higher energy tautomeric state is believed to result in mutations (Löwdin
mechanism (351)).  It is evident that using a laser at the
wavelength of 1.8 μ we may transfer a proton to an excited
state and thereby stimulate its tunneling to a higher energy
minimum.  This possibility has been discussed in (349) and
with the progress in picosecond tunable IR lasers must become
a subject of experimental studies.  Proton excitation must
also show itself in the electronic absorption spectrum and
thus can be used in two-step selective processes by the scheme:
"selective IR excitation of a proton + uv excitation of an

electron."

Temporal selectivity has been shown by Hearst and co-workers (352) to be particularly valuable in biological photochemistry. The planar tricyclic psoralen dye molecules are photochemically reactive at both ends (353). The $C_4$-cycloaddition reaction of psoralen with a DNA base at each end is shown in Fig. 57(a, b). The psoralen molecule slips in between the adjacent layers of base pairs of the DNA double helix, Fig. 57(c). Absorption of a single photon, $\lambda = 350$ nm, by psoralen causes bonding to a base on one of the DNA strands. The absorption of a second photon then causes the opposite end of the dye molecule to react with an adjacent base on the complimentary DNA strand. The DNA double helix is thus cross-linked by a covalent chemical bond. When the hydrogen bonds are broken, the two strands do not separate. A sufficient number of cross-links prevents genetic reproduction. A normal continuous photolysis produces statistical amounts of the mono-addition and cross-linking reactions; both are activated at the same wavelengths. Excitation with a 20 nsec 100 mJ pulse from a frequency-doubled ruby laser, 347 nm, produced a large yield of monoadduct without any detectable cross-linking. A second pulse cross-linked most of the molecules. Clearly a delay time of more than 20 nsec between the absorption of the first photon and the second is required for the second bond to form. This technique when fully developed will allow many valuable experiments on the structure and function of DNA molecules (353).

D.   Separation of Radioactive Isotopes and Nuclear Isomers

Selective laser photophysics methods have an extremely high selectivity and allow manipulation of atoms by means of optical and electrical fields without direct contact. Therefore, in principle, it is possible to apply these methods to the problem of radioactive waste management in the nuclear industry.

A laser approach to the treatment of radioactive wastes has potentially very important advantages over the conventional approaches. First it is possible to realize a remote treatment by means of laser radiation without direct contact with the main parts of the separator. Thus the radiation hazards may be reduced markedly. Second the laser treatment may, basically, provide a multi-purpose separation of chemical elements, isotopes and nuclear isomers simultaneously. In consequence it is possible to isolate radioactive elements which cannot practically be separated using physical and chemical properties. Isolation of the most dangerous radioactive elements would then be possible. Third, the laser

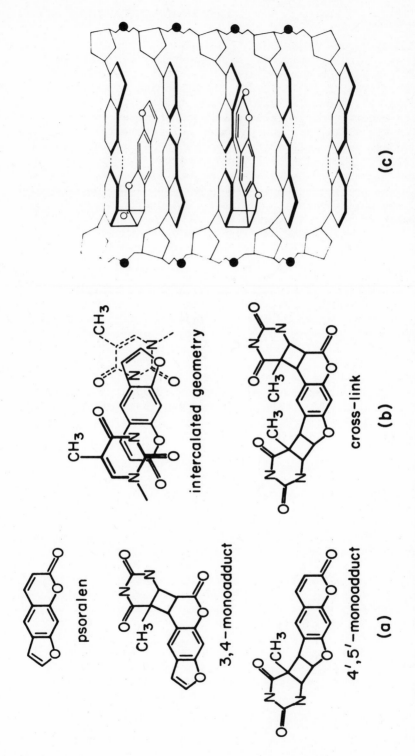

*Fig. 57. The photochemistry of psoralen intercalated in a DNA double helix. (a) Psoralen and monoadducts with the DNA base thymine. (b) Geometrical configuration psoralen relative to reacting thymines and structure of cross-linking adduct. (c) Monoadduct-DNA configuration.*

138

treatment can provide in principle a higher degree of concentration of any radioactive element.   Thus it is possible either to facilitate appropriate storage or further nuclear processing for example by destruction with intense neutron radiation, or by ejection into space.   It goes without saying that
these advantages can be realized fully only during multi-
purpose treatment connected with atomization of all elements
in the radioactive waste.   Of course, taking into account the
energy cost it is more profitable to make a preliminary chemical treatment possibly including laser treatment (photochemical treatment of molecular radioactive compounds) and then to
perform the more costly separation of difficult groups of
radioactive elements.

The separation of radioactive isotopes and especially of
nuclear isomers is important not only for nuclear technology
but also for preparation of inverted nuclear level populations.   This could be achieved by laser separation of excited
(isomeric) and unexcited nuclei (354).   An important contribution of selective laser photophysics to the creation of a
γ-ray laser may thus be hoped for.   The status of this problem
has been considered in refs. (355 - 357).

The estimates given in work (357) show that to make a
γ-ray laser in this way, we should prepare a *concentrate* of
excited nuclei during a fairly *short time* ($10^{-2}$ - $10^{2}$ sec).
Though the absolute amount of radioactive substance may be as
small as $10^{-8}$ gr, the necessity of preparing a concentrate of
excited nuclei forces us to give up schemes that use direct
excitation of the sample.   For direct excitation it is difficult to obtain simultaneously nuclear level population inver-
sion and excitation of a significant fraction (several percent)
of one sort of nucleus relative to all the other sorts in the
target.   It seems expedient to select nuclei in the desired
excited level.

Figure 58 shows a conceptual scheme for a γ-ray laser
which complies with the stated requirements.   When the target
is bombarded by a stream of neutrons, a relatively small
number of excited nuclei is produced.   The excited nuclei are
extracted from the target by, for example, rapidly vaporizing
the target with a pulse of laser radiation.   The next problem
is how to extract rapidly nuclei in the desired excited state.
For this purpose, the method of selective photoionization of
atoms is, in our opinion, suitable.   The existence of isomeric
structure in atomic spectra allows selective laser excitation
of only those atoms containing the desired excited nucleus.
For this purpose the wavelength $\lambda_1$ of the laser should coincide with the optical absorption line $\lambda_0$* belonging to the
atom with an excited nucleus M* in the desired state.   The
subsequent separation of the excited atoms $\tilde{A}$ (M*) is accom-

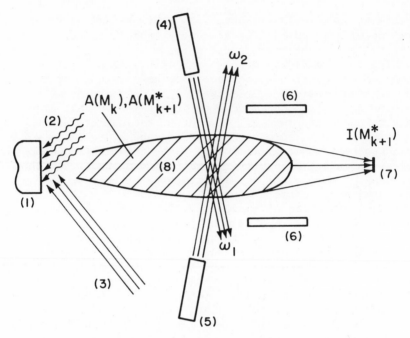

*Fig. 58.  Possible scheme of a γ-ray laser on high decay rate nuclear isomers employing laser sorting of excited nuclei $M_{k+1}$:  1 - target of $M_k$ atoms;  2 - beam of slow neutrons;  3 - laser beam for evaporation of surface layer of target;  4 - laser beam for excitation of $A(M_{k+1}^*)$ atoms; 5 - laser beam for ionization of excited atoms $A(M_{k+1}^*)$;  6 - collecting system;  7 - filament of atoms with excited nuclei $M_{k+1}^*$;  8 - stream of evaporated atoms (357).*

plished by photoionizing them with a second laser pulse. After simultaneous irradiation of the products of vaporization of the target by laser radiation at two wavelengths $\lambda_1$ and $\lambda_2$, only ions I(M*) of a given isomer M* are produced, gathered by the electromagnetic focusing system and deposited on a charged filament.

The limitation on gas density due to resonant charge exchange leads to a limitation on the maximum number of atoms which can be sorted per unit time.  The estimates presented in work (357) show that the scheme of Fig. 58 with laser sorting of excited nuclei through selective photoionization is quite realizable with realistic physical parameters.

E.    Selective Detection of Atoms, Molecules and Molecular Bonds

The methods of selective laser photophysics solve the problem of physical extraction of a particular atom or molecule from mixtures of constituents with very similar chemical properties. The first and relatively simpler part of this problem is selective detection of single atoms and molecules. Selective two- (or multi-) step photoionization of atoms and molecules is best suited to this purpose. Attention was given to this fact even in the first work (21) and was then stressed in (2).

*Detection of atoms.* The method is particularly simple in the case of atoms. As shown in Sec. III, selective excitation and subsequent photoionization of each particular atom in the mixture is guaranteed by the selection of proper frequencies and intensities of the laser beams. This method, in principle, may provide detection of single atoms, but, unlike that of fluorescence, it is a "destructive" method of detection. It is advisable to use it, for instance, to detect atoms in metastable states for which the probability of fluorescence is small and for which a relatively low power is needed for photoionization (especially with an external electric field present). It goes without saying that the photoionization method of selective atomic detection has a very important advantage over all other methods: that is its ability to extract a detected atom with the aid of external electric and magnetic fields.

After the first experiment on selective ionization of Rb atoms had been conducted in work (20), Sec. III, the method was successfully applied in work (358) to detect Na atoms in a beam. Clearly the development of selective photoionization to separate isotopes and purify substances will lead to vigorous scientific exploitation of spectroscopy by the detection and extraction of particular atoms from mixtures.

*Detection of complex molecules.* Selective detection of trace amounts of complex polyatomic molecules is a very hard problem. Mass spectral analysis is now a common method for detection and identification of complex molecules but its sensitivity is inadequate and there is practically no selectivity for complex molecules which differ only in spatial structure. Therefore the development of new detection methods is an urgent problem today.

The method of selective molecular ionization by laser radiation may be used as the basis for the so-called laser mass-spectrometer (359) illustrated schematically in Fig. 59. A laser with the tunable frequency $\omega_1$ excites selectively a vibrational (or electronic for some molecules) state of the

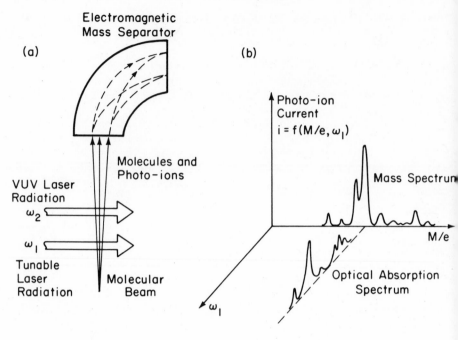

(a)     (b)

Molecules and
Photo-ions

VUV Laser
Radiation
$\omega_2$

$\omega_1$

Tunable
Laser          Molecular
Radiation      Beam

Photo-ion
Current
$i = f(M/e, \omega_l)$

Mass Spectrum

M/e

Optical Absorption
Spectrum

$\omega_l$

*Fig. 59.  Possible laser two-dimensional optical-mass-spectrometer (359).*

molecules.  The excitation shifts the edge of the molecular
photoionization band, usuallly in the vuv region, by a small
amount.  The second, vuv laser causes molecular photoionization, and its frequency is set at the point of maximum slope
of the photoionization band edge.  The selective excitation of
the molecules at $\omega_1$ even by a comparatively small amount
$\hbar\omega_1 = E_{exc} \simeq 0.1 - 0.5$ eV results in a marked change of the
photoionization cross section (0.01 - 0.1% when the molecular
distribution over the rotational states is accounted for),
that is, a change in the photocurrent.  Photoions are sent then
into a common mass-spectrometer which measures the mass-
spectrum, i.e. i = f(M/e).  With this version of mass-spectrometer we can also measure the photocurrent amplitude for a
fixed value of M/e as a function of the tunable laser frequency $\omega_1$.  In this case the IR spectrum of trace amounts of
complex molecules is measured since, with the $\omega_1$ frequency and
that of molecular absorption coincident, the molecules pass
into the excited state and hence the ionic photocurrent amplitude varies.  Thus the laser mass-spectrometer with selective
molecular ionization, instead of the usual nonselective ion-
nization by an electron beam (or by continuous vuv radiation), will produce at the same time the optical (IR and
visible) absorption spectrum and the mass-spectrum.  This

method enables one to obtain information on the spatial structure of molecules with the same mass, etc.

The systematic studies of this method of molecular detection were started after the simple vuv $H_2$-laser had been elaborated (360). First, the one-step photoionization of dimethylaniline and methylaniline molecules by $H_2$-laser radiation in the 1600 Å region (361) as well as NO molecules in the 1200 Å region have been carried out. Recently the two-step photoionization of $H_2CO$ molecules by joint action of radiation pulses of an $N_2$-laser with $\lambda_1$ = 3371 Å exciting the $^1A_2$ state and an $H_2$-laser at $\lambda_2$ = 1600 Å photoionizing the excited molecules have been realized (84). The experiments on molecular beams with a tunable laser are started also (348). In particular, in work (362) the one-step photoionization of dimethylaniline and other molecules in a molecular beam at the entrance of a mass-spectrometer by an $H_2$-laser at 1600 Å has been demonstrated. The magnitude of photoion current was increased by several orders. With the progress in the technique of the generation of tunable vuv radiation (see, for example, reports (363 - 365) and of excimer lasers (113) the possibilities of selective photoionization of molecules have increased drastically.

The two-step photoionization mass-spectrometer may become a universal, highly selective and sensitive detector of complex molecules for solving many scientific and applied problems. Indeed, since the detection of excited molecules through their photoionization is highly sensitive, we may hope to measure the IR spectrum of quantities of substances much smaller than possible with the best classical and laser IR spectrometers. It is also possible to realize at the same time an extremely high spectral resolution which is limited, in principle, only by the residual Doppler broadening because of the angular divergence of the molecular beam. A high efficiency, state-specific detector for molecular beam research should be possible. It may even be possible to achieve the sensitivity and selectivity necessary to detect and identify trace quantities of organic mixtures with the perfection that can be attained by the organs of smell of human beings and animals.

*Spatial localization of molecular bonds.* In principle the selective action of laser radiation on molecular bonds opens up the possibility of spatial localization of particular molecular bonds, and thus for macromolecular "mapping." The idea of this approach can be understood from the so-called photoelectronic (photoionic) laser microscope (336) which is schematically shown in Fig. 60. In the common field-electron or field-ion Müller projectors (366), an electron or an ion is pulled out of the material nonselectively by field-induced ionization in a strong electric field. Here ionization is

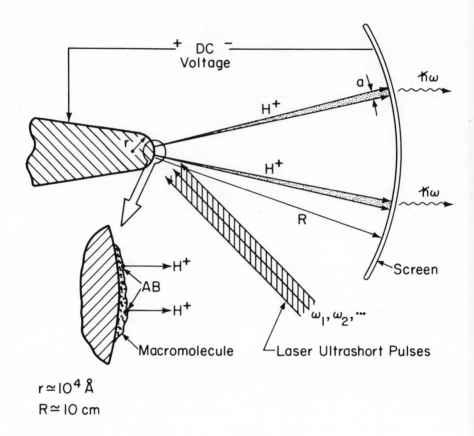

$r \simeq 10^4$ Å
$R \simeq 10$ cm

Fig. 60.  Possible laser-ion microscope for spatial localization of molecular bonds (336).

induced selectively by laser radiation. The only function of the d.c. electric field is to transport electrons or ions along radial paths to the projector screen. Selective photoionization near certain molecular bonds or chromophores in a macromolecule located on the point of the projector might be accomplished by the multistep scheme using several picosecond laser pulses at selected frequencies.

In case of selective ejection of an electron it is possible to attain a resolution of about 25 Å which is limited fundamentally by: the tangential velocity component of the emitted electron and by the uncertainty principle (de Broglie wavelength). In some cases after electron emission, the resultant positive molecular ion becomes unstable and spontaneously emits a proton. If we change the polarity in the projector, it is possible to accelerate protons instead of electrons to the screen. The spatial localization of the ionization point for protons must be $(m_p/m_e)^{\frac{1}{2}} \simeq 40$ times smaller than for photoelectrons.  Such a laser photoion- ic microscope may have a resolution quite sufficient to re- solve atomic details in molecular structures.  Further increase in the resolution of the laser photoemission micro- scope might also be possible by selective photoionization to heavier molecular ions.

The idea of this new approach to atomic-resolution micro- scopy is based on the combination of two important properties which usually belong to quite different methods.  Corpuscular (electron, for instance) microscopy ensures a high spatial resolution if the particle has a high energy.  But in this case the "contrast" of the image is lost automatically because the change in particle energy due to interaction with the object observed is difficult to resolve from the spread in the high initial energy.  Conversely, as the particle energy drops, the image "contrast" may increase considerably and energy resolution permits observation of image details, but the spatial resolution decreases.  The method of selective photoionization makes it possible to combine the high selec- tivity (or contrast) of the "optical channel" with the high spatial resolution of the "corpuscular (ionic) channel."

Determination of the nucleotide sequence in a DNA mole- cule which carries the hereditary information for every indi- vidual organism would be one of the most important applica- tions of the laser ion microscope.  This problem consists in selective ejection of a proton (or heavier molecular ion) either from the base pair A - T or G - C.  Of course, we may try to realize the method based on two-step selective IR-uv excitation of an electronic state through an intermediate vibrational state of a particular nucleotide and subsequent ionization of the excited electronic state.  It is necessary that the selectivity of this process be sufficient for good distinction between the pair of bonds A - T and G - C.  Parti- cular attention should be drawn to the conceptual possibility of preliminary chemical "marking" of either base pair to simplify selective excitation and ionization.  For instance, it is known (367) that one of the bases of DNA (cytosine) selectively reacts with hydroxylamine.  The reaction is followed by changes in the uv spectrum of the cytosine absorp- tion, the magnitude of which is much larger than that of the typical difference between the spectra of the bases.  In this case selective two-step photoionization through the electron- ically excited state of the cytosine-hydroxylamine reaction product may suffice.

When the magnification coefficient of a laser ionic pro- jector is $M = 3 \times 10^5$ and the screen dimension is $R = 10$ cm, a section of linear DNA chain with about $10^3$ nucleotides will be imaged.  Adjacent nucleotides are spaced 3.3 Å apart.  To record the sequence of nucleotides in long DNA chains, we shall need, of course, a sequence of overlapping projections.

Successful realization of a laser ionic microscope for observing biological molecules requires investigation of a number of rather difficult problems: 1) selective excitation of macromolecules adsorbed on surfaces; 2) search for schemes of selective dissociative ionization resulting in emission of protons or heavier molecular ions; 3) spatial scanning of the microscope needle point along the chain of a macromolecule, etc. The solution of the problem of direct observation of nucleotide sequences in genes, including those of man, would open up many possibilities in the study and control of heredity. Such a laser ionic microscope would be of immense value in many fields and certainly merits the very substantial effort which its development will require.

We are deeply grateful to our collaborators in Berkeley and Moscow for their contributions to all aspects of our research on laser isotope separation. We thank our colleagues from around the world who have eagerly shared their research results and ideas with us on many occasions. Our work has depended upon much appreciated support from the Academy of Sciences of the U.S.S.R., the University of California, and the National Science Foundation, the Army Research Office and the Energy Research and Development Administration of the U.S.A.

XIII. REFERENCES


1.  V.S. Letokhov, Science 180, 451 (1973).
2.  C.B. Moore, Acc. Chem. Res. 6, 323 (1973).
3.  V.S. Letokhov and C.B. Moore, Kvant. Elektron. (Moscow)
    3, 248 - 287 and 485 - 516 (1976). [Sov. J. Quantum
    Electron. 6, 129 - 150 and 259 - 276 (1976)].
4.  S.R. Leone, in Physics of Quantum Electronics Vol. 4,
    Laser Photochemistry, Tunable Lasers and other Topics.
    Eds. S.F. Jacobs et al. Addison-Wesley 1976, p. 17.
5.  J.P. Aldridge, J.H. Birely, C.D. Cantrell and D.C. Cart-
    wright, in Physics of Quantum Electronics Vol. 4,  Laser
    Photochemistry, Tunable Lasers and other Topics.    Eds.
    S.F. Jacobs et al. Addison-Wesley 1976, p. 57.
6.  T.R. Merton and H. Hartley, Nature (London) 105, 2630
    (1920).
7.  H. Hartley, A.O. Ponder, E.J. Bowen, and T.R. Merton,
    Philos. Mag. 43, 430 (1922).
8.  W. Kuhn and H. Martin, Naturwiss. 20, 772 (1932);  Z.
    Phys. Chem. Abt. B 21, 93 (1933).
9.  S. Mrozowski, Z. Phys. 78, 826 (1932);  78, 844 (1932).
10. K. Zuber, Nature (London) 136, 796 (1935);  Helv. Phys.
    Acta 8, 487 (1935).
11. K. Zuber, Helv. Phys. Acta 9, 285 (1936).
12. C.C. McDonald and H.E. Gunning, J. Chem. Phys. 20, 1817
    (1952).
13. B.H. Billings, W.J. Hitchcock, and M. Zelikoff, J. Chem.
    Phys. 21, 1762 (1953).
14. J.P. Morand and G. Nief, J. Chim. Phys. 65, 2058 (1968).
15. J.P. Morand, M. Wacongne,  and E. Roth, Energ. Nucl.
    (Paris) 10, 362 (1968).
16. W.B. Tiffany, H.W. Moos, and A.L. Schawlow, Science 157,
    40 (1967).
17. W.B. Tiffany, J. Chem. Phys. 48, 3019 (1968).
18. S.W. Mayer, M.A. Kwok, R.W.F. Gross, and D.J. Spencer,
    Appl. Phys. Lett. 17, 516 (1970).
19. J. Robieux and J.M. Auclair, French Patent No. 1391738,
    appl. October 21, 1963, publ. February 1, 1965;  USA
    Patent No. 3443087, appl. September 16, 1964, publ. May
    6, 1969.
20. R.V. Ambartzumian, V.P. Kalinin, and V.S. Letokhov,
    ZhETF Pis'ma Red. 13, 305 (1971) [JETP Lett. 13, 217
    (1971)].
21. V.S. Letokhov and R.V. Ambartzumian, IEEE J. Quantum
    Electron. QE-7, 305 (1971);  R.V. Ambartzumian and V.S.
    Letokhov, Appl. Opt. 11, 354 (1972).

22.  V.S. Letokhov, G.N. Makarov, and A.V. Puretzkii, ZhETF
     Pis'ma Red. 15, 709 (1972) [JETP Lett. 15, 501 (1972)]
     and R.V. Ambartzumian, V.S. Letokhov, G.N. Makarov, and
     A.V. Puretzkii, ZhETF Pis'ma Red. 17, 91 (1973) [JETP
     Lett. 17, 63 (1973)].

23.  E.S. Yeung and C.B. Moore, Appl. Phys. Lett. 21, 109
     (1972).

24.  V.S. Letokhov, Chem. Phys. Lett. 15, 221 (1972).

25.  R.V. Ambartzumian, V.S. Letokhov, E.A. Ryabov, and N.V.
     Chekalin, ZhETF Pis'ma Red. 20, 597 (1974) [JETP Lett.
     20, 273 (1974)].

26.  R.V. Ambartzumian, Yu. A. Gorokhov, V.S. Letokhov, and
     G.N. Makarov, ZhETF Pis'ma Red. 21, 375 (1975) [JETP
     Lett. 21, 171 (1975)].

27.  H. Dubost, L. Abouaf-Marguin, F. Legay, Phys. Rev. Lett.,
     29, 145 (1972).

28.  E.M. Belenov, E.P. Markin, A.N. Oraevskii, V.I. Romanenko,
     ZhETF Pis'ma Red. 18, 196 (1973) [JETP Lett. 18, 116
     (1973)].

29.  V.S. Letokhov, A.A. Makarov, J. Photochem. 3, 249 (1974).

30.  T.J. Manuccia and M.D. Clark, Appl. Phys. Lett. 28, 372
     (1976).

31.  I.I. Sobelman, Introduction to the Theory of Atomic Spec-
     tra, Pergamon Press, Oxford (1973).

32.  H.G. Kuhn, Atomic Spectra, 2nd ed., Academic Press, New
     York (1970), Ch. VI.

33.  Yu. B. Zel'dovich and I.I. Sobelman, ZhETF Pis'ma Red.
     21, 368 (1975) [JETP Lett. 21, 168 (1975)].

34.  G. Herzberg, Molecular Spectra and Molecular Structure,
     Vol. 1,  Spectra of Diatomic Molecules, Van Nostrand,
     New York 1950.

35.  G. Herzberg, Molecular Spectra and Molecular Structure,
     Vol. 2,  Infrared and Raman Spectra of Polyatomic Mole-
     cules, Van Nostrand, New York 1945.

36.  G. Herzberg, Molecular Spectra and Molecular Structure,
     Vol. 3,  Electronic Spectra and Electronic Structure of
     Polyatomic Molecules, Van Nostrand, New York 1966.

37.  E.D. Hinkley, App. Phys. Lett. 16, 351 (1970);  M.W.
     Goldberg and R. Yusek, Appl. Phys. Lett. 17, 349 (1970).

38.  R.J. Jensen, J.G. Marinuzzi, C.P. Robinson and S.D. Rock-
     wood, Laser Focus 12(5), 51 (1976).

39.  O.N. Kompanets, A.R. Kukudzhanov, V.S. Letokhov, V.G.
     Minogin, and E.L. Mikhailov, Zh. Eksp. Teor. Fiz. 69,
     32 (1975) [Sov. Phys. - JETP 42, 15 (1975)].

40.  J.W. Eerkens, Appl. Physics. 10, 15 (1976).

41.  C.G. Stevens and R.N. Zare, J. Mol. Spectrosc. 56, 167
     (1975).

42.  M.S. Sorem, T.W. Hänsch and A.L. Schawlow, Chem. Phys.
     Lett. 17, 300 (1972).

43. A.C.G. Mitchell and M.W. Zemansky, Resonance Radiation and Excited Atoms, Cambridge University Press (1934), pp. 97-103.

44. L. Andrews, Annual Rev. Phys. Chem. 22, 109 (1971); B. Meyer, Low Temperature Spectroscopy, American Elsevier, New York (1971).

45. G.H. Dieke and A.B.F. Duncan, Spectroscopic Properties of Uranium Compounds, McGraw-Hill, New York (1949), p 77.

46. H. Dubost, Thèse d'Etat, Univ. de Paris-Sud, 1975; and F. Legay, in Chemical and Biochemical Applications of Lasers, Vol. II. Ed. C.B. Moore. Academic Press, New York, 1977, p. 43.

47. V.V. Bertsev, T.D. Kolomiitseva and N.M. Tsyganenko, Opt. Spektrosk. 37, 463 (1974) [Opt. and Spectrosc. 37, 263 (1974)].

48. L.S. Vasilenko, V.P. Chebotaev, and A.V. Shishaev, ZhETF Pis'ma Red. 12, 161 (1970) [JETP Lett. 12, 113 (1979)] and V.P. Chebotaev, A.L. Golger, and V.S. Letokhov. Chem. Phys. 7, 316 (1975).

49. P.L. Kelley, H. Kildal and H.R. Schlossberg, Chem. Phys. Lett. 27, 62 (1974).

50. K. Shimoda, Appl. Phys. 9, 239 (1976).

51. S. Ezekiel and R. Weiss, Phys. Rev. Lett. 20, 91 (1968).

52. U. Brinkmann, W. Hartig, H. Telle, and H. Walther, Appl. Phys. 5, 109 (1974).

53. A. Bernhardt, D. Duerre, J. Simpson, L. Wood, J. Goldsborough, and C. Jones, IEEE J. Quantum Electron. QE-10, 789 (1974).

54. D.I. Kataev and A.A. Mal'tsev, Zh. Eksp. Teor. Fiz. 64, 1527 (1973) [Sov. Phys. - JETP 37, 772 (1973)].

55. S.E. Novick, P. Davies, S.J. Harris, and W. Klemperer, J. Chem. Phys. 59, 2273 (1973).

56. R.E. Smalley, L. Wharton and D.H. Levy, J. Chem. Phys. 63, 4977 (1976) and 64, 3266 (1976).

57. D.H. Levy, L. Wharton and R.E. Smalley, Chemical and Biochemical Applications of Lasers, Vol. II. Ed. C.B. Moore. Academic Press, New York 1977, p. 1.

58. Y.T. Lee, J.D. McDonald, P.R. LeBreton, and D.R. Herschbach, Rev. Sci. Instrum. 40, 1402 (1969).

59. V.K. Konyukhov and A.M. Prokhorov, ZhETF Pis'ma Red. 3, 436 (1966) [JETP Lett. 3, 286 (1966)].

60. N.G. Basov, A.N. Oraevskii, and V.A. Shcheglov, Zh. Tekh. Fiz. 37, 339 (1967) [Sov. Phys. - Tech. Phys. 12, 243 (1967)].

61. E.T. Gerry, Laser Focus 6, No. 12, 27 (1970).

62. L.A. Pakhomicheva, E.A. Sviridenkov, A.F. Suchkov, L.A. Titova, and S.S. Churilov, ZhETF Pis'ma Red. 12, 60 (1970) [JETP Lett. 12, 43 (1970)].

63. N.C. Peterson, M.J. Kurylo, W. Braun, A.M. Bass, and R.A. Keller, J. Opt. Soc. Am. 61, 746 (1971).

64. R.V. Ambartzumian, V.S. Letokhov, S.A. Maksimov, V.I. Mishin, and N.P. Furzikov, Kvantovaya Elektron. (Moscow) 2, 1851 (1975) [Sov. J. Quantum Electron. 5, 1018 (1975)].

65. S. Datta, R.W. Anderson and R.N. Zare, J. Chem. Phys. 63, 5503 (1976).

66. I. Nebenzahl and M. Levin, West German Patent No. 2312194, appl. March 19, 1972, publ. October 4, 1973.

67. R.H. Levy and G.S. Janes, USA Patent No. 3772519, appl. March 25, 1970, publ. November 13, 1973.

68. L.N. Ivanov and V.S. Letokhov, Kvantovaya Elektron. (Moscow) 2, 585 (1975) [Sov. J. Quantum Electron. 5, 329 (1975)].

69. R.V. Ambartzumian and V.S. Letokhov, Report on Gordon Research Conf. on Physics and Chemistry of Isotopes, Asilomar, California, 1974.

70. S.E. Harris and D.B. Lidow, Appl. Phys. Lett. 26, 104 (1975).

71. R.V. Ambartzumian, V.M. Apatin, V.S. Letokhov, A.A. Makarov, V.I. Mishin, A.A. Puretzkii, and N.P. Furzikov, Zh. Eksp. Teor. Fiz. 70, 1660(1976) [Sov. Phys. - JETP (1976)].

72. B.M. Smirnov, Atomic Collisions and Elementary Processes in Plasma, Atomizdat, Moscow, 1968.

73. S.A. Tuccio, J.W. Dubrin, O.G. Peterson, and B.B. Snavely, IEEE J. Quantum Electron QE-10, 790 (1974).

74. B.B. Snavely, R.W. Solarz, and S.A. Tuccio, in Lecture Notes in Physics Vol. 43, Laser Spectroscopy. Eds. S. Haroche et al. Proceedings of the Second International Conference, Megeve, France. Springer-Verlag 1975, p. 268.

75. G.S. Janes, I. Itzkan, C.T. Pike, R.H. Levy, and L. Levin, in Lecture Notes in Physics Vol. 43, Laser Spectroscopy. Eds. S. Haroche et al. Proceedings of the Second International Conference, Megeve, France. Springer-Verlag 1975, p. 454, and H.K. Forsen, G.S. Janes and R.H. Levy, A.N.S. Transactions 22, 312 (1975).

76. Press release of "Exxon Nuclear Company, Inc.," March 5, 1976 (courtesy of Dr. H.K. Forsen).

77. N.V. Karlov, B.B. Krynetsky, V.A. Mishin, A.M. Prokhorov, A.D. Saveliev and V.V. Smirnov, Kvantovaya Elektron. (Moscow) 3, 2486 (1976).

78. D.J. Bradley, P. Ewart, J.V. Nicholas, J.R.D. Shaw, and D.G. Thompson, Phys. Rev. Lett. 31, 263 (1973).

79. R.V. Ambartzumian, G.I. Bekov, V.S. Letokhov, and V.I. Mishin, ZhETF Pis'ma Red. 21, 595 (1975) [JETP Lett. 21, 279 (1975)].

80. T.W. Ducas, M.G. Littman, R.R. Freeman, D. Kleppner, Phys. Rev. Lett., 35, 366 (1975).

81.  M.G. Littman, M.L. Zimmerman, D. Kleppner, Phys. Rev.
     Lett., 37, 486 (1976).
82.  R.W. Solarz, C.A. May, L.R. Carlson, E.F. Worden, S.A.
     Johnson, J.A. Paisner, L.J. Radziemski, Phys. Rev. A.
     14, 1129 (1976).
83.  G.V. Karachevtsev, V.K. Potapov, and V.V. Sorokin, Dokl.
     Akad. Nauk SSSR 206, 104 (1972).
84.  S.V. Andreev, V.S. Antonov, I.N. Knyazev, V.S. Letokhov,
     Chem. Phys. Lett. 45, 166 (1977).
85.  S.D. Rockwood, in Optical Sciences Vol. 3, Tunable Lasers
     and Applications. Eds. A. Mooradian et al. Proceedings
     of the Loen Conference, Norway. Springer-Verlag 1976,
     p. 140.
86.  R.V. Ambartzumian, V.S. Letokhov, G.N. Makarov, A.A.
     Puretzkii, in Laser Spectroscopy. Eds. R. Brewer and
     A. Mooradian. Proceedings of the First Laser Spectro-
     scopy Conference, Vail, Colorado, 1973. Plenum Press,
     New York 1974, p. 611.
88.  R.V. Ambartzumian, V.S. Letokhov, G.N. Makarov, A.G.
     Platova, A.A. Puretzkii, O.A. Tumanov, Zh. Eksp. Teor.
     Fiz., 64, 770 (1973) [Sov. Phys. - JETP 37, 392 (1973)].
89.  R.V. Ambartzumian, V.S. Letokhov, G.N. Makarov, and A.A.
     Puretzkii, Zh. Eksp. Teor. Fiz. 68, 1736 (1975) [Sov.
     Phys. - JETP 41, 871 (1975)].
90.  P.F. Zittel and C.B. Moore, J. Chem. Phys. 58, 2004
     (1973).
91.  R.D. Bates, Jr., J.T. Knudtson, G.W. Flynn and A.M. Ronn,
     Chem. Phys. Letts. 8, 103 (1971).
92.  R.C.L. Yuan and G.W. Flynn, J. Chem. Phys. 57, 1316
     (1972).
93.  R.V. Ambartzumian, V.M. Apatin, and V.S. Letokhov,
     ZhETF Pis'ma Red. 15, 336 (1972) [JETP. Lett. 15, 237
     (1972)].
94.  R.V. Ambartzumian, Yu.A. Gorokhov, V.S. Letokhov, G.N.
     Makarov, A.A. Puretzkii, N.P. Furzikov, ZhETF Pis'ma
     Red. 23, 217 (1976) [JETP Lett. 23, 194 (1976)].
95.  V.S. Letokhov and A.A. Makarov, Zh. Eksp. Teor. Fiz. 63,
     2064 (1972) [Sov. Phys. - JETP 36, 1091 (1973)].
96.  R.E. Scurby, J.R. Lacher, and J.D. Park, J. Chem. Phys.
     19, 386 (1951).
97.  N.V. Karlov, G.P. Kuz'min, Yu.N. Petrov, and A.M. Prokho-
     rov, ZhETF Pis'ma Red. 7, 174 (1968) [JETP Lett. 7, 134
     (1968)].
98.  P. Lavigne and J.L. Lachambre, Appl. Phys. Lett. 19, 176
     (1971).
99.  V.S. Letokhov, A.A. Makarov, and E.A. Ryabov, Dokl. Akad.
     Nauk SSSR 212, 75 (1973) [Sov. Phys. - Dokl. 18, 603
     (1972)].

100. E.E. Stark, Jr., W.H. Reichelt, G.T. Schappert, and T.F. Stratton, Appl. Phys. Lett. $\underline{23}$, 322 (1973).

101. Y.S. Liu, Appl. Opt. $\underline{13}$, 2505 (1974) and J. Dupré, F. Meyer-Bourbonneux, C. Meyer and P. Barchewitz, Infrared Physics $\underline{16}$, 475 (1976).

102. H.M. Mould, W.C. Price, and G.R. Wilkinson, Spectrochim. Acta $\underline{15}$, 313 (1959) and F.O. Shimizu and T. Shimizu, J. Mol. Spectrosc. $\underline{36}$, 94 (1970).

103. A.D. Walsh and P.A. Warsop, Trans. Faraday Soc. $\underline{57}$, 345 (1961).

104. J.R. McNesby and H. Okabe, Adv. Photochem. $\underline{3}$, 157 (1964).

105. R.V. Ambartzumian, V.S. Letokhov, G.N. Makarov, and A. A. Puretzkii, Dokl. Akad. Nauk SSSR $\underline{211}$, 365 (1973).

106. N. Noguchi and Y. Izawa, Progress Report X, Osaka University (June, 1974), p. 63 and Annual Progress Rept. on Laser Fusion Program 1976, Institute of Laser Engineering Osaka University, p. 80.

107. S. Rockwood and S.W. Rabideau, IEEE J. Quantum Electron. $\underline{QE-10}$, 789 (1974).

108. S.R. Leone and C.B. Moore, Chem. Phys. Lett. $\underline{19}$, 340 (1973) and C.B. Moore and P.F. Zittel, Science $\underline{182}$, 541 (1973).

109. P.L. Houston, A.V. Nowak and J.I. Steinfeld, J. Chem. Phys. $\underline{58}$, 3373 (1973).

110. V.S. Letokhov and A.A. Makarov, J. Photochem. $\underline{2}$, 421 (1974).

111. R.V. Ambartzumian, Yu.A. Gorokhov, V.S. Letokhov, G.N. Makarov. ZhETF Pis'ma Red. $\underline{22}$, 96 (1975) [JETP Lett. $\underline{22}$, 43 (1975)].

112. M.L. Bhaumik, Laser Focus $\underline{12}$, (2) 54 (1976).

113. J.J. Ewing, Chemical and Biochemical Applications of Lasers, Vol. II. Ed. C.B. Moore. Academic Press, New York 1977, p. 241.

114. C.B. Moore and E.S. Yeung, U.S.A. Patent No. 3,983,020 appl. Jan. 26, 1973, publ. Sept. 28, 1976.

115. J.I. Steinfeld and A.N. Schweid, J. Chem. Phys. $\underline{53}$, 3304 (1970).

116. R.B. Kurzel and J.I. Steinfeld, J. Chem. Phys. $\underline{53}$, 3293 (1970) ;   J.I. Steinfeld and A.N. Schweid, J. Chem. Phys. $\underline{53}$, 3304 (1970) and G.D. Chapman and P.R. Bunker, J. Chem. Phys. $\underline{57}$, 2951 (1972).

117. V.S. Letokhov and Yu.E. Lozovik, Kvantovaya Elektron. (Moscow) $\underline{1}$, 2496 (1974) [Sov. J. Quantum Electron. $\underline{4}$, 1391 (1975)].

118. Y.T. Lee, private communication.

119. S.J. Harris, K.C. Janda, S.E. Novick and W. Klemperer, J. Chem. Phys. $\underline{63}$, 881 (1975).

120. S.R. Leone and C.B. Moore, Phys. Rev. Lett. $\underline{33}$, 269 (1974).

121.  F. Zaraga, N.S. Nogar and C.B. Moore, J. Mol. Spectrosc. 63, 564 (1976).

122.  K. Suzuki, P.H. Kim, K. Taki, and S. Namba, Japan J. Appl. Phys. 15, 2255 (1976).

123.  F. Zaraga, S.R. Leone and C.B. Moore, Chem. Phys. Lett. 42, 275 (1976) and K. Bergmann, S.R. Leone, and C.B. Moore, J. Chem. Phys. 63, 4161 (1975).

124.  S.A. Bazhutin, V.S. Letokhov, A.A. Makarov, and V.A. Semchishen, ZhETF Pis'ma Red. 18, 515 (1973) [JETP Lett. 18, 303 (1973)].

125.  V.S. Letokhov and V.A. Semchishen, Dokl. Akad. Nauk SSSR 222, 1071 (1975) [Sov. Phys. - Dokl. 20, 423 (1975)].

126.  V.I. Balikin, V.S. Letokhov, V.I. Mishin, V.A. Semchishen, Chem. Phys. 17, 111 (1976).

127.  M. Broyer, J. Vigue and J.-C. Lehmann, J. Chem. Phys. 64, 4793 (1976) and op. cited therein.

128.  C.D. Olson, K.K. Innes, J. Chem. Phys. 64, 2405 (1976).

129.  R.D. Hudson and V.L. Carter, Can. J. Chem. 47, 1840 (1969). The preparation of $^{18}O$ enriched $O_3$ by photolysis of $O_2$ with a filtered ArF laser at 193 nm has been reported. R.K. Sander, T.R. Loree, S.D. Rockwood, and S.M. Freund, Appl. Phys. Lett. 30, 151 (1977).

130.  H.B. Palmer and D.W. Naegeli, J. Mol. Spectrosc. 28, 417 (1968).

131.  T.R. Govers, C.A. van de Runstraat and F.J. de Heer, J. Phys. B6, 173 (1973).

132.  E.S. Yeung and C.B. Moore, J. Chem. Phys. 58, 3988 (1973).

133.  E.S. Yeung and C.B. Moore, J. Chem. Phys. 60, 2139 (1974).

134.  E.S. Yeung and C.B. Moore, in Fundamental and Applied Laser Physics. Eds. M. Feld et al. Proceedings of Esfahan Symposium, 1971. Wiley-Interscience, New York 1973, p. 223.

135.  R.V. Ambartzumian, V.M. Apatin, V.S. Letokhov, and V.I. Mishin, Kvantovaya Elektron. (Moscow) 2, 337 (1975) [Sov. J. Quantum Electron. 5, 191 (1975)].

136.  N.M. Bazhin, G.I. Skubnevskaya, N.I. Sorokin, and Yu.N. Molin, ZhETF Pis'ma Red. 20, 41 (1974) [JETP Lett. 20, 18 (1974)].

137.  J.H. Clark, Y. Haas, P.L. Houston, and C.B. Moore, Chem. Phys. Lett. 35, 82 (1975).

138.  A.P. Baronavski, J.H. Clark, Y. Haas, P.L. Houston, and C.B. Moore in Lecture Notes in Physics Vol. 43, Laser Spectroscopy. Eds. S. Haroche et al. Proceedings of the Second International Conference, Megeve, France. Springer-Verlag 1975, p. 259.

139.  J.B. Marling, Chem. Phys. Lett. $\underline{34}$, 84 (1975).
140.  A.P. Baronavski, A. Cabello, J.H. Clark, Y. Haas, P.L. Houston, A.H. Kung, C.B. Moore, J. Reilly, J.C. Weisshaar and M.B. Zughul in Optical Sciences Vol. 3, Tunable Lasers and Applications. Eds. A. Mooradian et al. Proceedings of the Loen Conference, Norway. Springer-Verlag 1976, p. 108.
141.  J.B. Marling, J. Chem. Phys $\underline{66}$,      (1977).
142.  J.H. Clark, J. Reilly, and C.B. Moore, J. Chem. Phys. (in press).
143.  J.H. Clark N.S. Nogar and C.B. Moore, J. Chem. Phys. (in press).
144.  R.D. McQuigg and J.G. Calvert, J. Am. Chem. Soc. $\underline{91}$, 1590 (1969).
145.  P.L. Houston and C.B. Moore, J. Chem. Phys. $\underline{65}$, 757 (1976).
146.  A.P. Baronavski, Ph.D. Thesis, University of California, Berkeley,1975.
147.  A. Cabello-Albala, J.C. Weisshaar and C.B. Moore, J. Chem. Phys. (in press).
148.  R.G. Miller and E.K.C. Lee, Chem. Phys. Lett. $\underline{27}$, 475 (1974).
149.  R.R. Karl and K.K. Innes, Chem. Phys. Lett. $\underline{36}$, 275 (1975).
150.  R.M. Hochstrasser and D.S. King, J. Am. Chem. Soc. $\underline{98}$, 5443 (1976).
151.  R.M. Hochstrasser and D.S. King, J. Am. Chem. Soc. $\underline{97}$, 4760 (1975).
152.  J. Langelaar, as quoted in Ref. 153.
153.  R.M. Hochstrasser, D.S. King and A.C. Nelson, Chem. Phys. Lett. $\underline{42}$, 8 (1976).
154.  D.S. King, C.T. Denny, R.M. Hochstrasser and A.B. Smith, III, J. Am. Chem. Soc. $\underline{99}$, 271 (1977).
155.  R.E. Smalley, D.A. Chandler, L. Wharton and D.H. Levy, J. Mol. Spectrosc. (in press).
156.  Z.B. Vukmirović and S.V. Ribnikov, J. Chem. Phys. $\underline{66}$, 7 (1977).
157.  M. Carlotti, G. DiLonardo, G. Galloni, and A. Trombetti, Trans. Faraday Soc. $\underline{67}$, 2852 (1971).
158.  R.V. Ambartzumian, V.S. Letokhov, G.N. Makarov, and A. A. Puretzkii, in Laser Spectroscopy. Eds. R. Brewer and A. Mooradian. Proceedings of the First Laser Spectroscopy Conference, Vail, Colorado, 1973. Plenum Press, New York 1974, p. 611.
159.  R.B. Woodward and R. Hoffmann, Conservation of Orbital Symmetry, Academic Press, New York 1971.
160.  C.B. Moore and G.C. Pimentel, J. Chem. Phys. $\underline{41}$, 3504 (1964).

161. N.R. Isenor, M.C. Richardson, Appl. Phys. Lett. 18, 224 (1971).
162. V.S. Letokhov, E.A. Ryabov, O.A. Tumanov, Opt. Commun. 5, 168 (1972).
163. V.S. Letokhov, E.A. Ryabov, and O.A. Tumanov, Zh. Eksp. Teor. Fiz. 63, 2025 (1972) [Sov. Phys. - JETP 36, 1069 (1973)].
164. R.V. Ambartzumian, N.V. Chekalin, V.S. Dolzhikov, V.S. Letokhov, E.A. Ryabov, Chem. Phys. Lett. 25, 515 (1974).
165. R.V. Ambartzumian, V.S. Dolzhikov, V.S. Letokhov, E.A. Ryabov, and N.V. Chekalin, Zh. Eksp. Teor. Fiz. 69, 72 (1975) [Sov. Phys. - JETP 42, 36 (1975)].
166. N.R. Isenor, V. Merchant, R.S. Hallsworth, M.S. Richardson, Can. J. Phys. 51, 1281 (1973).
167. R.V. Ambartzumian, Yu. A. Gorokhov, V.S. Letokhov, G. N. Makarov, E.A. Ryabov, and N.V. Chekalin, Kvantovaya Elektron. (Moscow) 2, 2197 (1975) [Sov. J. Quantum Electron. 5, 1196 (1975)].
168. R.V. Ambartzumian, Yu.A. Gorokhov, V.S. Letokhov, and G.N. Makarov, Zh. Eksp. Teor. Fiz. 69, 1956 (1975) [Sov. Phys. - JETP 42, 993 (1975)].
169. R.V. Ambartzumian, N.V. Chekalin, Yu.A. Gorokhov, V.S. Letokhov, G.N. Makarov, and E.A. Ryabov, in Lecture Notes in Physics Vol. 43, Laser Spectroscopy. Eds. S. Haroche et al. Proceedings of the Second International Conference, Megeve, France. Springer-Verlag 1975, p. 121, and Proceedings of the Fourth Vavilov Conference on Nonlinear Optics, Novosibirsk 1975 in Kvantovaya Elektron. (Moscow) 3, 802 (1976) [Sov. J. Quantum Electron. 6, 437 (1976)].
170. J.L. Lyman, R.J. Jensen, J. Rink, C.P. Robinson, S.D. Rockwood, Appl. Phys. Lett. 27, 87 (1975).
171. J.L. Lyman, S.D. Rockwood, J. Appl. Phys. 47, 595 (1976).
172. A. Yogev and R.M.J. Benmair, J. Am. Chem. Soc. 97, 4430 (1975).
173. N.V. Chekalin, V.S. Dolzhikov, Yu.R. Kolomiisky, V.N. Lokhman, V.S. Letokhov, and E.A. Ryabov, Phys. Lett. 59A, 243 (1976).
174. G. Koren, U.P. Oppenheim, D. Tal, M. Okon and R. Weil, Appl. Phys. Lett. 29, 40 (1976).
175. S.M. Freund and J.L. Lyman, postdeadline paper at Tunable Lasers and Applications Conference, Loen, Norway 1976.
176. R.V. Ambartzumian, Yu.A. Gorokhov, V.S. Letokhov, G.N. Makarov, and A.A. Puretzkii, ZhETF Pis'ma Red. 23, 26 (1976) [JETP Lett. 23, 22 (1976)].
177. R.V. Ambartzumian, Yu.A. Gorokhov, V.S. Letokhov, G.N. Makarov, A.A. Puretzkii, and N.P. Furzikov, Zh. Eksp. Teor. Fiz. 71, N2(8) (1976) [Sov. Phys. - JETP (1976)].

178.  R.V. Ambartzumian, N.P. Furzikov, Yu.A. Gorokhov, V.S.
      Letokhov, G.N. Makarov, and A.A. Puretzkii, Opt.
      Commun. 18, 517 (1976).
179.  V.N. Bagratashvili, I.N. Knyazev, Yu.R. Kolomiisky, V.S.
      Letokhov, V.V. Lobko, and E.A. Ryabov, Kvantovaya
      Elektron. (Moscow) [Sov. J. Quantum Electron.] (in
      press).
180.  R.V. Ambartzumian, Yu.A. Gorokhov, V.S. Letokhov, G.N.
      Makarov, and A.A. Puretzkii, ZhETF Pis'ma Red. 22,
      374 (1975) [JETP Lett. 22, 177 (1975)].
181.  R.V. Ambartzumian, Yu.A. Gorokhov, V.S. Letokhov, G.N.
      Makarov, and A.A. Puretzkii, Phys. Lett. 56A, 183
      (1976).
182.  K.L. Kompa, in Optical Sciences Vol. 3, Tunable Lasers
      and Applications. Eds. A. Mooradian et al. Proceed-
      ings of the Loen Conference, Norway. Springer-Verlag
      1976, p. 177.
183.  V.N. Bagratashvili, I.N. Knyazev, V.S. Letokhov and V.V.
      Lobko, Opt. Commun. 14, 426 (1975).
184.  B.H. Mahan, Acc. Chem. Res. 8, 55 (1975).
185.  J.G. Calvert and J.N. Pitts, Jr., Photochemistry, Wiley,
      New York 1966.
186.  J.P. Morand, M. Wacongne, and E. Roth, Energ. Nucl.
      (Paris) 10, 362 (1968).
187.  F. Botter, D. Leger, and R. Darras, B.I.S.T. de Comm. à
      l'Energie Atomique de France 183, 25 (1973).
188.  R. Pertel and H.E. Gunning, Can. J. Chem. 37, 35 (1959).
189.  C.C. McDonald and H.E. Gunning, J. Chem. Phys. 20, 1817
      (1952).
190.  R. Pertel and H.E. Gunning, J. Chem. Phys. 26, 219
      (1957).
191.  K.R. Osborn, C.C. McDonald, and H.E. Gunning, J. Chem.
      Phys. 26, 124 (1957).
192.  H.E. Gunning, Can. J. Chem. 36, 89 (1958).
193.  C.C. McDonald, J.R. McDowell, and H.E. Gunning, Can. J.
      Chem. 37, 930 (1959).
194.  K.R. Osborn and H.E. Gunning, Can. J. Chem. 37, 1315
      (1959).
195.  J.R. McDowell, C.C. McDonald, and H.E. Gunning, Can. J.
      Chem. 37, 1432 (1959).
196.  A.G. Sherwood and H.E. Gunning, Can. J. Chem. 38, 466
      (1960).
197.  H.E. Gunning, J. Chim. Phys. Phys. - Chim. Biol. 60, 197
      (1963).
198.  H.E. Gunning and O.P. Strausz, Adv. Photochem. 1, 209
      (1963).
199.  M. Desnoyer, G. Nief, and E. Roth, J. Chim. Phys. 60, 14
      (1963).
200.  J. P. Morand and G. Nief, J. Chim. Phys. 65, 2058 (1968).

201. C.F. Schmidt, R.R. Reeves, Jr., and P. Harteck, Ber. Bunsenges. Phys. Chem. 72, 129 (1968).
202. O. Dunn, P. Harteck, and S. Dondes, J. Phys. Chem. 77, 878 (1973).
203. V.S. Letokhov and V.A. Semchishen, Spectrosc. Lett. 8, 263 (1975).
204. N.N. Semenov, Some Problems in Chemical Kinetics and Reactivity (in Russisan), Izd. AN SSSR, Moscow 1958, p. 331.
205. R.M. Badger and J.W. Urmston, Proc. Nat. Acad. Sci. U.S.A. 16, 808 (1930).
206. A.G. Harris and J.E. Willard, J. Am. Chem. Soc. 76, 4678 (1954) and T.A. Gover and J.E. Willard, J. Am. Chem. Soc. 82, 3816 (1960).
207. S.V. Filseth and J.E. Willard, J. Am. Chem. Soc. 84, 3806 (1962).
208. L.C. Glasgow and J.E. Willard, J. Phys. Chem. 77, 1585 (1973).
209. D.D. Liu, S. Datta, and R.N. Zare, J. Am. Chem. Soc. 97, 2557 (1975).
210. D.M. Brenner, S. Datta, and R.N. Zare, J. Am. Chem. Soc. 99, (1977).
211. E. Hulthen, N. Johansson, and U. Pilsäter, Ark. Fys. 14, Paper 3, 31 (1958) and E. Hulthen, N. Järlsäter, and L. Koffman, Ark. Fys. 18, Paper 35, 479 (1969).
212. M. Stuke and F.P. Schäfer, Chem. Phys. Lett. (1977).
213. S.J. Harris, J. Am. Chem. Soc. (submitted).
214. M. Lamotte, H.J. Dewey, R.A. Keller, and J.J. Ritter, Chem. Phys. Lett. 30, 165 (1975); see also H. Okabe, J. Chem. Phys. 66, 2058 (1977).
215. R. Gibert, J. Chim. Phys. 60, 205 (1963).
216. K. Gürs, West German Patent No. 1 959 767, appl. November 28, 1969, publ. June 3, 1971; Addendum: West German Patent No. 2 150 232, appl. October 8, 1971, publ. July 13, 1972.
217. R.D. Levine and A. Ben-Shaul, Chemical and Biochemical Applications of Lasers, Vol. II. Ed. C.B. Moore. Academic Press, New York 1977, p. 145.
218. R.B. Bernstein and R.D. Levine, Adv. At. Mol. Phys. 11, 215 (1975).
219. R.D. Levine and R.B. Bernstein, in Modern Theoretical Chemistry, Vol. 3, Dynamics of Molecular Collisions. Ed. W.H. Miller. Plenum Press, New York 1975.
220. K.G. Anlauf, D.H. Maylotte, J.C. Polanyi, and R.B. Bernstein, J. Chem. Phys. 51, 5716 (1969).
221. T.J. Odiorne, P.R. Brooks, and J.V.V. Kasper, J. Chem. Phys. 55, 1980 (1971).

222.  A.M. Ding, L.J. Kirsch, D.S. Perry, J.C. Polanyi, and
      J.L. Schreiber, Faraday Discuss. Chem. Soc. 55, 252
      (1973).
223.  D.J. Douglas, J.C. Polanyi, and J.J. Sloan, J. Chem.
      Phys. 59, 6679 (1973).
224.  D. Arnoldi and J. Wolfrum, Chem. Phys. Lett. 24, 234
      (1974).
225.  R.D. Brown and I.W.M. Smith, Int. J. Chem. Kinet. 7, 301
      (1975).
226.  Z. Karny, B. Katz, and A. Szöke, Chem. Phys. Lett. 35,
      100 (1975).
227.  R.G. Macdonald and C.B. Moore, J. Chem. Phys. (in
      press).
228.  D.Arnoldi, K. Kaufmann, and J. Wolfrum, Phys. Rev. Lett.
      34, 1597 (1975).
229.  S.R. Leone, R.G. Macdonald, and C.B. Moore, J. Chem.
      Phys. 63, 4735 (1975).
230.  J.H. Birely, J.V.V. Kasper, F. Hai, and L.A. Darnton,
      Chem. Phys. Lett. 31, 220 (1975).
231.  J.L. Lyman and R.J. Jensen, Chem. Phys. Lett. 13, 421
      (1972).
232.  S.H. Bauer, D.M. Lederman, E.L. Resler, Jr., and E.R.
      Fischer, Int. J. Chem. Kinet. 5, 93 (1973).
233.  R.J. Gordon and M.C. Lin, Chem. Phys. Lett. 22, 262
      (1973) and J. Chem. Phys. 64, 1058 (1976).
234.  M.J. Kurylo, W. Braun, A. Kaldor, S.M. Freund, and R.
      P. Wayne, J. Photochem. 3, 71 (1974).
235.  M.J. Kurylo, W. Braun, C.N. Xuan, and A. Kaldor, J. Chem.
      Phys. 62, 2065 (1975).
236.  J.C. Stephenson and S.M. Freund, J. Chem. Phys. 65, 4303
      (1976).
237.  S.M. Freund and J.C. Stephenson, Chem. Phys. Lett. 41,
      157 (1976).
238.  E.R. Lory, S.H. Bauer, and T. Manuccia, J. Phys. Chem.
      79, 545 (1975).
239.  K.-R. Chien and S.H. Bauer, J. Phys. Chem. 80, 1405
      (1976).
240.  A. Yogev, R.M.J. Loewenstein, and D. Amar, J. Am. Chem.
      Soc. 94, 1091 (1972).
241.  N.V. Karlov, Yu.N. Petrov, A.M. Prokhorov, and O.M.
      Stel'makh, ZhETF Pis'ma Red. 11, 220 (1970) [JETP Lett.
      11, 135 (1970)].
242.  N.G. Basov, E.P. Markin, A.N. Oraevskii, A.V. Pankratov,
      and A.N. Skachkov, ZhETF Pis'ma Red. 14, 251 (1971)
      [JETP Lett. 14, 165 (1971)].
243.  N.G. Basov, E.P. Markin, A.N. Oraevskii, and A.V. Pan-
      kratov, Dokl. Akad. Nauk SSSR 198, 1043 (1971) [Sov.
      Phys. - Dokl. 16, 445 (1971)].

244. N.G. Basov, E.M. Belenov, L.K. Gavrilina, V.A. Isakov, E.P. Markin, A.N. Oraevskii, V.I. Romanenko, and N.B. Ferapontov, ZhETF Pis'ma Red. 20, 607 (1974) [JETP Lett. 20, 277 (1974)].

245. N.G. Basov, E.M. Belenov, V.A. Isakov, E.P. Markin, A.N. Oraevskii, V.I. Romanenko, N.B. Ferapontov, Kvantovaya Elektron. (Moscow) 2, 938 (1975) [Sov. J. Quantum Electron. 5, 510 (1975)].

246. J.L. Lyman and R.J. Jensen, J. Phys. Chem. 77, 883 (1973).

247. S.D. Rockwood and J.W. Hudson, Chem. Phys. Lett. 34, 542 (1975).

248. A.K. Petrov, A.N. Mikheev, V.N. Sinel'nikov, and Yu. N. Molin, Dokl. Akad. Nauk SSSR 212, 915 (1973) and V.I. Gritsai, L.N. Krasnoperov, and V.N. Panfilov, Dokl. Akad. Nauk SSSR 212, 1368 (1973).

249. V.P. Strunin, N.K. Serdyuk, E.N. Chesnokov, and V.N. Panfilov, React. Kin. and Catal. Lett. 3, 131 (1975).

250. H.R. Bachmann, H. Nöth, R. Rinck, and K.L. Kompa, Chem. Phys. Lett. 29, 627 (1974) and 33, 261 (1975).

251. T.J. Manuccia, M.D. Clark, and E.R. Lory, Opt. Commun. 18, 219 (1976).

252. R.N. Zitter, R.A. Lau, and K.S. Wills, J. Am. Chem. Soc. 97, 2578 (1975).

253. D.F. Dever and E. Grunwald, J. Am. Chem. Soc. 98, 5055 (1976).

254. R.G. Bray, W. Henke, S.K. Liu, R.V. Reddy, and M.J. Berry, Chem Phys. Lett. (in press).

255. A.V. Eletskii, V.D. Klimov, and V.A. Legasov, Khimiya Vysokikh Energii 10, 126 (1976) [High Energy Chem. 10, 110 (1976)].

256. J.H. Birely and J.L. Lyman, J. Photochem. 4, 269 (1975) and S.H. Bauer, Ann. Rev. Phys. Chem. (in press) (1977).

257. N.G. Basov, A.N. Oraevskii, A.V. Pankratov, in Chemical and Biochemical Applications of Lasers, Vol. I. Ed. C.B. Moore. Academic Press, New York 1974, p. 203.

258. K. Bergmann, S.R. Leone, R.G. Macdonald, and C.B. Moore, Isr. J. Chem. 14, 105 (1975) [Twenty-fifth Intern. Union of Pure and Applied Chemistry Congress, Jerusalem, (1975)].

259. N.D. Artamonova, V.T. Platonenko, and R.V. Khokhlov, Zh. Eksp. Teor. Fiz. 58, 2195 (1970) [Sov. Phys. JETP 31, 1185 (1970)].

260. E.M. Belenov, E.P. Markin, A.N. Oraevskii, and V.I. Romanenko, ZhETF Pis'ma Red. 18, 196 (1973) [JETP Lett. 18, 116 (1973)].

261. N.G. Basov, E.M. Belenov. E.P. Markin, A.N. Oraevskii, and A.V. Pankratov, Zh. Eksp. Teor. Fiz. 64, 485 (1973) [Sov. Phys. - JETP 37, 247 (1973)].

262.  N.G. Basov, E.M. Belenov, E.P. Markin, A.N. Oraevskii, and A.V. Pankratov, in Fundamental and Applied Laser Physics. Eds. M. Feld et al. Proceedings of Esfahan Symposium, 1971. Wiley-Interscience, New York 1973, p. 239.

263.  V.S. Letokhov, A.A. Makarov, J. Photochem. 3, 249 (1974).

264.  A.V. Eletskii and A.N. Starostin, Fiz. Plazmy 1, 684 (1975) [Sov. J. Plasma Phys. 1, 377 (1975)].

265.  S.M. Freund and J.J. Ritter, Chem. Phys. Lett. 32, 255 (1975).

266.  I. Shamah and G. Flynn, J. Am. Chem. Soc. 99, (1977).

267.  S. Mukamel and J. Ross, J. Chem. Phys. 66,        (1977).

268.  R.M. Osgood, Jr., P.B. Sackett, and A. Javan, Appl. Phys. Lett. 22, 254 (1973).

269.  R. Frey, J. Lukasik, and J. Ducuing, Chem. Phys. Lett. 14, 514 (1972).

270.  S. Kimel, A. Ron, and S. Speiser, Chem. Phys. Lett. 28, 109 (1974).

271.  C. Cohen, C. Bordé, and L. Henry, Comptes Rend. Acad. Sci. Paris 265B, 267 (1967).

272.  J.W. Robinson, P. Moses, and N. Katayama, Spectrosc. Lett. 5, 333 (1972).

273.  R.V. Ambartzumian, V.S. Letokhov, G.N. Makarov, A.G. Platova, A.A. Puretzkii, and O.A. Tumanov, Zh. Eksp. Teor. Fiz. 64, 771 (1973) [Sov. Phys. - JETP 37, 392 (1973)].

274.  O.N. Kompanets, V.S. Letokhov, and V.G. Minogin, Kvantovaya Elektron. (Moscow) 2, 370 (1975) [Sov. J. Quantum Electron. 5, 211 (1975)].

275.  K. Bergmann and C.B. Moore, J. Chem. Phys. 63, 643 (1975).

276.  C.C. Badcock, W.C. Hwang, and J.F. Kalsch, Chem. Phys. Lett. (in press).

277.  R.D. Deslattes, M. Lamotte, H.J. Dewey, R.A. Keller, in Lecture Notes in Physics Vol. 43, Laser Spectroscopy Eds. S. Haroche et al. Proceedings of the Second International Conference, Megeve, France. Springer-Verlag 1975, p. 296.

278.  C. Willis, R.A. Back, R. Corkum, R.D. McAlpine, and F.K. McClusky, Chem. Phys. Lett. 38, 336 (1976).

279.  J.W. Eerkens, Opt. Commun. 18, 32 (1976).

280.  P. N. Lebedev, Ann. Phys. 32, 441 (1910) and P.N. Lebedev, Selected Papers (in Russian), Gostekhizdat, Moscow-Leningrad (1949).

281.  R. Frisch, Z. Phys. 86, 42 (1933).

282.  V.S. Letokhov, Comm. on Atomic and Molecular Physics 6, 119 (1977).

283.  J.-L. Picque, J.-L. Vialle, Opt. Commun. 5, 402 (1972).

284. A.Yu. Usikov, V.N. Kontorovich, E.A. Kaner, and P.V. Bliokh, Author's Certificate No. 174 432 (in Russian), appl. January 22, 1963, publ. August 28, 1965; Ukr. Fiz. Zh. 17, 1245 (1972).

285. A. Ashkin, Phys. Rev. Lett. 24, 156 (1970) and 25, 1321 (1970).

286. J. Pressman, U.S.A. Patent No. 3 558 877, appl. December 19, 1966, publ. January 26, 1971.

287. A.F. Bernhardt, D.E. Duerre, J.R. Simpson, L.L. Wood, Appl. Phys. Lett. 25, 617 (1974), Opt. Commun. 16, 169 (1976), J. Opt. Soc. Am. 66, 416 and 420 (1976), and A.F. Bernhardt, Appl. Phys. 9, 19 (1976).

288. O.I. Sumbaev, A.F. Mezentsev, V.I. Marushenko, A.S. Rylnikov, and G.A. Ivanov, Yad. Fiz. 9, 906 (1969) [Sov. J. Nucl. Phys. 9, 529 (1969)].

289. J.E. Bjorkholm, A. Ashkin, D.B. Pearson, Appl. Phys. Lett. 27, 534 (1975).

290. I. Nebenzahl, A. Szöke, Appl. Phys. Lett. 25, 327 (1974).

291. E. Courtens, in Laser Handbook, Vol. 2. Eds. F.T. Arecchi et al. North-Holland Publishing Company 1972, p. 1259.

292. E.B. Treacy, Phys. Lett. A27, 421 (1968).

293. A.P. Kazantzev, Zh. Eksp. Teor. Fiz. 63, 1628 (1972) and 66, 1599 (1974) [Sov. Phys. - JETP 63, 861 (1973) and 66, 784 (1974)].

294. S.S. Alimpiev, N.V. Karlov, A.M. Prokhorov, and B.G. Sartakov, ZhETF Pis'ma Red. 21, 257 (1975) [JETP Lett. 21, 117 (1975)].

295. M.R. Aliev, ZhETF Pis'ma Red. 22, 165 (1975) [JETP Lett. 22, 76 (1975)].

296. V.I. Mishin, Kvantovaya Elektron. 4,    (1977) [Sov. J. Quantum Electron.    (1977)].

297. C. Fabré and S. Haroche, Opt. Commun. 15, 254 (1975).

298. T.F. Gallagher, S.A. Edelstein, and R.M. Hill, Phys. Rev. A11, 1504 (1975).

299. C.J. Collins and N.S. Bowman, in Isotope Effects in Chemical Reactions, ACS Monograph N 167, Van Nostrand Reinhold, New York 1970.

300. Separation of Isotopes. Ed. H. London. London George Newnes Ltd., London 1961.

301. C.E. Treanor, J.W. Rich, R.G.I. Rehm, J. Chem. Phys. 48, 1798 (1968).

302. J.D. Teare, R.L. Taylor, and C.W. von Rosenberg, Jr., Nature 225, 240 (1970).

303. N.D. Artamonova, V.G. Platonenko, and R.V. Khokhlov, Zh. Eksp. Teor. Fiz. 58, 2195 (1970) [Sov. Phys. - JETP 31, 1185 (1970)].

304. N.G. Basov, E.M. Belenov, L.K. Gavrilina, V.A. Isakov, E.P. Markin, A.N. Oraevskii, V.I. Romanenko, and N.B.

Ferapontov, ZhETF Pis'ma Red. 20, 607 (1974) [JETP Lett. 20, 277 (1974)].

305.  N.G. Basov, E.M. Belenov, V.A. Isakov, E.P. Markin, A.N. Oraevskii, V.I. Romanenko, and N.B. Ferapontov, Kvantovaya Elektron. 2, 938 (1975) [Sov. J. Quantum Electron. 5, 510 (1975)].

306.  N.G. Basov, E.M. Belenov, L.K. Gavrilina, V.A. Isakov, E.P. Markin, A.N. Oraevskii, V.I. Romanenko, and N.B. Ferapontov, ZhETF Pis'ma Red. 19, 336 (1974) [JETP Lett. 19, 190 (1974)].

307.  T.J. Manuccia and M.D. Clark, Appl. Phys. Lett. 28, 372 (1976).

308.  G.I. Tkeshelashvili et al. Proceedings of the First National Winter School "Laser Isotope Separation," Bakuriani, USSR, March 1976. (in press).

309.  N.G. Basov, E.M. Belenov, V.A. Isakov, Yu.S. Leonov, E.P. Markin, A.N. Oraevskii, and V.I. Romanenko, ZhETF Pis'ma Red. 22, 221 (1975) [JETP Lett. 22, 102 (1975)].

310.  M.A. El-Sayed, M.T. Wauk, and G.W. Robinson, Mol. Phys. 5, 205 (1962).

311.  G.C. Nieman and G.W. Robinson, J. Chem. Phys. 37, 2150 (1962).

312.  R.Z. Sagdeev, S.V. Kamishan, A.A. Obinochnii, and Yu.N. Molin, ZhETF Pis'ma Red. 22, 584 (1975) [JETP Lett. 22, 287 (1975)].

313.  Wash 1139(74) "Nuclear Power Growth 1974-2000 (1974). and L. Cox, Nuclear Power, Uranium Enrichment and Ore Projections through Year 2000, Lawrence Livermore Laboratory, U.S. Energy Research and Development Administration (1975).

314.  J.C. Vanderleeden, Laser Focus (1977). (in press).

315.  B.M. Casper, Bull. Atomic Scientists, January 1977, p. 29.

316.  A.S. Krass, Science 196, 721 (1977).

317.  L.O. Love, Science 185, 343 (1974).

318.  W. Spindel, Am. Chem. Soc. Symposium Series 11, 77 (1975).

319.  R.D. Hauck, J. Environ. Quality 2 , 317 (1973).

320.  A.P. Edwards and R.D. Hauck, Soil Sci. Soc. Amer. Proc. 38, 765 (1974).

321.  A.L. Allen, F.J. Stevenson, and L.T. Kurtz, J. Environ. Quality 2, 120 (1973).

322.  A.P. Edwards, J. Environ. Quality 2, 382 (1973).

323.  A.S. Gomenyuk, V.P. Zharov, V.S. Letokhov, and E.A. Ryabov, Kvantovaya Elektron. (Moscow) 3, 369 (1976) [Sov. J. Quantum Electron. 6, 195 (1976)].

324.  A.S. Provorov and V.P. Chebotaev, Dokl. Akad. Nauk. SSSR 208, 318 (1973) [Sov. Phys. - Dokl. 18, 56 (1973)].

325.  A.L. Golger and V.S. Letokhov, Kvantovaya Elektron.
      (Moscow) 2, 1508 (1975) [Sov. J. Quantum Electron. 5,
      811 (1975)].
326.  R.W.F. Gross, Opt. Eng. 13, 506 (1974).
327.  S. Haroche, J.C. Pebay-Peyroula, T.W. Hänsch, and S.E.
      Harris, Eds. Lecture Notes in Physics Vol. 43, Laser
      Spectroscopy. Proceedings of the Second International
      Conference, Megeve, France. Springer-Verlag 1975.
328.  A. Mooradian, T. Jaeger, and P. Stokseth, Eds. Optical
      Sciences Vol. 3, Tunable Lasers and Applications.
      Proceedings of the Loen Conference, Norway. Springer-
      Verlag 1976.
329.  Proceedings of the Fourth Vavilov Conference on Non-
      linear Optics, Novosibirsk 1975, Kvantovaya Elektron.
      3, N4, (1976) [Sov. J. Quantum Electron. 6, 379 (1976)].
330.  Laser Induced Chemistry Conference, February 1976,
      Steamboat Springs, Colorado, USA (Laser Focus, April
      1976, p. 12).
331.  Stephen F. Jacobs, Murray Sargent, III, Marlan O. Scully,
      and Charles T. Walker, Eds. Physics of Quantum Elec-
      tronics Vol. 4, Laser Photochemistry, Tunable Lasers,
      and Other Topics. Addison-Wesley 1976.
332.  Materials of USSR National School "Laser Isotope Separa-
      tion," March 1976, Tvilisi, USSR. (in press).
333.  V.S. Letokhov, J. Photochemistry 4, 185 (1975).
334.  K.S. Gochelashvili, N.V. Karlov, A.N. Orlov, R.P. Petrov,
      Yu.N. Petrov, and A.M. Prokhorov, ZhETF Pis'ma Red. 21,
      640 (1975) [JETP Lett. 21, 302 (1975)].
335.  M.S. Djidjoev, R.V. Khokhlov, A.V. Kiselev, V.I. Lygin,
      V.A. Namiot, A.I. Osipov, V.I. Panchenko, and B.I.
      Provotorov, in Optical Sciences Vol. 3, Tunable Lasers
      and Applications. Eds. A. Mooradian et al. Proceed-
      ings of the Loen Conference, Norway. Springer-Verlag
      1976, p. 100.
336.  V.S. Letokhov, Kvantovaya Elektron. (Moscow) 2, 930
      (1975) [Sov. J. Quantum Electron. 2, 506 (1975)], and
      Phys. Lett. 51A, 231 (1975).
337.  V.S. Letokhov, Spectrosc. Lett. 8, 697 (1975).
338.  V.S. Letokhov, V.I. Mishin, and A.A. Puretzkii, in
      Chemistry of Plasma (in Russian), Atomizdat, Moscow
      1977. (in press).
339.  R.V. Ambartzumian, Yu.A. Gorokhov, S.L. Grigorovich,
      V.S. Letokhov, G.N. Makarov, Yu.A. Malinin, A.A.
      Puretzkii, E.P. Filippov, and M.P. Furzikov, Kvantovaya
      Elektron. (Moscow) 4, 171 (1977) [Sov. J. Quantum
      Electron. 7,    (1977)].
340.  H.R. Bachman, H. Nöth, R. Rinck, and K.L. Kompa, Chem.
      Phys. Lett. 45, 169 (1977), Ber. Bunsenges. Phys. Chem.
      81,    (1977), and references cited therein for many

laser induced reactions of boron compounds.

341.  W.K. Swainson, British Patent No. 1 243 043, appl. July
      8, 1968, publ. August 8, 1971.

342.  R.H. Barnes, C.E. Moeller, J.F. Kircher, and C.M. Verber,
      in Development of Gaseous Media for 3-Dimensional
      Displays, Report of Battelle Columbus Laboratories
      1975, and Appl. Phys. Lett. $\underline{24}$, 610 (1974).

343.  G.L. DePoorter and C.K. Rofer-DePoorter, LA-5630-MS
      Report of Los Alamos Scientific Laboratory (1976).

344.  T. Donohue, Chem. Phys. Lett. (in press).

345.  A. Laubereau and W. Kaiser, in Chemical and Biochemical
      Applications of Lasers, Vol. II. Ed. C.B. Moore.
      Academic Press, New York 1977, p. 87.

346.  T.G. Spiro, in Chemical and Biochemical Applications of
      Lasers, Vol. I. Ed. C.B. Moore. Academic Press, New
      York 1974, p. 29, and Accts. Chem. Res. $\underline{7}$, 339 (1974).

347.  R. Mathies, A.R. Oseroff, T.B. Freedman, and L. Stryer,
      in Optical Sciences Vol. 3, Tunable Lasers and
      Applications. Eds. A. Mooradian et al. Proceedings of
      the Loen Conference, Norway. Springer-Verlag 1976, p.
      294.

348.  V.S. Letokhov, in Optical Sciences Vol. 3, Tunable
      Lasers and Applications. Eds. A. Mooradian et al.
      Proceedings of the Loen Conference, Norway. Springer-
      Verlag 1976, p. 122.

349.  Janos Ladik, in Quantenbiochemie für Chemiker und
      Biologen. Akademiai Kiado, Budapest 1972.

350.  H. Susi, in Structure and Stability of Biological Macro-
      molecules. Eds. S.N. Timasheff et al. Marcel Dekker,
      New York 1969.

351.  P.O. Löwdin, Adv. Quantum Chem. $\underline{2}$, 213 (1968) and in
      Electronic Aspects of Biochemistry. Ed. B. Pullman.
      Academic Press, New York 1964, p. 167.

352.  B.H. Johnston, M.A. Johnson, C.B. Moore, and J.E. Hearst,
      Science. (in press).

353.  C.-K.J. Shen and J.E. Hearst, Proc. Natl. Acad. Sci.
      U.S.A. $\underline{73}$, 2649 (1976) and C.V. Hanson, C.-K.J. Shen,
      and J.E. Hearst, Science $\underline{193}$, 62 (1976).

354.  V.S. Letokhov, Opt. Commun. $\underline{7}$, 59 (1973)

355.  G. Baldwin, and R.V. Khokhlov, Phys. Today $\underline{28}$, 32 (1975).

356.  V.I. Gol'danskii, and Yu.M. Kagan, Zh. Eksp. Teor. Fiz.
      $\underline{64}$, 90 (1973) [Sov. Phys. - JETP $\underline{37}$, 49 (1973)].

357.  V.S. Letokhov, Zh. Eksp. Teor. Fiz. $\underline{64}$, 1555 (1973)
      [Sov. Phys. - JETP $\underline{37}$, 787 (1973)].

358.  H.T. Duong, P. Jacquinot, S. Liberman, J. Pinard, and
      J.-L. Vialle, C. R. Acad. Sci. (Paris) $\underline{206B}$, 909 (1973).

359.  V.S. Letokhov, Usp. Fiz. Nauk $\underline{118}$, 199 (1976) [Sov.
      Phys. - Usp. $\underline{19}$, 109 (1976)].

360.  I.N. Knyazev, V.S. Letokhov, and V.G. Movshev, IEEE J. Quantum Electron. QE-11, 805 (1975).

361.  S.V. Andreeev, V.S. Antonov, I.N. Knyazev, V.S. Letokhov, and V.G. Movshev, Phys. Lett. 54A, 91 (1975).

362.  V.S. Letokhov, V.G. Movshev, and V.K. Potapov, Kvanto-vaya Elektron. (in press)

363.  S.E. Harris, J.F. Young, A.H. Kung, D.M. Bloom, and G.C. Bjorklund, in Laser Spectroscopy.  Eds. R.G. Brewer and A. Mooradian.  Plenum, New York 1973, p. 59.

364.  P.P. Sorokin, J.A. Armstrong, R.W. Dreyfus, R.T. Hodgson, J.R. Lankard, L.H. Manganaro, and J.J. Wynne, in Lecture Notes in Physics Vol. 43, Laser Spectroscopy Eds. S. Haroche et al.  Proceedings of the Second International Conference, Megeve, France.  Springer-Verlag 1975, p.46.

365.  B.P. Stoicheff, and S.C. Wallace, in Optical Sciences Vol. 3, Tunable Lasers and Applications.  Eds. A. Mooradian et al.  Proceedings of the Loen Conference, Norway.  Springer-Verlag 1976, p. 1.

366.  E.W. Müller and Tien Tzon Tzong, in Field Ion Microscopy. American Elsevier, New York 1969.

367.  D.W. Werwoerd, H. Kohlhage, and W. Zillig, Nature 192, 1038 (1961).

# 2

# Multiple Photon Infrared Laser Photochemistry

*R. V. AMBARTZUMIAN and V. S. LETOKHOV*

INSTITUTE OF SPECTROSCOPY OF THE SOVIET ACADEMY OF SCIENCES

## I.   INTRODUCTION

The photochemical effect of visible and ultraviolet light based on excitation of atomic and molecular electronic states is well known (see, for example, (1)).  The photochemical effect of infrared light based on the excitation of molecular vibrational levels has been discovered relatively recently (2).  The intensity of conventional sources of monochromatic infrared radiation was too weak to allow study of the infrared photochemical effect in detail.  The situation changed fundamentally with the advent of lasers, which are practically ideal sources of radiation from the vacuum ultraviolet to the far infrared region for very many applications, photochemistry included.  The development of tunable lasers gave impetus to works on selective photochemistry, photochemical isotope separation in particular (see (3) and chapter I of the present volume).  The unique combination of the many properties (frequency tuning, high power, short duration, spatial coherence, monochromaticity and temporal coherence) of a laser light beam has made it possible not only to realize the known photochemical methods but also to elaborate a number of basically new photochemical approaches which could not be realized with common light sources.  The approach based on multiple photon excitation of the vibrational (and probably electronic) states of polyatomic molecules in a strong infrared field discussed here is a radically new photochemical approach.

To understand the place of this new approach among those known for a long time or discovered recently, it is useful to classify the selective photochemical methods by type of excitation and type of photochemical conversion of excitation, as well as to compare their pros and cons.  In these cases we are dealing with potentially selective photoprocesses in the main.  The term "selectivity" in laser photochemistry has two meanings.  First, it may deal with selective photochemical

167

conversion of a particular molecule in a mixture with others. Such selectivity may be called "intermolecular." We always mean it, for instance, when speaking about photochemical iso- tope separation.  Secondly, it may deal, in principle, with selective photoexcitation of any one molecular bond.  If we ensure a chemical reaction of such a molecule before excita- tion transfer to many molecular bonds, we may hope that the photochemical reaction will be selectively controlled.  Such selectivity can be conveniently called "intramolecular."  It may be realized only in a chemical reaction with an acceptor sensitive not only to energy but also to the type of molecu- lar excitation.  This potentiality of selective laser photo- chemistry has not been studied experimentally yet.  Therefore, when using below the term "selectivity" we always mean "inter- molecular selectivity."

## A.    Classification of Laser Photochemistry Methods

### 1.    Type of Photoexcitation
In Fig. 1 various types of selective molecular photoexci-

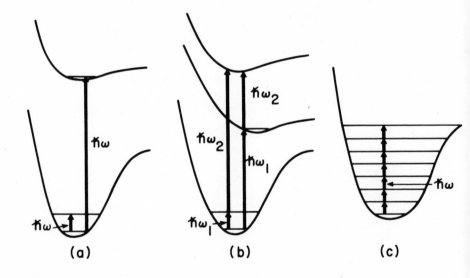

Fig. 1.  Types of selective molecular photoexcitation: a)  single-step excitation of electronic or vibrational states;  b)  two-step excitation of electronic states through intermediate vibrational or electronic states;  c)  multiple photon excitation by IR radiation.

tation are shown in a very simple form.  The classical (pre-
laser) photochemical method is based on one-step excitation
of an electronic state of an atom or a molecule.  This type of
molecular excitation has a serious disadvantage for selective
photochemistry.  Most molecules, especially polyatomic, have
comparatively wide structureless bands of electronic absorption at normal temperature.  So this scheme can not be used
for, say, isotopically selective excitation of molecules.
Only a limited number of simple, in the main two- or three-
atom, molecules have narrow lines of electronic absorption
suitable for isotopically selective excitation.  On the other
hand, the excitation of electronic states is beneficial because of a high quantum yield of the photochemical reaction.

One-step excitation of a molecular vibrational state
(photochemistry in the ground electronic state) features
rather high excitation selectivity both for simple and complex
molecules.  The main disadvantage of this method is the fast
relaxation of vibrational excitation to heat and hence a low
quantum yield of the subsequent photochemical process.  Besides, the method can be used only for photochemical reactions
with low activation energy.

Two-step excitation of a molecular electronic state
through an intermediate vibrational state by joint action of
IR and UV radiation (Fig. 1b) combines the advantages of one-
step IR and UV excitation processes and removes their disadvantages (4).  In two-step photoexcitation by a two-frequency (IR + UV) laser field it is possible to separate the
functions of selective excitation, when a molecule derives
rather low energy (IR photon), and absorption of much larger
energy (UV photon) by the selectively-excited molecule.  This
type of two-step photoexcitation has high enough selectivity
together with a high quantum yield of photochemistry.
The merits of two-step (IR + UV) excitation must vividly show
themselves in the case of condensed media where the discrepancy between the high selectivity and the high quantum yield
at normal temperatures becomes unsolvable.

Two-step excitation of molecules through an intermediate
electronic state (Fig. 1b) is not so universal as IR + UV
excitation.  Its only advantage over one-step excitation of
electronic states is the strong possibility of exciting states
with specific properties and high-lying states without VUV
radiation.

For polyatomic molecules it is possible to excite selectively high vibrational and even excited electronic states
under the action of sufficiently powerful IR radiation alone
(5).  Due to multiple absorption of IR photons of the same
frequency a molecule derives an energy comparable to the typical energy of electronic excitation (Fig. 1c).  Therefore,
we can realize at the same time the excitation selectivity

which suffices to separate isotopes and a rather high quantum yield of the subsequent photochemical process. In this case it is also possible to separate the functions of selective excitation and subsequent absorption of high energy by a selectively excited molecule in a two-frequency IR field (6) which provides enhancement in the selectivity of the process. A limitation of the method of multiple photon IR radiation absorption is that it can be applied only to polyatomic molecules having a high density of excited vibrational levels in the ground electronic state.

The two last methods of selective photochemistry illustrated in Fig. 1b and 1c have been realized only by means of laser radiation since in this case high population of intermediate quantum levels is required. Basically, this cannot be realized with any efficiency using conventional incoherent light sources because of the low temperature of their radiation. Conventional light sources are applicable to one-step processes only, with higher efficiency for electronic states than for vibrational ones.

Table 1 gives a cumulative list of pros and cons for each photochemical method mentioned. It is quite possible, of course, to realize those schemes of selective excitation which are intermediate with respect to the ones considered. For example, even simple molecules excited successively by a multifrequency IR field can reach high vibrational levels. Or we may excite vibrational levels by two-frequency visible laser radiation by Raman-type processes. So there may be no sharp boundaries between different methods, and the classification of the methods as well as of their pros and cons is conditional to some extent.

*Type of photochemical process.* An excited molecule can participate in subsequent photochemical processes. Several different mechanisms of this participation may be classified in the most general and simplified form thus: 1) photochemical reaction of an (electronically or vibrationally) excited molecule with a proper acceptor; 2) photodissociation (or photopredissociation) of an excited molecule; 3) photoisomerization, i.e. rearrangement of the spatial structure of an excited molecule. All these types of photochemical conversions are well known in photochemistry (1). But as far as the attainment of high selectivity in the photochemical process and its universality are concerned, these mechanisms are far from being equivalent. Each of them has its pros and cons for selective photochemistry.

The first process (chemical reaction of excited molecules) is potentially rather universal. Really, any molecule has an excited electronic state with increased reactivity which may be selectively excited by laser radiation with the use of a proper scheme (Fig. 1). This process requires

TABLE 1

*Comparison Between Various Methods of Selective Molecular Photoexcitation*

| Method | Advantage | Disadvantage |
|---|---|---|
| A) One-step excitation of: | | |
|   a) electronic state | 1) High excitation energy <br> 2) Little role of thermal effects | 1) Low selectivity for polyatomic molecules |
|   b) Vibrational state | 1) High selectivity | 1) Low excitation energy <br> 2) Loss of selectivity because of thermal heating |
| B) Two-step excitation of electronic state through intermediate state: | | |
|   a) vibrational state | 1) High excitation energy <br> 2) High selectivity <br> 3) Little role of thermal effects | |
|   b) electronic state | 1) High excitation energy <br> 2) Little role of thermal effects <br> 3) Universality in exciting high-energy states | 1) Low selectivity for polyatomic molecules |
| C) Multiple photon excitation of high vibrational levels | 1) High excitation energy <br> 2) High selectivity <br> 3) Little role of thermal effects | 1) Unapplicable for simple molecules |

however, a suitable acceptor which would react at a higher
rate with excited molecules rather than with unexcited ones.
The rate at which excited molecules react with an acceptor
must be much higher than that of excitation transfer during
their collisions with unwanted molecules as well as  the rate
of excitation relaxation (see Sec. I of Chapter I). Another
fundamental requirement consists in conserving the photochemical reaction selectivity in the inevitable subsequent secondary photochemical reactions. These are rather rigid requirements which can be, as a rule, complied with non-trivially.
It suffices to say that quite recently high isotopic selectivity of photochemical reactions has been demonstrated (an
electronically excited molecule, ICl, reacted with the $C_2H_2Br_2$
acceptor molecule) (7). All of these requirements are very
difficult for vibrationally excited molecules when the difference in the reaction rates of excited and unexcited molecules
is relatively small.

Molecular photodissociation is as universal a process as
the photochemical reaction of excited molecules. Photodissociation can be realized either by exciting an unstable electronic state (repulsive term) of the molecule or within the
ground electronic state through strong vibrational excitation.
Until very recently, only photodissociation through an excited
electronic state was known, and only the creation of high-power pulsed IR lasers allowed the second possibility to be
realized. These two methods of selective molecular photodissociation differ greatly from one another.

Every molecule has excited unstable electronic states
suitable for molecular photodissociation. Since a molecule on
a repulsive electronic surface dissociates very quickly
(about $10^{-13}$ to $10^{-14}$ sec), relaxation and excitation transfer in such a short time are of no importance, of course. Because the band of electronic absorption at the transition to a
dissociative state is wide, excitation selectivity may be en-
sured only by means of a two- (or multi-) step process (Fig.
1b). So the requirements of small losses of excitation because of relaxation and transfer of excitation apply to the
intermediate excited state. These requirements can always be
met by choice of the intensity and duration of the second-step
UV radiation (4). Since dissociation products are always radicals, it is necessary to use scavengers which react with the
radicals without affecting the initial molecules. The choice
of such scavengers which do not result in significant losses
of selectivity through secondary photochemical processes is a
simpler task than for chemical reactions of excited molecules.

Sometimes electronically excited molecules dissociate not
due to the repulsive nature of the electronic state, but due
to the intersection of the stable and unstable states. The
dissociation in this case, called photopredissociation, occurs

more slowly, say, in $10^{-6}$ to $10^{-12}$ sec, for different molecules in excited electronic-vibrational-rotational states.  By
virtue of the fact that the absorption lines of the transition
to a predissociated state are narrow, we can make the high
selectivity of excitation useful for one-step photoexcitation
(8, 9).  This gives a new advantage to the photodissociation
method, though at the expense of universality because of the
limited number of molecules which exhibit the photopredissociation phenomenon.

Photodissociation in the ground electronic state (Fig.
1c) even has advantages over that through excited electronic
states.  First, it requires less energy;  secondly, it produces, in principle, lower energy radicals.  But it was impossible to carry out this process with its advantages for
lack of a proper method of molecular photoexcitation.  The
question is seemingly only one of the excitation of molecular
transitions.  However, this would require the use of multifrequency IR laser radiation for photoexcitation up to the
dissociation limit since the series of vibration-rotation
transitions are not equidistant.  This is possible in principle but the present level of tunable laser engineering does
not enable us to do it for the time being.  The situation
became much easier after the effect of collisionless dissociation of polyatomic molecules in intense IR laser fields had
been revealed (10, 11).  A rich structure of the vibration-
rotation transitions in polyatomic molecules allows absorption of a very large number of IR photons if the laser field
is sufficiently intense.  And what is more, such a process of
photodissociation in a single-frequency, intense IR field
is so selective that it enables isotope separation (5, 12).
Thus, photodissociation in the ground electronic state in an
intense IR field proves to be a simpler process but again at
the expense of some universality.  (It cannot be applied to
simple, two- or three-atom molecules).  Its other pros and
cons are common to any photodissociation method.

Molecular photoisomerization, like photodissociation, is
a unimolecular photochemical process which requires no collisions with other particles.  This is an advantage of both
methods over the chemical reaction of excited molecules which
necessitates collisions with all the ensuing limitations
(loss of excitation and selectivity).  But photoisomerization,
unlike the first two methods, does not need any acceptor, and
so there is almost no selectivity loss in secondary photochemical processes for this method.  The only limitation of this
method is the relatively small number of the molecules which
exhibit this effect.  Since the final state of the molecular
phototransition is stable, selective photoisomerization in the
case of a narrow absorption line may be done by one-step pho-

TABLE 2

*Comparison Between Various Photochemical Molecular Processes*

| Process | Advantage | Disadvantage |
|---|---|---|
| A) Chemical reaction of excited molecule | 1) Universality | 1) Special choice of a proper acceptor<br>2) Loss of selectivity in secondary photochemical processes<br>3) Loss of excitation because of relaxation<br>4) Loss of selectivity because of excitation transfer |
| B) Photodissociation (photopredissociation) | 1) Universality (except for photopredissociation)<br>2) Low losses of excitation because of relaxation<br>3) Low losses of selectivity because of excitation transfer | 1) Necessity of two- (multi-)step photoexcitation (except for photopredissociation)<br>2) Special choice of an acceptor to scavenge radicals<br>3) Loss of selectivity in secondary photochemical processes |
| C) Photoisomerization | 1) Low losses of selectivity because of excitation transfer<br>2) No necessity of using an acceptor<br>3) No losses of selectivity in secondary photochemical processes | 1) No universality |

toexcitation (13). Otherwise two-step photoexcitation can be applied.

Table 2 summarizes the pros and cons of the photochemical processes considered. This table is not complete because we have not considered such molecular processes as photoionization and dissociative photoionization resulting in charged particles (electrons, positive and negative ions). Despite the fact that some proposals on selective photoionization of molecules by two-step photoexcitation have been made (14) and that a first successful experiment has recently been carried out (15), we doubt whether these processes have any advantages over photodissociation for selective photochemistry. Their application will rather be limited to the area of selective laser detection of complex molecules and molecular bonds, where it is of principal importance to produce charged particles after laser selective action (16). (see the conclusion of Chapter I).

The task of our article is to review infrared laser photochemistry based on multiple photon absorption of intense IR radiation by polyatomic molecules. Thus, we are going to consider only the third mentioned method (Fig. 1c) of selective photoexcitation. As to photochemical processes, we shall discuss all the mentioned processes (Table 2) though most experimental results are concerned with molecular photodissociation. Before we turn to the detailed discussion of the problem, it is advisable to give a short review of the works which preceded the discovery of the effect of isotopically selective molecular photodissociation in a strong IR field (5), and to list the basic effects observed.

B.    Early Works.  Basic Effects.

The creation of high-power IR lasers opened the way for studying the resonance interaction of an intense IR field with molecular systems. Bordé et al. (17) observed the visible luminescence of ammonia and its decomposition products under the irradiation of a CW high-power $CO_2$ laser the frequency of which coincided with the vibration-rotation absorption line of the $\nu_3$ band. This effect is explained (17, 18) as thermal heating of $NH_3$ by laser radiation to high temperatures. Similar experiments were conducted later with the $CO_2$ laser on $BCl_3$ molecules (19). Though the authors of this work directed their attention to the possibility of nonthermal dissociation of $BCl_3$ in their experiment, now we know that under CW laser IR radiation with power from $10^2$ to $10^3$ W/cm$^2$ and at a gas pressure over 1 torr, molecular dissociation may only be thermal in nature.

The creation of high-power, pulsed TEA $CO_2$ lasers (20, 21) made it possible to begin studying the action of a more intense ($10^6$-$10^9$ W/cm$^2$) pulsed IR field on molecules. In work (22) the authors observed an optical breakdown and visible fluorescence of some molecular gases ($NH_3$, $SiF_4$, $BF_3$, $CCl_2F_2$, etc.) as a $CO_2$ laser pulse was focused on to them. They found that visible fluorescence occurred at intensities much less than the optical breakdown threshold in the focus. The fluorescence intensity depended drastically on laser power. Studies of the fluorescence spectrum revealed that it belonged to electronically excited molecular dissociation products. In (23, 24) more detailed quantitative measurements of the kinetics and spectra of fluorescence of $NH_3$ and $C_2F_3Cl$ molecules under the action of $CO_2$ laser pulses were taken. The investigation of fluorescence temporal characteristics showed that it appeared without delay (measurement accuracy of 30 nsec) with respect to the leading edge of the laser pulse. Ultraviolet radiation was observed too, which suggested that the high electronic states of the $NH_3$ molecule were excited under the action of a high-power IR radiation pulse. The studies of the fluorescence intensity of $NH_3$ as a function of $CO_2$-laser frequency showed that the fluorescence was associated with resonance absorption of radiation, because it occurred only when the radiation frequency was tuned to one of the absorption lines. One more experiment was carried out in work (25) where, also under the action of a $CO_2$-laser pulse, the $N_2F_4$ molecule dissociated at a rate higher than that of thermal relaxation of the absorbed energy.

The data obtained in (22 - 25) unambiguously pointed to the nonthermal character of molecular gas fluorescence in an intense pulsed IR field. The discussions of potential dissociation mechanisms (23, 24) disclosed at once that it would be non-trivial to explain the experimental data obtained. The first assumption on nonthermal dissociation mechanism relied on the intense IR field, aided by V-V transfer of excitation, to "heat" up the vibrational degrees of freedom of a molecular gas and to cause dissociation with the formation of electronically excited products. But on the basis of a more detailed analysis of the kinetics of molecular vibrational excitation by IR radiation done in work (26) it was concluded in (24) that such a mechanism was not responsible for molecular gas luminescence under experimental conditions (22 - 24). Actually, according to work (26), the rate of vibrational energy absorption into a molecular system as it interacts with monochromatic IR radiation complies with the condition

$$\frac{dE_{vib}}{dt} < \hbar\omega_o \frac{q}{2\tau_{rot}} ,$$ (I.1)

where $\hbar\omega_o$ is the vibrational quantum energy, $q$ is the relative
portion of the molecules at the lower rotational-vibrational
molecular transition 0 - 1 being in resonance with the field,
$\tau_{rot}$ is the time of rotational relaxation.   The estimate (I.1)
results from the "bottleneck" effect since the molecules are
distributed over many rotational sublevels.   It holds true,
whatever the mechanism of further excitation of molecules to
high vibrational levels, provided that a molecule can absorb
only one IR photon.   It is clear that the limitation on the
rate of vibrational excitation (I.1) rules out the role of
the mechanism of V-V exchange in explaining almost instant
dissociation of polyatomic molecules.   As it was stressed in
work (24), the only possibility to go beyond the limits of
(I.1) is direct radiation excitation at all successive rotational-vibrational transitions without rotational relaxation
at the intermediate levels.   That is excitation by the following scheme

$$|v = 0, \; J = J_1 > \to \; |1, \; J_2 > \to \; |2, \; J_3 > \to \; , \; etc. \qquad (I.2)$$

In this case a molecule being in resonance with the field
would absorb a great number of IR photons.   This mechanism was
discarded however in (24) because of the vibrational anharmonicity which makes it impossible to attain a multistep resonance of the molecule with a great number of vibrational
transitions.   To get rid of the contradiction, in work (24) an
hypothesis was made on electronic level excitation and molecular dissociation due to a small number of free electrons
accumulating energy in the IR field and colliding then with
the molecules.   But this mechanism was also rather tenuous because it was difficult to explain the resonance nature of
fluorescence excitation.   An attempt to modify somehow the
electronic mechanism involving considerations about the resonance transport properties of molecular gases was made
as well (27).

   To clarify the mechanism of molecular dissociation and,
in particular, to test two mutually exclusive hypotheses
(direct radiation excitation and excitation by hot electrons),
in 1972 our laboratory conducted some experiments on isotopically selective dissociation of $^{15}NH_3$ molecules in a mixture
with $^{14}NH_3$ by focused  $CO_2$-laser pulses.   These experiments
yielded negative results.   Afterwards we decided to carry out
a detailed investigation into the fluorescence and dissociation effects on a more suitable molecule, which was $BCl_3$.   The
goal of the experiments was to explore further the minor role
of collisions observed in (23, 24).   For this purpose
the experiments on $BCl_3$, as opposed to those in (22 - 25),
were tried at low pressures (to 0.03 torr).   The results of
the research were published in (11).   We found both collisionless dissociation of $BCl_3$ molecules in a strong IR field and a

collisional contribution to the dissociation of highly excited molecules. Analogous results had been obtained in (10) on the $SiF_4$ molecule. The experiments at low pressures demonstrated that an explanation of the effect must be sought in the collisionless interaction of a polyatomic molecule with an intense IR field. For this reason, in (10) an assumption was made on a possible role of a large number of highly excited vibrational levels forming a "vibrational quasi-continuum" in polyatomic molecules. This effect, upon which no particular emphasis was placed in early works, removed the problem of field frequency detuning from vibrational transitions between high excited levels. The problem of anharmonicity for excitation of lower vibrational levels still remained to be solved. The broadening of vibrational transitions because of dynamic Stark effect, the subject of discussions in (10, 11), could not solve the problem since the broadening by field was small compared to anharmonicity.

Subsequent experiments were aimed at observing the effect of isotopically selective dissociation of $BCl_3$ molecules. After many lame attempts to find this effect through direct observation of changes in the isotopic composition of $BCl_3$ some experiments were conducted to study the isotope effect in the BO-radical chemiluminescence. This radical is formed by reaction of the dissociation products of the $BCl_3$ molecule and the $O_2$ acceptor molecule. Work (5) discovered the effect of isotopically selective dissociation of the $BCl_3$ molecule. This was of vital importance from two standpoints. First, it became clear at last that the dissociation of an isolated molecule by a strong IR field with no collisions involved was really practicable. Secondly, another efficient and simple method was elaborated for isotope separation within the frameworks of our photodissociative approach to laser isotope separation. Approximately at that time work (28) was published where small (15%) selectivity of dissociation of trans-2-butene molecules in a mixture with cis-2-butene was done by pulsed $CO_2$-laser radiation. The experimental data obtained were very limited however; so it was impossible to decide among several potential mechanisms including those with collisions involved.

After the experiments on isotopically selective dissociation of $BCl_3$ molecules carried out in 1974 (5) it became evident that the effect was general in character and must be observed in many polyatomic molecules. Indeed, some successful experiments on isotopically selective dissociation of $SF_6$ molecules and on macroscopic enrichment of sulphur isotopes were tried at once (12). These experiments were repeated in work (29) sent to press after work (12) had been published. Isotopically selective dissociation, with a lower enrichment factor though, was demonstrated even on the $OsO_4$ molecule which had a very small isotope shift (30). In the same work it was possible to observe directly the excitation of the high vibrational levels of $OsO_4$ from variations in the UV absorption spectrum. The excitation of such levels below the dissociation limit can probably explain the results of the experiments (31) in which an isotopically selective reaction of $BCl_3$ molecules with $H_2S$ (and $D_2S$) took place under $CO_2$-laser pulses.

The successful experiments on isotopically selective dissociation of polyatomic $BCl_3$ (5) and $SF_6$ (12) molecules generated considerable interest both among experimentalists studying laser isotope separation and theorists dealing with the non-trivial effect of collisionless molecular dissociation in an intense IR field. But further considerable progress in the interpretation was made after some new features of this effect had been found. In work (32) the effect of isotopically selective dissociation of molecules ($SF_6$ and $CCl_4$) was found by excitation of weak overtone absorption bands. The threshold intensity of radiation for dissociation through overtone bands turned out to coincide with the threshold for fundamental bands. In work (33) the frequency dependence of the dissociation rate of $SF_6$ molecules was studied, and also it was found that the maximum was shifted to long wavelengths by an amount approximately equal to the anharmonic shift. On this basis a mechanism was proposed to compensate for the anharmonic shift in the transitions between several lower vibrational levels by changes in the rotational energy of the molecule. This made it possible to synthesize the hypothesis of high level excitation due to a vibrational quasi-continuum (10) with the effect of isotopically selective excitation of several lower levels and thereby to explain qualitatively the effect. Some direct experiments on selective molecular dissociation in a two-frequency IR field provided support for this mechanism (6).

This year selective photochemistry of molecules in intense IR laser fields has become the subject of numerous papers and an urgent topic at laser conferences. The detailed analysis of all these papers and the present-day situation are given below.

II.  MULTIPLE PHOTON ABSORPTION OF IR RADIATION:  EXPERIMENT

As stated in the preceding section, the basic phenomenon in all the processes considered is collision-free absorption of a large number of IR photons by a single molecule. The principal question here is how many IR photons are absorbed by a molecule which has interacted with a field of a given inten-

sity and frequency. This question is of interest when we deal
not only with the interaction between high-intensity radiation
and molecular vibrational levels, but also with chemical kinetics, reactivity of highly vibrationally excited molecules,
description of the dissociation processes, intramolecular
vibration-vibration relaxation and many other topics. In this
section we shall consider in detail the methods used in studies of multiple photon excitations as well as the results
obtained from the studies of this interesting effect.


A.   Experimental Technique

So far multiple photon absorption has been studied quantitatively by at least three methods (Fig. 2) (30, 34 - 36), and
it should be emphasized that all three methods give the same
quantitative results if all the errors of the experiments are
taken into account.

*Direct calorimetry.* In this method (34) the energy incident on a cell containing the radiation-absorbing molecular
gas and the energy transmitted through the cell are measured
simultaneously (Fig. 2a). The cell length is chosen so that
not more than 10% of the incident radiation could be absorbed
in the gas. Because of this it is possible to assume that the
intensity is the same along the entire length of the cell.
This condition simplifies interpretation of the measurements
but gives data of minimal precision.

Although this type of measurement would seem to be absolute, it may turn out that the measured reduction of energy in
the beam is related not to energy absorption by the molecules
but rather to scattering of some type. For example, Raman
scattering, conversion of exciting radiation into radiation
at different frequencies which can be absorbed by the cell
windows, harmonic generation, etc. may occur. If such competing processes are efficient enough, the measurements will
overestimate the energy directly absorbed by molecules. The
processes mentioned above are quite possible when the resonance character of the radiation with the molecular vibrations
is taken into account. A further disadvantage of this method
is that it is impossible to take measurements above the optical damage threshold of the window materials.

*Optoacoustic Detection.* This technique (Fig. 2b) is
based on the measurement of the intensity of an acoustic wave
in an absorbing gas which originates due to the conversion of
the vibrational energy of the molecules to translational
motion (V-T relaxation). The acoustic wave amplitude is
usually registered by a capacitance microphone. The choice of
a special construction of optoacoustic detectors allows one
to make measurements at low power in weak fields, where the

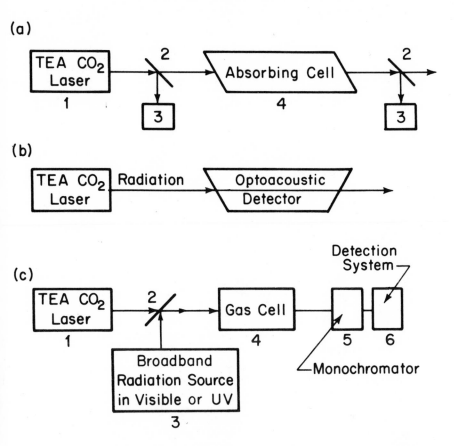

Fig. 2. *Techniques used in the study of multiple-photon absorption:  a)   direct calorimetry of absorbed energy (34): 1 - TEA $CO_2$ laser, 2 - beam splitters, 3 - calorimeters, 4 - cell with an absorber;  b)   optoacoustic detector:   for details see Fig. 3;  c)   infrared-ultraviolet (IR-UV) double resonance (30, 37):   1 - TEA $CO_2$ laser, 2 - beam splitters, 3- broadband source in the visible or UV, 4 - absorber, 5 - grating monochromator, 6 - detection system.*

absorption is quite linear, as well as at power levels at which optical breakdown takes place.   This is an important advantage of the method.   Figure 3 represents schematically an optical microphone which registers the optoacoustic signal from a spatially-defined region which coincides with the waist of the focused laser beam (35).

The signal from other regions of the cell, as seen in Fig. 3 is cut off by a set of irises.   In work (36) the micro-

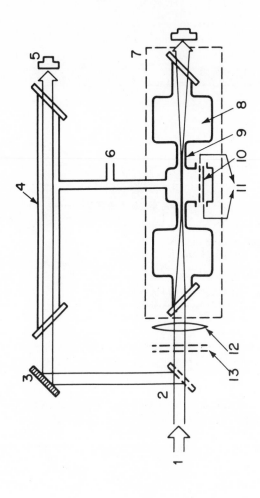

*Fig. 3. The detailed design of an optoacoustic detector (35). 1 — incident radiation from a laser, 2 — NaCl beam splitter, 3 — mirror, 4 — 2-meter long cell with the absorber used for calibration, 5 — calibrated thermopile, 6 — gas inlet, 7 — outer shield of the detector, 8 — buffer volume, 9 — iris, 10 — capacitance microphone, 11 — output to the recording system, 12 — focusing lens of NaCl, 13 — attenuators. The cell 4 and the detector itself are drawn to different scales.*

phone was placed in a long side arm located opposite to the zone of intense field. This also allows the signal from the spatially-defined zone of intense field, the beam waist defined by the lens, to be observed clearly.

Attention must be drawn to the shortcomings of this measuring technique. First, the acoustic signal amplitude should be calibrated to some given absorbed energy. In work (35) this calibration was done by simultaneous measurement of the signal from the optoacoustic detector and the absorbed energy by simple calorimetry in the fashion discussed earlier. Thus, the measuring error which entered into the calibration was automatically extended to all subsequent measurements. Secondly, when taking measurements in gases for which the V-T relaxation rate is slow enough to be comparable to the volume diffusion rate of excited molecules, it is necessary to perform absolute calibrations at each pressure because the shape and hence the amplitude of the acoustic pulses may vary. This last circumstance is of particular importance since all measurements of multiple photon absorption are taken at very low pressures (from 0.05 to 0.25 torr) of absorbing gas so as to avoid the influence of collisional processes. At such pressures the above considerations should be taken into account. One more shortcoming of this technique is that it is impossible to differentiate between the signal arising from vibrational excitation relaxation and that resulting from the heat evolution in exothermic or endothermic reactions which can be observed at high enough intensities of exciting radiation.

It must be mentioned that the measurements of multiple photon absorption taken by the first (34) and second (35) methods gave practically the same results in the region of overlap. Both of these methods, as a matter of fact, are calorimetric and differ only in the types of detectors used.

*IR-UV Double Resonance.* This technique (30) for observing the population of vibrational levels of excited molecules does not give precise quantitative data on the amount of energy absorbed. But, together with the calorimetric measurements it yields a lot of information on times characteristic of vibrational excitation energy redistribution. This technique was suggested in (37) and successfully applied in (30) to investigate the multiple photon excitation of $OsO_4$ molecules by an intense IR field. In this method the time-resolved red shift in the electronic absorption is studied (Fig. 2c). The red shift is caused by transitions from vibrationally excited levels of the ground state to levels of the excited electronic state. The transitions which are studied are schematically drawn in Fig. 4. The value of the red shift in molecular electronic transitions is proportional to the number of vibrational levels at which absorption occurs since in this case we record transitions corresponding to those from

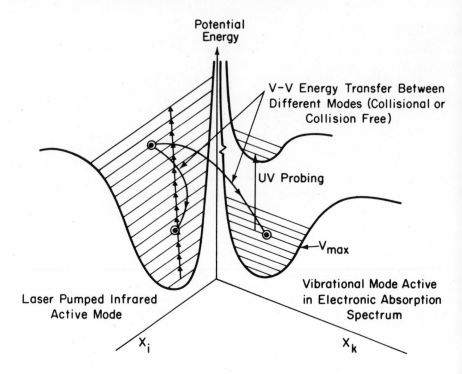

*Fig. 4.   The transition scheme in IR-UV double resonance. The transition  from $v_{max}$ to $v = 0$ in the electronically excited state gives the maximum value of red shift in the VIS-UV absorption bands.*

the level $v \gg 1$ of ground electronic state to $v = 0$ of excited electronic state.   The maximum red shift of photoabsorption gives the $v_{max}$.   It is apparent that in some cases $v_{max}$ derived from this measurement may be affected by the sensitivity of the apparatus and too low a value may be observed. The pulse shape of induced absorption can give information on the population mechanism of a level observed:   direct radiative population, population by V-V exchange, or thermal population.   Also, from observing the decay of vibrational level population one may obtain additional information on the time of V-T relaxation.   In (38) these questions are covered in detail.

Unfortunately, this method does not give direct quantitative data on the amount of the IR field energy absorbed by molecules because one should know the Franck-Condon factors for high-lying vibrational levels and one must also make some assumptions about the distribution of the population over the levels.   It is also necessary that a molecule being investigated should have an absorption spectrum in the UV or visible ranges and the absorption in this transition should be much higher than in the transitions that are pumped by infrared. In this case one can realize the highest sensitivity.

*Comparison of the Methods.* The first two methods give quantitative data on absorbed energy while the third method gives a qualitative energetic picture and an exact picture of the time dependence of processes. Thus, although these methods have certain differences, they complement each other.

It is evident that monochromatic laser radiation interacts only with a part of molecules. This is associated with molecular distribution over rotational levels. The portion of interacting molecules depends, of course, on the spectral width of the exciting radiation. Also it depends on exciting radiation intensity since the dynamic Stark effect broadens the transition. Thus, with increasing intensity more and more molecules are involved in the absorption process.

Direct calorimetry simply yields the value of absorbed energy $E_{abs}$. The real number of radiation quanta absorbed by each molecule which has interacted with the radiation field is expressed as:

$$n_{real} = \langle n \rangle \frac{\Delta \nu_{p.br.}}{\delta \nu_{branch}} \tag{II.1}$$

where $\langle n \rangle$ is given by the expression

$$\langle n \rangle = \frac{E_{abs}}{\hbar \omega N_o} \tag{II.2}$$

$\hbar \omega$ is the energy of the pumping radiation quantum, $N_o$ is the total number of molecules in a volume under irradiation (34), $\delta \nu_{branch}$ is the half-width of the absorption band branch, $\Delta \nu_{p.br.}$ is the value of power broadening of transitions in the pumping field.

The validity of such a method for evaluating the number of absorbed quanta has recently been proved experimentally (39, 40) by measuring the width $\Delta \nu_{hole}$ of the "hole" burned by the radiation of a high-power laser in the distribution over rotational levels of the absorbing molecules. A "hole" in the absorption spectrum is formed due to migration of the molecules to upper vibrational levels. It has been demonstrated in (39, 40) that the width of a "hole" in the molecular distribution over rotational levels within the limits of experimental error agrees with the value

$$\Delta \nu_{hole} = \Delta \nu_{p.br.} = \frac{\mu_{01} E_{laser}}{h}, \tag{II.3}$$

where $\mu_{01}$ is the matrix element of pumped transition, $E_{laser}$ is the intensity of the light wave.

## B.    Experimental Data

Of particular interest are the following items:   a) dependence of the $n_{real}$ and $<n>$ introduced above on exciting radiation intensity;   b) frequency dependence of these quantities;   c)   transformation of a linear absorption spectrum to a substantially nonlinear one, as well as the influence of collisions.  All these points are considered below.

1.  *Dependence of $n_{real}$ and $<n>$ on Radiation Intensity.*

The dependence of $<n>$ on laser field intensity, $I_{laser}$, was first measured in (34), and then such measurements were performed again in (35, 41).  The experiments on measuring $<n>$ were carried out in low-pressure gases where there were no gas-kinetic collisions in the time of pulse duration $\tau_p$. Usually, the half-width of a pulse from a Transversly Excited Atmospheric pressure (TEA) $CO_2$-laser with a nitrogen-free mixture, used in all experiments is 90 to 100 nsec.  It is enough to have an absorbing gas pressure less than 0.25 torr, since the time between gas-kinetic collisions is, as a rule, greater than 50 nsec at one torr.  The investigations have shown that the dependence of $<n>$ on $I_{laser}$ depends significantly on the excitation frequency.  The results typical of $OsO_4$ and $SF_6$ molecules are given in Fig. 5 (34).  It can be seen that, when the $OsO_4$ molecules are pumped through the P-branch of the $\nu_3$ vibrational mode, $<n>$ increases linearly to a definite value with increasing intensity of the pumping laser.  When the same molecules are pumped through the R-branch of the $\nu_3$ transition, the dependence $n(I_{laser})$ is different and $<n>$ tends to saturate.  The same picture is valid for other polyatomic molecules as well (34, 35).  It is evident that, since the number of molecules interacting with the field increases in proportion to $I^{\frac{1}{2}}_{laser}$, the real number of quanta absorbed by a molecule $n_{real}$ increases also in proportion to $I^{\frac{1}{2}}$.

It can be seen at once that 1)  the molecules absorb an energy $\hbar\omega n_{real}$ that is much higher than the energy of molecular dissociation $D_0$;  2)   even in low intensity (from $10^5$ W/cm$^2$) fields a molecule can absorb large energies which suggests that field broadening is of no significance in the process of energy absorption;  3)   at a certain pumping intensity the dependence of $<n>$ changes and becomes proportional to $I^{\frac{1}{2}}$.  This points to the fact that the process of energy absorption by a molecule is over and an increase in $<n>$ is caused through involvement of new molecules in the process of absorption due to power broadening.  It is not unlikely that

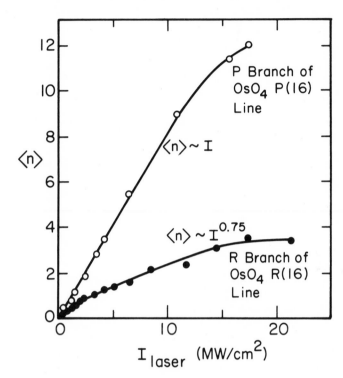

*Fig. 5a.   The number of photons averaged over all the mo-
lecules in the irradiation volume absorbed in $OsO_4$, <n>, versus pump intensity.   The pumping was performed by the $CO_2$
lines indicated in the 10.6 μ band (41, 42).*

these molecules participating in absorption also have some
characteristic energy distribution within $n_{real}$ but, unfortunately, for the time being there are no techniques available
to investigate this distribution.

The experimental results obtained for the $SF_6$ molecule
show that at 23 MW/cm$^2$ the real number of quanta of $CO_2$ radiation absorbed by one molecule $n_{real} \approx 100 \pm 15$.   This
means that each molecule accumulates energy of about 13 eV
which is from 3 to 3.5 times higher than the energy needed to
break the molecular bond, $SF_6 \rightarrow SF_5 + F$ - that is much higher
than $D_0$.   It can be seen from the results presented in the
next paragraph that the $SF_6$ molecule dissociates in the IR
laser field through a channel which is energetically not favorable.

At intensities of over 23 MW/cm$^2$, $n_{real}$ no longer increases.   This value of intensity agrees with the experimental
value of molecular dissociation threshold.

When pumped through the R-branch of the vibrational tran-

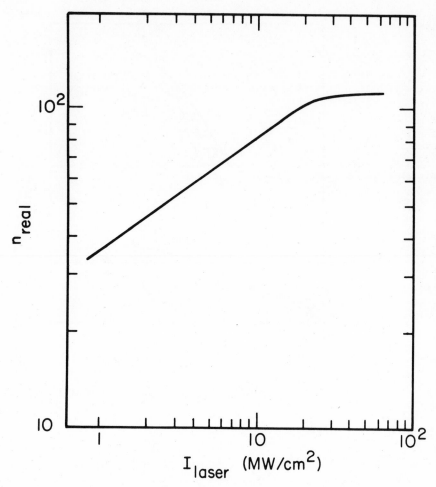

Fig. 5b. *The dependence of $n_{real}$, real number of infra-red photons absorbed by a molecule which has interacted with the radiation field, versus pump intensity. The data is for 0.1 torr of $SF_6$, P(16) line of $CO_2$ laser (34).*

sition the molecule also absorbs energy with $n_{real} \gg 1$, which is very important for estimating the validity of the multiple photon absorption model. In the experiments on double IR-UV resonance (30) the levels $v \leq 10$ were populated after the energy stored in the $\nu_3$ mode had been redistributed over other modes. The molecules in that case were pumped through the R-branch. These results suggest that before the redistribution of the energy from the laser-pumped $\nu_3$ mode over the vibrational manifold, the vibrational levels of $\nu_3$ were populated up to $v = 40 - 90$ in a field of 10 MW/cm².

But the direct calorimetric measurements taken under the same conditions gave $<n>$ = 2.6.   This fact also proved that $n_{real} >> <n>$.

Some experiments on multiple photon absorption under the conditions when rotational relaxation can be realized have shown that $<n>$ increases as the buffer gas pressure builds up. The dependence of $<n>$ on buffer gas pressure measured for $SF_6$ is given in Fig. 6.   One can see that an increase of hydrogen pressure by only 16 torr, all other things being equal, doubles $<n>$.   This means that in a pulse time (100 nsec) rotational relaxation fills the "hole" burned in the original molecular distribution over rotational levels by the pumping laser.   Similar results - sharp increase of $<n>$ with a buffer

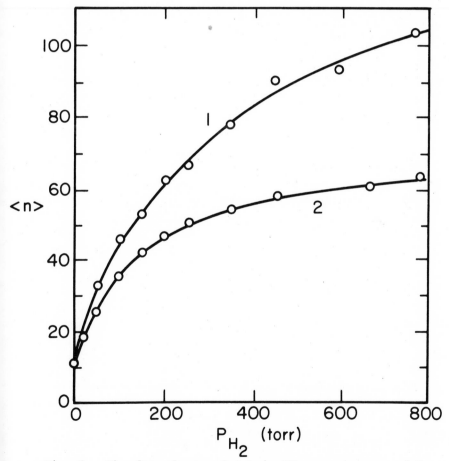

Fig. 6.   The dependence of $<n>$ in $SF_6$ on pressure of $H_2$ buffer gas.   The pump intensity 10 MW/cm$^2$.   The curves 1 and 2 correspond to $SF_6$ pressures 0.5 and 1 torr    (34).

gas added into the cell with the absorbing gas $C_2H_2Cl_2$ - were obtained in (44), where 30 torr of Xe added increased the absorbed energy by an order of magnitude. Such a sharp increase of absorbed energy $<n>\hbar\omega$ is explained by the fact that due to a large value for the rotational constant of the $C_2H_2Cl_2$ molecule the portion of molecules taking part in IR radiation absorption is smaller compared to $SF_6$. Therefore the influence of rotational relaxation on $<n>$ for light molecules is much more important than for heavy ones.

2.    *Resonance Characteristics of Multiple Photon Absorption.*

    It is of particular interest to study the resonance characteristics of multiple photon absorption because this enables us to have an idea about the structure of the vibrational levels with v > 1 which is difficult to do now by the methods of classical spectroscopy. In addition, detection of narrow peaks in multiple photon absorption spectra will allow high-selectivity excitation in isotopic molecules, and especially in those where the isotope shift is small compared to the absorption bandwidth of the transition v = 0 → v = 1 directly pumped by laser. It is also evident that, as the pumping laser intensity increases, resonances narrower than $\Delta\nu = \mu E/h$ cannot be detected because in this case they become masked by power broadening..

    Figure 7 shows the dependence of $<n>$ in $SF_6$ on laser fre-

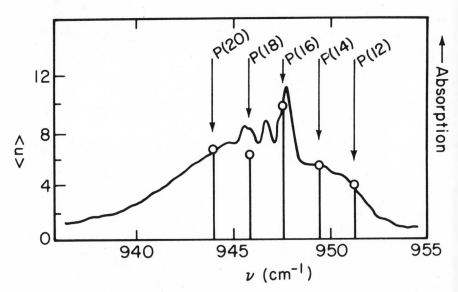

*Fig. 7.    The dependence of $<n>$ in $SF_6$ on pump frequency. The intensity on different laser lines was kept constant. The linear absorption spectrum of $SF_6$ is shown for illustration (34).*

quency.  The laser intensity at all frequencies is held
constant, 8 MW/cm$^2$.  There is a minimum observed at the P(18)
line of the $CO_2$-laser, and already at P(20) multiple photon
absorption builds up.  Similar results were obtained in work
(36).  It is also shown in (36) that with increasing power the
absorption at P(20), P(22) and other lines of the longer wave
region of the spectrum increases and is about twice as much
as that at P(16).  That is, change in the multiple photon
absorption spectrum with $I_{laser}$ takes place.

In work (35) $<n>$ in $SF_6$ was measured as a function of
laser frequency at various intensities.  The results are presented in Fig. 8.  As is seen, the spectrum of multiple photon
absorption is deformed.  Poor accuracy of the experimental
measurements did not make it possible however to observe the
structure of multiple photon absorption as was done in (34,
36).  Figure 9 shows the spectrum of multiquantum absorption
in $C_2H_4$ (35).  Its structure becomes blurred as the exciting
laser intensity increases.  This is quite understandable since
field broadening makes it impossible to detect narrow resonances.

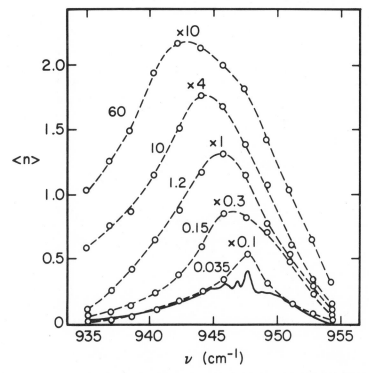

Fig. 8.  The dependence of $<n>$ in $SF_6$ measured by an
optoacoustic detector (35).  The parameter of the curves is
the laser intensity (MW/cm$^2$) (35).

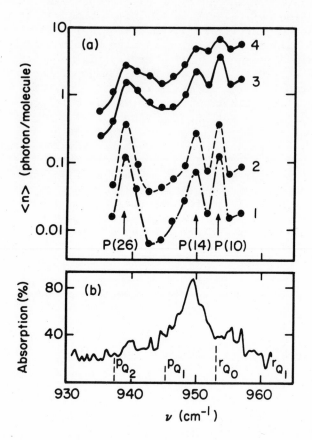

*Fig. 9.* (a) *Multiple photon absorption spectrum of ethylene* ($P_{C_2H_4}$ = *0.7 torr); laser intensity for each curve is 1 -* $I_{las}$ = *7 MW/cm$^2$, 2 -* $I_{las}$ = *70 MW/cm$^2$, 3 -* $I_{las}$ = *700 MW/cm$^2$ and 4 -* $I_{las}$ = *1800 MW/cm$^2$.* (b) *Linear absorption spectrum of* $C_2H_4$ *(*$P_{C_2H_4}$= *40 torr) taken with 0.4 cm$^{-1}$ resolution.*

Narrow resonances in multiple photon absorption were observed in heavy polyatomic molecules, such as $OsO_4$ (41) by examining the visible luminescence in $OsO_4$ which accompanied the dissociation process. The dependence of visible lumines-cence intensity in $OsO_4$ on $CO_2$-laser frequency is drawn in Fig. 10. The field intensity in the area of luminescence observation is 60 MW/cm$^2$. Such variations in intensity can apparently be explained by the structure of multiple photon transitions in $OsO_4$, for all other parameters were thoroughly controlled and were the same for all frequencies of the $CO_2$-laser. As will be seen in Sec. VI below, separation of

Os isotopes was realized only by virtue of such a structure of multiple photon transitions.

In principle, because of differences in the spectra of multiple photon transitions for isotopically different molecules, we may use for pumping transitions of molecules which have no isotope shift (42, 43). These experiments will be discussed in detail in Sec. VI.

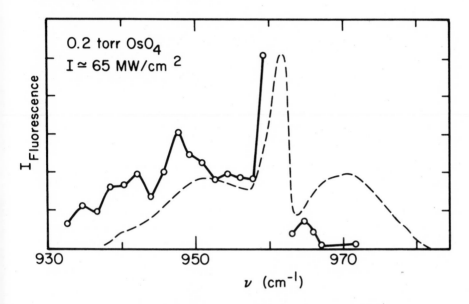

Fig. 10.  The frequency dependence of multi-photon induced visible luminescence intensity in $OsO_4$ (arbitrary units). The linear absorption of $OsO_4$ (----) is also shown (41).

III. MULTIPLE PHOTON DISSOCIATION OF POLYATOMIC MOLECULES

A.   Research Methods

The multiple photon dissociation of polyatomic molecules by an intense infrared laser field is explored by a variety of methods. The simplest of them is following the concentrations of stable products in a reaction vessel, both starting material (absorber molecules) and dissociation products. The apparatus used may vary in each specific case. This method was found extremely convenient for investigation of those molecules which undergo dissociation without any noticeable reverse reactions. The $SF_6$ molecule is rather the best example (6, 12, 32 - 34, 45 - 47, 89).  In the absence of reverse reactions this method provided correct quantitative data on dissociation rate and so on.

This method cannot be applied for investigation of the dissociation process for those molecules such as $BCl_3$, and $SiF_4$ (2, 51, 53) whose dissociation products tend to recombine in reverse reactions, unless a scavenger is added in the reaction vessel.  It is rather obvious that in the latter case the data on dissociation rates, quantum yield, etc. is strongly dependent on the choice of the scavenger.

The dissociation kinetics sometimes is studied by observing the so-called, visible luminescence, i.e. the radiation of electronically excited fragments of the molecules which are formed in dissociation processes.  It was found (41, 44, 90) that the integral of the luminescence signal is directly proportional to the amount of the absorbing gas dissociated. This observation allows us to connect the measurements made in the course of visible luminescence studies with studies of the dissociation processes.  As it was shown in (52) and also (90), the electronically excited fragments are not the major products of the multiple photon dissociation;  most of the radicals are formed in the ground state.  The application of tunable dye lasers for the excitation of fluorescence from the radicals proved to be very fruitful as this permits detection of the quantum states of the primary radicals.  This kind of information cannot be obtained by any other method.

Very recently IR dissociation has become the subject of studies in molecular beam arrangements with direct mass spectrometer studies of dissociation products (48, 53, 118, 119).

A combination of laser excited fluorescence and the molecular beam technique would be an ideal method for these studies, as it would provide the most complete data on dissociation products.

In the following sections we shall review the experimental data on multiple photon dissociation of polyatomic molecules, and it will be quite clear that the observed phenomena are general in character.

B.    Threshold Characteristics of the Dissociation Rate

The dissociation rate $W_{dis}$ or, in other words, the fraction of molecules dissociating per pulse, has been investigated in papers (6, 33, 46, 55, 89).  Dissociation rate $W_{dis}$ is defined as

$$W_{dis} = n_p \ ln(N_o/N) \qquad \text{(III.1)}$$

where N, $N_o$ are current and initial concentrations of the ori-

ginal molecules, and $n_p$ is the number of irradiation pulses.
As the experiments have shown (46), $W_{dis}$ is a function of
exciting radiation intensity, and pressure of both absorbing
and buffer gases.

The most careful measurements of the dissociation rate of
the $SF_6$ molecule (without adding any scavenger) were done in
(46). The experiments were carried out with well collimated
laser beams without focusing the radiation into the cells
which is a governing factor for similar experiments. The cell
length was selected so that the beam intensity was reduced
because of the absorption not more than 3 - 5%. The results
of such measurements are shown in Fig. 11. The lower scale
is given in MW/cm$^2$. The pulse half-width was 90 nsec. The
energy of a pulse tail, if there was any, was less than 5%

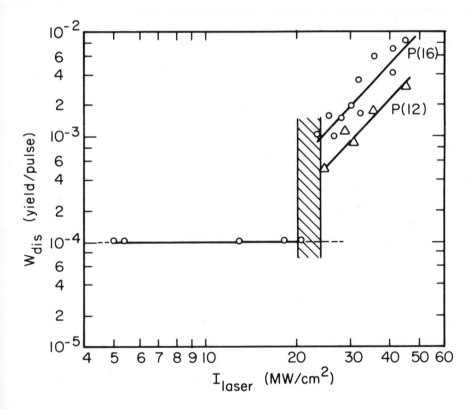

Fig. 11. The dependence of the dissociation rate $W_{dis}$
in $SF_6$ on laser intensity. Direct propagating beam measurements, no scavenger added. The laser pulse length was always
the same - 90 nsec FWHM, the $SF_6$ pressure was 0.2 torr (46).

of the total pulse energy which was achieved by not using nitrogen in the gas mixture in the TEA $CO_2$-laser. The measurements were done at various frequencies, the laser lines P(16) and P(12). It is evident that in both cases the dependence of dissociation rate on intensity is the same, $W_{dis} \propto (I_{las})^3$. The error in the exponent is within ± 0.5.

Comprehensive measuring of the threshold characteristics for dissociation of $SF_6$ done in (49) shows that the threshold is a function of the laser frequency. It was found that the dissociation threshold in $SF_6$ has its maximum value in the region of the R-branch of the $\nu_3$ transition of $SF_6$ (27 MW/cm$^2$) and falls down monotonically to 8 MW/cm$^2$ when the laser frequency is shifted to the red 13 cm$^{-1}$ from the Q-branch of the $\nu_3$ mode. The error limits for the threshold intensity are shown by the shadowed region on Fig. 11.

In a number of papers different dependencies for the dissociation rate versus laser intensity were obtained. In (54) the dissociation rate was proportional to $(I_{las})^{1.5}$ and in (41, 55) before the intensity region where $W_{dis} \propto (I_{las})^3$, the dependence $W_{dis} \propto (I_{las})^7$ was also observed. As it was shown in (50) the dependence $W_{dis} \propto (I_{las})^{1.5}$ arises from the dependence $W_{dis} \propto (I_{las})^3$ when the experiments are performed with the focused-beam geometry.

The dependence $W_{dis} \propto (I_{las})^7$ which precedes the intensity region $W_{dis} \propto (I_{las})^3$ (Fig. 12) cannot be explained easily. However, the following possibilities should be considered: the character of the dependence $W_{dis}$ versus $I_{las}$ should greatly depend on laser-beam properties such as: a) transverse mode structure, b) self-mode-locking which is usually observed in TEA lasers, c) presence of a long tail which follows the main laser spike and so on. Another possibility is the occurance of chemical reactions between molecules which are very highly excited and impurities or scavenger gas in the reaction vessel.

In favor of the latter assumption are the data on dissociation of $SF_6$ in a molecular beam arrangement (48) where sharp threshold dependences were also observed. It is clear that in an extremely intense laser field the probability of dissociation per pulse will tend to unity i.e. the $W_{dis}(I_{las})$ dependence will show saturation.

Comparing the experimental data on multiple photon absorption (Fig. 5) with the data on dissociation rate one can see the coincidence of the observed intensity threshold and the intensity region where the $<n>$ begin to be proportional to $(I_{las})^{\frac{1}{2}}$.

To investigate the threshold dependence of the dissociation rate of ethylene molecules on laser radiation intensity

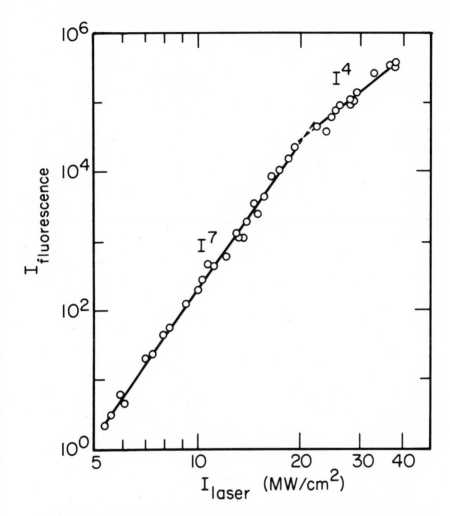

Fig. 12. The dependence of the visible luminescence intensity (arbitrary units) in $OsO_4$ versus pump intensity. The
laser pulse length was 100 nsec FWHM, the $OsO_4$ pressure 0.15
torr (41).

(90), the resultant primary radicals $C_2$ in the electronic
ground state $a^3\Pi_u$ were excited by pulsed dye laser radiation.
The experimental diagram is given in Fig. 13.

The great sensitivity of the method of laser-excited luminescence permitted, as for the $OsO_4$ molecule, accurate measurements of the threshold dependences of the advent of primary
dissociation product radicals, $C_2$. The results of these
measurements are given in Fig. 14. The measurements show (90)
that in the range 150 - 250 MW/cm$^2$ the primary product yield

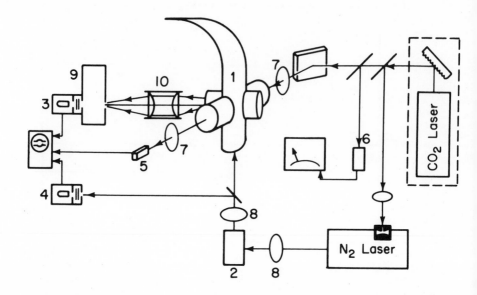

*Fig. 13. Experimental setup for detection of dissocia-
tion products in ground electronic states (90). 1 - cell with
absorber (ethylene), 2 - $N_2$-pumped dye laser, 3, 4 - photo-
multipliers, 5 - photon-drag detector, 6 - thermopile, 7 -
Ge-lenses, 8 - cylindrical lenses, 9 - monochromator and 10 -
focusing lenses.*

or the dissociation rate $W_{dis}$ depends on $(I_{las})^{14}$ and over
250 MW/cm$^2$ the dependence is changed    to $(I_{las})^{5/2}$.  The
measurements were taken in focused beams of $CO_2$ laser radiation and this, probably, resulted in such a dependence.

Figure 14 shows that the data obtained from probing the
unexcited $C_2$ radicals by laser radiation correlates with the
data obtained from the studies of the visible luminescence of
electronically excited radicals formed by dissociation in the
IR field.  In the same work the presence of intensity fluctuation of laser radiation due to its multimode nature has been
proposed as one possible interpretation of the dependence
$W_{dis} \propto (I_{las})^{14}$

C.    Resonance Characteristics of the Dissociation Rate

It is quite evident that to understand the processes of
multiple photon dissociation of molecules it is very important

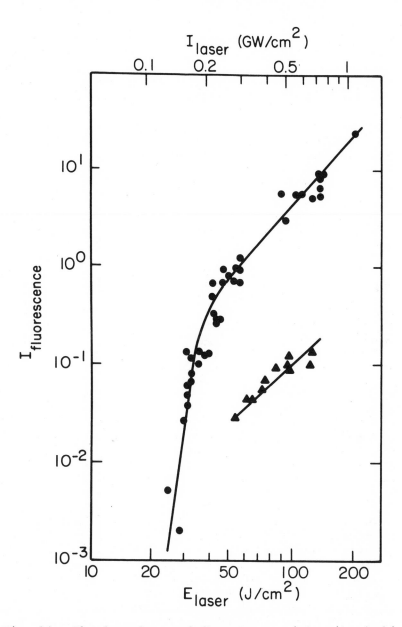

Fig. 14.   The dependence of fluorescence intensity (arbitrary units) from $C_2$ radicals on $CO_2$ pump laser intensity.
$C_2$ is produced in $C_2H_4$ during the dissociation of $C_2H_4$ by an
IR field.   The circles correspond to the fluorescence intensity excited by a dye laser;   The triangles correspond to the
visible luminescence intensity without a dye laser (90).

to obtain information on resonance characteristics of the
dissociation process.  This point is of importance for the
problem of isotope separation by this method which will be
considered in detail in Sec. VI.  These characteristics are
of great interest from the spectroscopic point of view as they
represent the spectra of transitions to those molecular levels
from which the dissociation takes place.  The resonance characteristics of multiple photon dissociation have been investigated for a number of molecules such as $SF_6$, $C_2H_4$, $NH_3$, $OsO_4$,
$BCl_3$, $SiF_4$, $CCl_4$.

The dependence of dissociation rate of the molecule $SF_6$
on exciting radiation frequency is given in Fig. 15.  In these

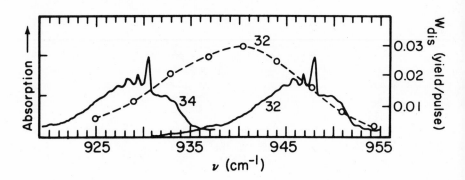

*Fig. 15.  The dependence of the dissociation rate $W_{dis}$
for $SF_6$ molecules on $CO_2$ laser frequency.  The $SF_6$ pressure
is 0.2 torr, the energy at each laser frequency is 1.5 J with
the beam focused.  The linear absorption of $SF_6$ in arbitrary
units is also shown (33).*

experiments the energy and the power were held constant at all
frequencies.  The pressure of $SF_6$ was 0.2 torr.  The linear
absorption spectrum is given in this figure for comparison.
Figure 15 shows that the maximum of the dissociation rate
$W_{dis}$ is shifted by approximately 7 $cm^{-1}$ to the red with respect to the Q-branch absorption of the mode $\nu_3$ (33, 46), and
it is much broader than the absorption band.

As the investigation has shown, the position of the maximum in dissociation rate $W_{dis}$ depends essentially on pressure
and shifts to the red as the absorber gas pressure increases.

At the minimum pressure under which the experiments were
conducted, 0.05 torr, the maximum is shifted by about 5 $cm^{-1}$.
As mentioned above (49), the frequency shift of the maximum

of dissociation rate also depends on laser intensity. So at low intensity, ~ 14 MW/cm$^2$, the maximum shift was 13 cm$^{-1}$ and fell to 5 cm$^{-1}$ at 40 MW/cm$^2$. The intensities indicated resulted from averaging the intensity over the entire irradiated volume.

From the analyses of this data it is clear that it would be a mistake, of course, to connect the value of this red shift with the magnitude of the anharmonicity of $\nu_3$. For example, in BCl$_3$ the maximum of $W_{dis}$ is red-shifted by 14 cm$^{-1}$, while the anharmonicity is 1.6 cm$^{-1}$.

The dependence $W_{dis}$ $(\nu)$ is a complex function of molecular distribution over rotational levels and cross sections of transitions up to the levels from which the molecule may dissociate. Later we will return to this question once more.

The dependence shown in Fig. 15 was obtained by measuring the dissociation rate of SF$_6$ in a focused beam arrangement. The intensity in the focal region, where the dissociation process goes most effectively, reached 200 - 300 MW/cm$^2$. At these intensities the power broadening reaches several wavenumbers, and this of course smoothes out the dissociation rate curve. Any possible resonance structure, which may be less than the power broadening, is washed out. The recent measurements of $W_{dis}$ in SF$_6$ versus laser frequency carried out just above the dissociation threshold (28 MW/cm$^2$) showed some structure, which correlated with the structure which is seen in Fig. 7. Collisions contribute as well, smoothing the dissociation curve and shifting its maximum to the red where the cross section of multiple photon absorption is larger. As in the case of the SF$_6$ molecule, the dissociation rate versus frequency for the molecules BCl$_3$ and SiF$_4$ exhibit the same behavior: the maximum is shifted to the lower frequencies and with increasing pressure the maximum shifts more and more.

As was shown in Sec. II (Fig. 10), the dependence of visible luminescence intensity of OsO$_4$ on laser frequency has sharp maxima and minima. By contrast the molecule NH$_3$ lacks entirely any dependence of $W_{dis}$ on laser frequency (55). It must be said that the example of the NH$_3$ molecule is not only atypical but rather unique. This is because the value of anharmonicity in NH$_3$ is $\simeq$ 130 cm$^{-1}$ which cannot be found deliberately in other polyatomic molecules. Taking into account an extremely low quantum yield of dissociation of NH$_3$ (under low pressures) it would be likely to suppose that the dissociation of NH$_3$ molecules takes place from the vibrational hot bands where the value of anharmonicity is far lower than for the ground state. Unfortunately, because of the shortage of experimental data on NH$_3$ dissociation, no unique conclusion can be drawn at the present time.

D.    The Role of Collisions in the Dissociation Process

As mentioned in Sec. I the dissociation of molecules
excited by multiple photon absorption takes place both instantaneously during a pulse and as a result of collisions. It is
necessary, however, to stress that the decay of excited molecules in the collisional phase can be caused by chemical reactions of overexcited particles both with molecules of the same
kind and with other molecules in the irradiated volume (reaction vessel). Collisional dissociation can be differentiated
from chemical reactions only by the reaction products.

Let us consider what influence may be produced on the
dissociation rate $W_{dis}$ by collisions.

Collisions of excited molecules with unexcited ones can
be of two types according to the nature of the colliding
particles: with energy transfer from an excited molecule to a
nonexcited one and without such transfer. Under the latter
conditions the distribution of excitation energy between modes
inside the excited molecule may be distinctly changed (intramolecular V - V transfer). As the energy inside the molecule
due to its interaction with the radiation field is stored
mainly in one mode, the redistribution of the stored energy
from this excited mode to other modes may influence the capa-
bility of the molecule both to dissociate from the excited
state and to undergo chemical reactions.

The influence of collisions on the dissociation rate is
investigated in most detail for the molecule $SF_6$ (46). It
has been found experimentally that in those pressure ranges
where the measurements were done, the dissociation rate decreases with increasing pressure of both buffer gas and pure
$SF_6$ according to the law $W_{dis}$

$$W_{dis} = exp \ (- \ P/P_o) \ ,$$

where $P_0$ is the characteristic pressure at which the dissociation rate is decreased by e times. The experimental data are
given in Fig. 16.

It is seen that the molecules having normal vibrational
modes resonant with $SF_6$, i.e. vibrations $\nu_2$ of $NH_3$ and $\nu_7$ of
$C_2H_4$, reduce the dissociation rate $W_{dis}$ much more quickly
than $O_2$ or Ar. This indicates that the dominant process
which suppresses dissociation is most likely the V - V exchange between $SF_6$ and $NH_3$ or $C_2H_4$. Addition of $O_2$ or Ar
exert an almost identical influence if the difference between
the gas kinetic cross sections of $O_2$ and Ar are taken into
account. It is hardly probable that addition of 9 torr of Ar
might cause full V - T relaxation in $SF_6$, although the exact

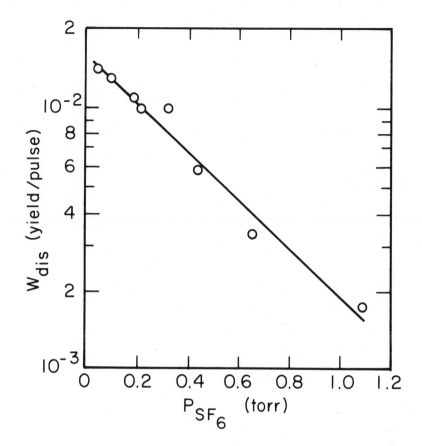

Fig. 16a.   The dependence of the dissociation rate $W_{dis}$ on $SF_6$ pressure (46).

data on the rate of V - T relaxation of $SF_6$ by Ar are known only for the level v = 1 (59) (for this pressure of Ar the time of V - T relaxation of $SF_6$ is ≃ 50 μsec).  The decrease in $W_{dis}$ with increasing pressure of an inert buffer gas is most likely caused by a collisional intramolecular V - V relaxation which populates modes unfavorable for dissociation.

The strong dependence of the dissociation rate $W_{dis}$ on collisions seems to contradict the observations made with the molecular beams (48), where 30% of molecules dissociate per pulse.  The dissociation of 30% of the molecules in the beam means that approximately 100% of the excited molecules dissociate since $\Delta\nu_{Rabi}$ for the experimental conditions of (48) was approximately $\Delta\nu_{(P-branch)}/3$.

From these two observations a very important conclusion may be drawn:  the superexcited molecule needs noticeable time

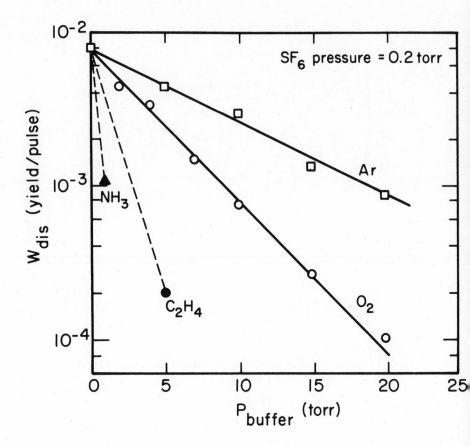

*Fig. 16b. The dependence of $W_{dis}$ in $SF_6$ on the pressure
of buffer gases, $NH_3$, $C_2H_4$, $O_2$, and $Ar$, added to the absorber
cell. The $SF_6$ pressure was constant at 0.2 torr. The buffer
gas pressure is given in torr (46).*

(3 - 5 μsec for $SF_6$) to dissociate in a collision free regime.
The last figure is based on observation (48) that $W_{dis}$ is independent of pressure when the time between successive collisions becomes greater than 5 μsec.

The former conclusion is valid on the assumption that the
molecule absorbs an equal amount of energy during the laser
pulse in both cases - in molecular beam and in absorption cell.
This assumption seems to be valid since the energy build-up
process in both cases is the same because the Rabi frequency
is larger than the reciprocal time of $SF_6$-$SF_6$ collisions in
the pressure range examined.

## E.  Molecular Dissociation by Overtone Excitation

In the experiments described above molecular excitation due to multiple photon absorption was realized under conditions when the laser frequency coincided with the absorption frequency of a fundamental molecular vibration.  The data obtained from excitation of weak combination molecular vibrations are of great interest for the interpretation of physical problems of molecular dissociation in intense IR fields.  Such experiments were performed in (32) and (57).

The $SF_6$ molecule has two composite vibrational bands in the region of 10 $\mu$ which are assigned to $\nu_2 + \nu_6$ and $\nu_3 + \nu_2 - \nu_5$.  The dissociation rate of $SF_6$ was studied in the case of excitation of these composite bands (32).  The experimental conditions of all experiments were the same:  i.e. the initial pressure of $SF_6$ in the cell, 0.2 torr, the geometry of the experiments, the power and energy emitted by the laser.

The experimental data may be summarized in the following way.  The dissociation rates in the case of excitation of various vibrational bands relate as:

$$W_{dis}^{\nu_3} : W_{dis}^{\nu_2 + \nu_6} : W_{dis}^{\nu_3 + \nu_2 - \nu_5} \quad as \quad \mu_{\nu_3} : \mu_{\nu_2 + \nu_6} : \mu_{\nu_3 + \nu_2 - \nu_5}$$

where $\mu_{\nu_k}$ is the dipole moment of the transition $\nu_k$.  It is seen from the data presented that the dissociation rates are proportional to $\mu_{\nu_k}$.  Consequently, the dissociation rate, other conditions equal, is determined mainly by the value $\Delta\omega_{Rabi}$, i.e. by the relative portion of the molecules which are directly excited by the laser radiation.

The main reason for the deviation from strict proportionality of $W_{dis}$ to the magnitude of the matrix element of the transition pumped is:  1) coarse selection of the laser frequencies for the pumping of the different bands; and 2) different transition cross sections in the vibrational quasi-continuum.  The latter factor would result in a reduction of dissociation rate as the exciting frequency shifts to the blue.

The latter reason should be taken into account if the molecule dissociates through the same vibrational quasi-continuum.  This seems to be the case since the threshold for dissociation was found to be the same for the $\nu_3$ mode pumping and for the $\nu_2 + \nu_6$ mode.  It may happen that the molecules dissociate not due to pumping the $\nu_2 + \nu_6$ mode but due to excitation through the high frequency wing of the $\nu_3$ band.

To check whether the dissociation of $SF_6$ in the case of pumping the $\nu_2 + \nu_6$ vibration is due to the vibration $\nu_2 + \nu_6$ and not due to $\nu_3$-band wing pumping, the mass-spectrometric analysis of the dissociation products as well as of the re-

maining $SF_6$ gas in the cell was done. The analysis has shown
that no enrichment of any sulphur isotope was observed when
the vibration $\nu_2 + \nu_6$ was pumped. This proves that the dissociation was not due to pumping the $\nu_3$ band. The analogous
results were obtained in work (47) where the authors observed
no enrichment in $SF_6$ when pumped in the frequency region of
970 $cm^{-1}$. Further, while pumping the triple combination
vibration with the vibration $\nu_3$, i.e. $\nu_3 + \nu_2 - \nu_5$, the mass
spectrometric analysis revealed the remaining $SF_6$ to be enriched with the isotope $^{34}S$ by an amount which corresponded
to the amount of dissociated $SF_6$. This indicates the high
selectivity of the dissociation process.

As stated above, the dissociation of molecules by IR radiation in pumping overtones should be more selective as
compared with pumping fundamental bands.

The dissociation of the molecule $CCl_4$ (57), as an example
proved this suggestion to be quite real. The molecule $CCl_4$
has a weak absorption band at 10 $\mu$. The identification of
vibrational transitions is given in Herzberg's book where it
is ascribed to the vibrations $\nu_3 + \nu_2$ and $\nu_4 + \nu_2 + \nu_1$ (82).

In work (57) the dependence of the dissociation rate $W_{dis}$
in $CCl_4$ on laser radiation frequency was investigated. The
laser power at all frequencies and the initial pressure of
$CCl_4$ were held constant. The experiments were carried out in
focused beams; the intensity in the focus reached 300 MW/$cm^2$.
The dissociation rate was measured by IR spectrophotometry
after the cell was irradiated by a series of pulses.

The dependence of the dissociation rate $W_{dis}$ in the $CCl_4$
molecule together with a linear absorption band are shown in
Fig. 17. It is immediately seen that there is no dissociation
in the high-frequency part of the absorption. The low-frequency part of the absorption band has a rather narrow peak
of disssociation rate, its half-width being much narrower than
the absorption band.

It is evident that even in spite of the large magnitude
of the intensity in the focal region, where the dissociation
of the molecule $CCl_4$ mainly takes place, the power broadening
is small compared to the band half-width. This causes the
curve $W_{dis}$ to become narrower even despite the fact that the
experiments were carried out at 0.2 torr when molecular collisions may play a role resulting, as mentioned in the previous
section, in the broadening the resonance curve $W_{dis}$.

F.   Dissociation by Two Pulses of IR Radiation

As follows from the results of the experiments described
in Sec. II and in the previous parts of this section, the pro-

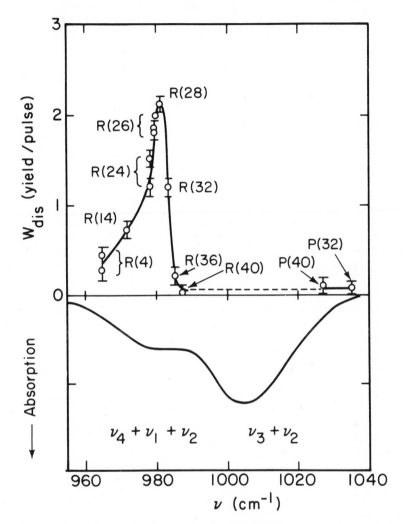

*Fig. 17. The dependence of the dissociation rate $W_{dis}$
on $CO_2$ pump frequency. The $CCl_4$ pressure is 0.2 torr, without
a scavenger. The absorption of $CCl_4$ at the pressure of 90
torr is also shown (57).*

cess of energy storage by a molecule comes about effectively
at intensities of about 1 $MW/cm^2$ and even lower.  The sharp
threshold of dissociation is connected, evidently with saturation of some transitions lying in the quasi-continuum range of
vibrational levels.

The independent behavior of two quantities - the number
of absorbed quanta of pumping radiation $<n>$ and the dissocia-

tion rate $W_{dis}$ - on exciting radiation intensity permits us to perform separately the functions of selective excitation and molecular dissociation with pulses at two different IR frequencies. A field at the frequency $\nu_1$ in resonance with the 0 - 1 vibrational transition, weak by comparison to the dissociation threshold intensity, excites the molecules to higher vibrational levels. At the same time an intense radiation field at frequency $\nu_2$, out of resonance with the 0 - 1 transition causes the dissociation of the molecules selectively excited by $\nu_1$.

Such a separation of the functions of excitation and dissociation between two pulses of different frequencies and intensities, permits us to avoid the effect of power broadening of vibrational transitions by simple reduction of the exciting field intensity at the frequency $\nu_1$. It means that the molecular dissociation selectivity increases, an effect of basic importance for molecules with a high threshold of dissociation.

The dependence of the dissociation rate $W_{dis}$ of a molecule of $SF_6$ in a two frequency IR laser field on the frequencies ($\nu_1$, $\nu_2$) and intensities ($I_1$, $I_2$) of time-synchronized pulsed $CO_2$-lasers have been experimentally studied. In the experiment TEA $CO_2$-lasers with frequency selective cavities were used. The maximum energy of the laser at the frequency $\nu_1$ was 0.5 J with a pulse duration of 250 nsec. The laser at the frequency $\nu_2$ emitted pulses with an energy of 3 J and a duration of 90 nsec. The synchronization was done with an accuracy of (1 ± 0.5) μsec. The radiation at the frequency $\nu_1$ was directed to the cell as an unfocused beam, and the radiation at the frequency $\nu_2$ was focused by a NaCl lens with f = 100 mm. The dependence of selective dissociation $W_{dis}$ ($I_2$) on the intensity $I_2$ of the non-resonant IR field at the frequency $\nu_2$ was measured in a beam just slightly collimated by a NaCl lens with f = 1000 mm. The cell filled with $SF_6$ gas was cooled to T = 193 ± 5°K. The amount of dissociated $SF_6$ gas in the cell after irradiation was determined by IR spectrophotometry. The experiments were conducted at an $SF_6$ pressure in the cell of 0.2 torr.

Figure 18 shows the dependences of dissociation rate on the frequency $\nu_1$ of the resonant IR laser field (selecting step) which coincides with the band of the $SF_6$ molecule. The power $I_1$ of the selecting-step radiation was held constant for all $CO_2$ laser lines and was 4 MW/cm$^2$. The average power of the second-step, focused radiation $I_2$ was 58 MW/cm$^2$. Resonance characteristics (1, 2) were obtained at 300°K and 193 ± 5°K respectively. Curve 3 relates to the dissociation of $SF_6$ in a single-frequency, powerful IR laser field (33), with the average power of a focused beam being 30 MW/cm$^2$. The frequency of non-resonant IR radiation was $\nu_2$ = 1084 cm$^{-1}$

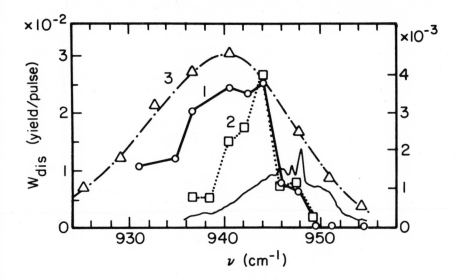

Fig. 18.   The dependence of the dissociation rate $W_{dis}$ in
$SF_6$ on the frequency $\nu_1$ which is in resonance with the absorption.   The frequency $\nu_2 = 1084$ $cm^{-1}$.   $I_1 - 4$ $MW/cm^2$, $I_2$ (averaged over the irradiation volume) - 60 $MW/cm^2$ (focused beam).
The curves 1 and 2    correspond to two different temperatures
of $SF_6$ - 300 and 190°K (right scale).   The linear absorption of
$SF_6$ and the dissociation rate dependence on frequency in the
single frequency case at T = 300°K are also shown (curve 3,
left   scale) (89).

(line P (30)).   The dissociation of $SF_6$ occured only during
simultaneous irradiation by two lasers.   The comparison of the
two resonance characteristics for the cases of dissociation in
a one frequency (curve 3) and a two-frequency (curve 1) IR
laser field shows that there is a sharp narrowing of the dissociation edge from the high-frequency side and nearly no dissociation in the excitation range of the R-branch of the vibration-rotation band in the two-frequency case.   The low-frequency edge of the dissociation contour was practically unchanged
compared to single frequency dissociation, but the cooling of
$SF_6$ gas to (193 ± 5)°K caused a sharp narrowing of this edge
of the $W_{dis}$ ($\nu_1$) contour.   The half-width of the resonance
curve $W_{dis}$ ($\nu_1$) is 5 $cm^{-1}$ which is approximately four times
less than that in a single frequency field.   The maximum of
the dependence $W_{dis}$ ($\nu_1$) is shifted relative to the maximum of
the Q-branch of the IR linear absorption spectrum by
(4.5 ± 0.5) $cm^{-1}$.   With single-frequency dissociation the
shift is (7 ± 0.5) $cm^{-1}$ (33).   The results of the comparison
between the resonance characteristics for the two methods of

$SF_6$ dissociation (single-frequency and two-frequency) permit us to conclude that the dissociation of $SF_6$ molecules in an intense single-frequency, IR-laser field in the R-branch of a vibration-rotation band is connected with a large value of laser intensity.  It is possible that in an intense laser field effective excitation of the molecule $SF_6$ takes place at the weak vibration-rotation transitions within the R-branch which cannot be observed in the linear IR absorption spectrum. The comparison between the resonance characteristics of $SF_6$ dissociation in a two-frequency IR laser field at 300°K and when cooled to $T = (193 \pm 5)$°K (curves 1, 2, Fig. 18) shows that the low-frequency edge of the dissociation contour $W_{dis}$ ($\nu_1$) depends on hot bands in the IR absorption spectrum.

Figure 19 (solid curve) shows the dependence of dissociation rate on radiation intensity for the selecting step $I_1$ at the fixed frequency $\nu_1 = 942.4$ cm$^{-1}$.  The power and frequency of the non-resonant laser field were $I_2 = 60$ MW/cm$^2$, and $\nu_2$ = 1048.6 cm$^{-1}$ respectively.  For comparison (dashed line) the dependence of the number of photons $<n>$ absorbed by a $SF_6$

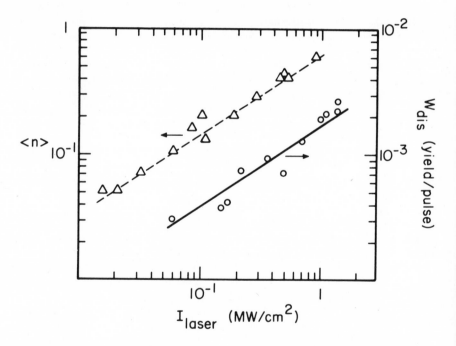

Fig. 19.  The dependence of the dissociation rate, $W_{dis}$, in $SF_6$ on intensity of selecting radiation $I_1$ ($\nu_1 = 942$ cm$^{-1}$), $I_2$ (averaged) = 60 MW/cm$^2$, $\nu_2 = 1041$ cm$^{-1}$ (89).  The dotted line shows the dependence of $<n>$ absorbed in $SF_6$ at the same intensities $I_1$ (89).

molecule in the single-frequency IR laser field on this field
intensity is shown. It is seen from Fig. 19 that both
straight lines have the same slope which determines the power
dependence for $<n>$ and $W_{dis}$

$$W_{dis}, \ <n> \ \sim \ I_1^{0.7} \qquad\qquad (III.2)$$

It is also worth noting that with two-frequency excitation the
dissociation occurs at a considerable rate in fields with $I_1$
= 60 kW/cm$^2$. The observed correlation of the dissociation
rate with the number of photons absorbed by a single SF$_6$ molecule points to the fact that the increase in the dissociation
rate with selecting step intensity $I_1$ is due to the increase
in the number of absorbed photons. The dependence of $W_{dis}$ on
$I$ close to $\sqrt{I}$ (III.2) shows that the main mechanism leading to
an increase in the number of photons absorbed by a single SF$_6$
molecule, as the laser radiation intensity rises, is power
broadening ($\Delta\nu_{p.br.} \sim I_{las}^{0.5}$) because of which new vibration-
rotation transitions become involved in the absorption process.
The dependence $W_{dis}$ $(I_2)$, Fig. 19, is the direct confirmation
of the fact that the dissociation threshold (33) is not
connected with overcoming anharmonicity by field broadening
of the first vibrational levels but is most likely determined
by the transitions in the vibrational quasi-continuum. Consequently, the dependence of dissociation rate on non-resonant
IR laser field intensity $W_{dis}$ $(I_2)$ must exhibit a threshold.
Such a dependence is given in Fig. 20 ($\nu_1$ = 942.4 cm$^{-1}$, $\nu_2$
= 979 cm$^{-1}$, $I_1$ = 4 MW/cm$^2$) and it is evident that its character is that of threshold and the threshold intensity value $I_2$
= 20 MW/cm$^2$ is very near the value $I$ = 23 MW/cm$^2$ obtained in
the single-frequency regime, and the cubic dependence on intensities is observed in both cases:

$$W_{dis} \ \sim \ I_2^3, \ I_{las}^3 \qquad\qquad (III.3)$$

In Fig. 21 the dependence of dissociation rate on the
frequency of intense non-resonant IR laser field is shown.
The frequency of the weak resonant field $\nu_1$ was fixed and
equal to 942.4 cm$^{-1}$ (line P (22)). The average power $I_2$ of
the non-resonant IR laser field was held constant for all frequencies and was equal to 58 MW/cm$^2$. The selecting-step laser
radiation power $I_1$ was 4 MW/cm$^2$. The maximum detuning of the
frequency from the center of the linear IR absorption spectrum
of SF$_6$ ($\nu_3$) was fixed by the CO$_2$ laser output range at 130
cm$^{-1}$. The frequency corresponding to the minimum detuning
equal to $\simeq$ 20 cm$^{-1}$ was selected so that the intense IR laser
field $I_2$ could not cause dissociation in the absence of a weak
resonance field at the frequency $\nu_1$. Figure 21 shows that the
dissociation rate increases as the frequency $\nu_2$ approaches the

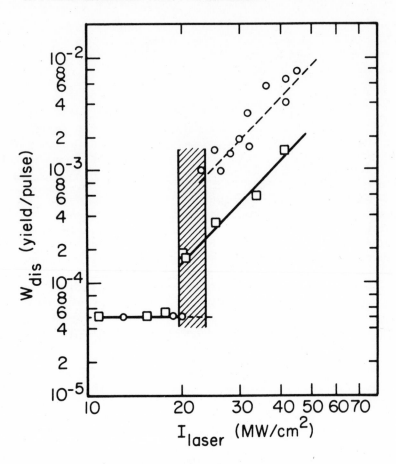

*Fig. 20.    The dependence of the dissociation rate, $W_{dis}$,
in $SF_6$ on intensity of the nonresonant field $I_2$ (solid line).
The dotted line shows the same dependence for the single frequency dissociation case.    The shadowed region shows the possible error of determination of the threshold.    The absolute
value of the error in determining the dissociation rate is
shown below.    The double log abscissa and the ordinate scales
are in 3:1 ratio (89).*

$\nu_3$ band.  As a result of the fact that the nonresonant IR laser radiation $\nu_2$ is absorbed in the vibrational quasi-continuum, the dependence $W_{dis}$ ($\nu_2$) in Fig. 21 is the dispersion
characteristic of vibrational quasi-continuum.  It is difficult to obtain the dissociation characteristic of the quasi-
continuum by this method in the vibrational band range as the
molecules dissociate from the field $I_2$ at the frequency $\nu_2$
irrespective of whether a weak field at the frequency $\nu_1$ is

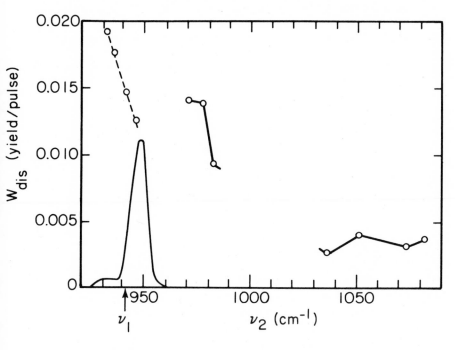

Fig. 21. *The dispersion characteristics of the vibrational quasi-continuum - the dependence of $W_{dis}$ on the frequency of the nonresonant field $\nu_2$; $\nu_1$ was fixed at 942.4 $cm^{-1}$ (89). The dotted line is the result of calculations from the data presented in Fig. 18 (89, 42).*

applied or not. But if we take into account that the characteristic $W_{dis}$ ($\nu_1$) at the fixed frequency $\nu_2$ also reflects the spectrum of transitions to higher levels, it is possible to obtain the dispersion characteristic of the quasi-continuum within the vibrational band $\nu_3$ proceeding from the data for $W_{dis}$ ($\nu_1$) as dissociation is done by an IR field of the same frequency $\nu_2$. For this purpose it is necessary to obtain the dissociation rate dependence on $\nu_1$ for a single molecule. It would be easy to do it taking into account that the number of molecules excited by a weak field is directly proportional to the linear absorption coefficient at the given frequency. In this case the quasi-continuum spectrum within the vibration $\nu_3$ in $SF_6$ looks like a sloped line, and the results of such construction coincide both for the data obtained at room temperature and at 196°K. These results are given in Fig. 21 (dashed line). These results may be added to those in (99) where absorption in $SF_6$ was investigated in double-frequency experiments. In (99) the absorption of the radiation, which in this work was of weak intensity and acted as probing radiation, at the frequency of 925 $cm^{-1}$ was higher by 1.6 times than at 907 $cm^{-1}$. All these data support the conclusion that the vibrational "quasi-continuum" in $SF_6$ represents a rather extremely wide resonance with its half-width 150 - 200 $cm^{-1}$.

The resonance properties of the quasi-continuum mean that due to the difference of absorption cross sections in the quasi-continuum it is possible to obtain some selectivity in molecular dissociation.

So, if we take a particular instance of the molecule $SF_6$, taking the value of isotopic shift and the width of the resonance curve of the quasi-continuum, it is possible to obtain 20 to 30% enrichment with heavier isotopes of sulphur even with full V - V exchange between molecules of different isotopic composition.

One possible conclusion may be drawn from the fact that at lower frequencies in respect to normal vibrations the absorption cross section in the vibrational quasi-continuum is much larger than it is at the frequency of normal modes. Say, in the first experiments of Isenor the dissociation of $SiF_4$ was performed at frequencies $\simeq$ 940 $cm^{-1}$ while the $\nu_3$ mode of $SiF_4$ lies at 1030 $cm^{-1}$. The process may start from a 3 to 4 quantum transition of a molecule and as it reaches the vibrational quasi-continuum, it dissociates effectively. This point of view, of course, must be checked experimentally.

It seems that in the absence of absorption the multiquantum transitions to the vibrational quasi-continuum would cause some dissociation in intense ($10^9 W/cm^2$) laser fields.

In this section the term "quasi-continuum" was used to identify the transitions of a molecule which are above the level $v > 1$ of the excited mode.

IV.    THEORETICAL INTERPRETATION OF MULTIPLE PHOTON ABSORPTION

AND DISSOCIATION OF POLYATOMIC MOLECULES IN AN INTENSE

IR FIELD

A.    Statement of the Problem.    Basic Models

It is rather difficult to describe theoretically the effect of multiple photon absorption and dissociation of polyatomic molecules in an intense IR field discussed in Secs. II and III. So far, in essence, there has been only an approximate, semi-quantitative theory developed for multiple photon absorption and only qualitative estimates have been done for dissociation. It should be emphasized that the effect of multiquantum excitation and dissociation of molecules by radiation with a frequency, $\omega$, which is much higher than the frequency of molecular vibrations $\omega_{vib}$ ($\omega \gg \omega_{vib}$), is rather evident. This process was first estimated by Askarian (60), and Bunkin et al. (61). In these works consideration was also given to molecular vibrational excitation by intense radiation with a frequency $\omega \simeq \omega_{vib}$ and with overtone frequencies. The radiation intensity for molecular dissociation (for the model of an anharmonic Morse oscillator) was estimated as about $10^{12}$ W/cm$^2$. The effects of strong excitation and dissociation of polyatomic molecules being discussed in this chapter have nothing to do with the problem covered in (60, 61) since they occur in a resonant laser field with $\omega \simeq \omega_{vib}$, with an intensity three to five orders of magnitude less than those required for multiquantum dissociation in the works mentioned. The effect considered in (60, 61) must be observed even in the case of diatomic molecules, whereas the effects under discussion are characteristic of polyatomic molecules only.

The main difficulty in explaining these effects is that strong excitation and dissociation of polyatomic molecules takes place in fields of only $10^6$ to $10^8$ W/cm$^2$ intensity, even though the molecular vibrations are anharmonic. In an intense resonant IR field, power broadening of the transition due to the dynamic Stark effect naturally takes place, and this broadening may in principle compensate for anharmonicity. Such a possibility for strong excitation of molecular vibrations was discussed as long ago as in (62). In this case, however, the intensity of IR radiation must also be extremely high. Actually, the magnitude of the level splitting or broadening in a strong field is given by

$$\Delta\omega_{p.br.} = \frac{\mu E}{\hbar} = \omega_R \qquad (IV.1)$$

where $\mu$ is the transition dipole matrix element or more correctly its projection along the field direction ($\mu = \mu_0/\sqrt{3}$, where $\mu_0$ is the absolute value of the transition dipole moment.[1]

---

[1]Here we use the notation of Rabi frequency $\omega_R = \mu E/\hbar$, which is to be found in the Rabi formula for the time dependent probability of occupation of a single quantum state $b$ of a two-level system under the resonant field $E \cos\omega_0 t$:

$$\left|b\right|^2 = sin^2\left[\frac{\omega_R}{2}(t-t_o)\right] = \frac{1}{2}\left[1 - cos\ \omega_R(t-t_o)\right]$$

Some authors use as a Rabi frequency of $\mu E/2\hbar$ and sometimes in the same paper both notations are used simultaneously. This fact must be taken into account by our readers during the

The anharmonic shift of the frequency of the $|v-1> \to |v>$ vibrational transition relative to the $|0> \to |1>$ frequency is

$$\Delta\omega_{anh} = 2\pi c\, B_{anh}(v-1), \tag{IV.2}$$

where $B_{anh} = 2X_{ii}v_i/c$ is the anharmonicity constant for the i-th vibrational mode. The power broadening and anharmonicity shift are equal at a laser field intensity of

$$I = \frac{c}{8\pi}\left(\frac{\hbar}{\mu}\Delta\omega_{anh}\right)^2. \tag{IV.3}$$

A molecule of $BCl_3$, for example, can dissociate when the vibration $v_3$ is excited by radiation. For this purpose, it must be excited up to the $v = 42$ level of $v_3$ because the dissociation limit is $39,000\ cm^{-1}$. When the value of anharmonicity $B_{anh} = 1.65\ cm^{-1}$ (19) and the dipole moment of transition is approximately, $\mu = 0.2$ Debye, the pulse intensity required must be about $4.5 \times 10^{11}\ W/cm^2$. The instantaneous luminescence phase of $BCl_3$ in our experiments (11, 51) appeared however at intensities of no higher than $10^9\ W/cm^2$. These simple considerations (51) showed that for the model of anharmonicity compensation by power broadening that a pulse of intensity about $10^9\ W/cm^2$ could excite only several of the lowest vibrational levels. Strong excitation and dissociation of $SF_6$ molecules could be observed at more moderate powers (from $10^6$ to $10^7\ W/cm^2$). At such intensities the power broadening is at least one order of magnitude less than the anharmonicity and cannot even explain the excitation of $v = 2$. Therefore the broadening of molecular transitions in an intense IR field is not by itself a large enough effect. It is still necessary to find an explanation for why the molecules overcome so easily the "anharmonicity barrier" at comparatively moderate light intensities.

Isenor et al. (10) and then Bloembergen (63), Akulin et al. (64) pointed out the possibility of molecular dissociation from highly excited levels located far below the dissociation limit, due to the fact that the spectrum of transitions between highly excited multiply degenerate molecular levels is almost continuous. This so-called "vibrational quasi-con-

---

numerical comparison of results from different works. The standard notations for frequency in

$$\omega[rad/sec] = 2\pi v\ [Hz] = 2\pi\frac{v}{c}\ [cm^{-1}],$$

where c is speed of light, should be kept in mind. In some works the Rabi frequency is multiplied by $2\pi$ because of improper use of the value $2\pi\mu E/\hbar$ as Rabi frequency or by $\pi$ times (employing the value $2\pi\mu E/2\hbar$).

tinuum" obviates the problem of detuning between the field
frequency and the vibrational transitions between highly
excited anharmonic levels.  The quasi-continuum may also ensure, in principle, fast molecular excitation to the dissociation limit.  When an intense field interacts with the continuum of overlapping vibrational levels, it is apparent that
a large number of normal vibrations must be excited at the
same time.  This can explain to some extent why the energy
absorbed by a molecule considerably exceeds the dissociation
energy of the molecular bond in resonance with the IR field.
Absorption by the transitions in the vibrational quasi-continuum exists only for polyatomic molecules; this also can
explain the absence of dissociation for two- and three-atomic
molecules being acted upon by an IR field with its intensity
of up to $10^9$ W/cm$^2$.  But this mechanism of absorption is nonselective and offers no explanation for the isotopically
selective excitation of lower nonoverlapping vibrational
levels.

Bloembergen combined (63) the ideas of compensating
anharmonicity for lower transitions through power broadening
along with those of vibrational quasi-continuum to explain
collisionless dissociation.  Using the classical analysis of
an anharmonic oscillator he estimated the threshold power for
exciting several lower vibrational levels (v = 3 - 4) to the
lower boundary of the vibrational quasi-continuum, to be about
$10^9$ W/cm$^2$.  This approach may be used to explain the dissociation of some molecules having rather high threshold intensity of dissociation (probably BCl3 and others) but it cannot
be used to explain molecular excitation and dissociation at
lower intensities, from $10^6$ to $10^7$ W/cm$^2$ (SF$_6$, OsO$_4$ and
others).

Another step was made in our work (33), where we proposed
a mechanism of "soft" compensation for anharmonicity of the
lower transitions by changes in the molecular rotational
energy.  Many scientists apparently considered this possibility but in our work (33) this mechanism resulted directly from
experiments on the frequency dependence of the dissociation
yield of SF$_6$ (Sec. III).  By virtue of a triple vibrational
resonance (PQR-resonance) a polyatomic molecule in a rather
weak field ($10^4$ to $10^5$ W/cm$^2$) may be excited to the vibrational level v = 3.  Further excitation of a molecule which is
characterized by a low boundary of the vibrational quasi-continuum is done by virtue of a direct transition to the lower
levels of the vibrational quasi-continuum and subsequent
transitions within it.  Such a case is perhaps realizable for
SF$_6$ molecules (Fig. 22(a)).  Later there was another mechanism
of "soft" compensation for anharmonic frequency shifts suggested in (66).  It was based on anharmonic splitting of the
excited degenerate vibrational levels of a polyatomic molecule,

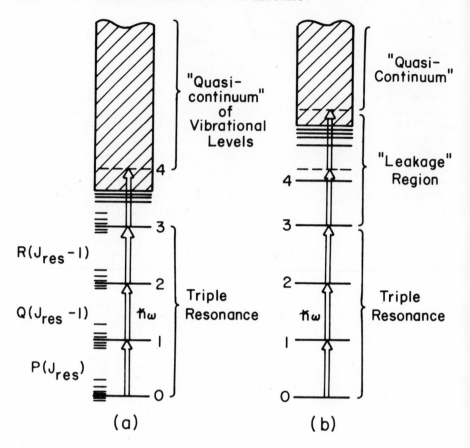

*Fig. 22. Models of quasi-resonance multiple photon absorption without anharmonicity compensation by power broadening: a) excitation of the level v = 3 by virtue of a "triple resonance" at successive vibrational-rotational transitions; b) "leakage" of molecules at detuned transitions from lower resonance levels into the vibrational "quasi-continuum" (65).*

$SF_6$ in particular.

For the molecules whose lower boundary of the vibrational quasi-continuum is above the level v = 4 multiple photon transitions may take place between v = 3 and the levels of the vibrational quasi-continuum.  In this case it is not obligatory at all that power broadening should fully compensate detuning the field frequency out of exact resonance with an intermediate level, v = 4, for instance.  At comparatively small power broadening the rate of the two-photon transitions such as v = 3 → v = 5, permits molecular "leakage" from the lower discrete levels v = 0, 1, 2, 3 into the vibrational quasi-continuum throughout a laser pulse of duration $\tau_p = 10^{-7}$ sec. This effect of "leakage" during excitation of a multi-level molecular system was studied in (67). Such a situation may occur, for example, in $BCl_3$ molecules (Fig. 22(b)).

The effect of the vibrational quasi-continuum in combination with rotational and (or) anharmonic compensation of anharmonicity and leakage in multiphoton transitions forms the basis for the theoretical interpretation of collisionless excitation and dissociation of polyatomic molecules. In the ensuing theoretical works much consideration is given to several models with due regard for these effects: 1) model of anharmonic oscillator and the multiphoton excitation resonance of its levels (68 - 70); 2) model of rotating anharmonic oscillator with regard to rotational compensation of anharmonicity on several lower transitions (71, 72); 3) coupling of the lower vibrational levels with those of the vibrational quasi-continuum (72 - 77). These basic models are considered in more detail below.

B.  Model of Anharmonic Oscillator

First we consider a simpler model, that of an anharmonic oscillator with frequency $\omega_0$ under the action of an external electromagnetic field with frequency $\omega$. The exact solution of this problem is well known (78). Note that an attempt to use this simplest possible model in explaining collisionless dissociation was made as long ago as in (51, 79).

The excitation probability of a v-th vibrational level in the "ladder" of equidistant levels is

$$P(v,t) = \frac{1}{v!} [w(t)]^V exp[-w(t)],  \qquad (IV.4)$$

where w(t) in the particular case of an electromagnetic field $E \cdot \sin\omega t$ has the form

$$w(t) = \left(\frac{\mu E}{\hbar}\right)^2 \frac{1-\cos(\omega-\omega_o)t}{2(\omega-\omega_o)^2}$$

$$= \left(\frac{\omega_R}{\omega-\omega_o}\right)^2 \sin^2 \frac{1}{2}(\omega-\omega_o)t,  \qquad (IV.5)$$

where $\omega_R = \mu E/\hbar$ is the Rabi oscillation frequency. The parameter w(t) means the average energy of the oscillator in units of the quantum energy $\hbar\omega$. Thus at any instant of interaction with the field a Poisson distribution exists for the excitation probability of harmonic oscillator levels. The average energy of the oscillator varies periodically with time at the detuning frequency $|\omega-\omega_0|$ between the levels $v_0 = 0$ and $v_{max}$ = $w_{max}$ = $[\omega_R/(\omega-\omega_0)]^2$. In the case of exact resonance the average number of photons absorbed by the oscillator in the finite time $\tau_p$ of interaction with a coherent light pulse is expressed as:

$$v_{max} = \frac{1}{4} (\omega_R \tau_p)^2 \qquad (IV.6)$$

For example, a harmonic oscillator with $\mu = 0.1$ Debye in a pulsed field of intensity $I = 10^7$ W/cm$^2$ and duration $\tau_p$ = $10^{-7}$ sec can be excited to the level $v_{max}$ $(\tau_p) = 2 \cdot 10^6$; that is, it absorbs from the field a huge energy of about $10^5$ eV (for $\hbar\omega = 0.1$ eV). In the case of finite detuning of the oscillator frequency $\omega_0$ and the field frequency $\omega$, the oscillator can absorb energy only if the pulsed field is switched off in a time shorter than the period of oscillations. If the field is switched in a time longer than $|\omega-\omega_0|^{-1}$, the oscillator returns to its initial state.

Unfortunately, to the extent that the harmonic oscillator easily absorbs a large number of photons, to just that extent is the model far from describing a real molecule. The anharmonicity in a molecule limits the excitation rate of even the lowest levels $v \geq 2$ if $\omega_R << \Delta\omega_{anh}$, that is, when the field broadening cannot compensate for the anharmonicity of vibrations.

Now we consider a slightly anharmonic oscillator which describes an i-th vibrational mode of the molecule. The energy levels of the anharmonic oscillator may be written in the form:

$$W(v_i) = v_i \hbar\omega^i + \hbar\omega^i X_{ii} v_i(v_i - 1). \qquad (IV.7)$$

In this expression $\hbar\omega^i$ corresponds to the fundamental frequency of the lowest transition $v_i = 0 \rightarrow v_i = 1$. The frequency of each subsequent n-th transition $v_i = (n-1) \rightarrow v_i = n$ decreases slightly because of anharmonicity:

$$\hbar\omega_n^i = \hbar\omega^i - (n - 1) 2|X_{ii}| \hbar\omega^i. \qquad (IV.8)$$

The next task is to solve the Schrödinger equation (or the appropriate equation for the density matrix):

$$[H_o + \vec{\mu} \cdot \vec{E} cos\omega t] \psi = i\hbar\dot\psi, \qquad (IV.9)$$

where $H_o$ is a Hamiltonian with a sequence of eigenvalues (IV.7); $\vec{\mu}$ is the operator of transition dipole moment; $\vec{E}$ and $\omega$ are the amplitude and oscillation frequency of the electric field. Yet, even such a straightforward problem for the simplest one-dimensional anharmonic oscillator has not been exactly solved. As a rule, in such cases the problem may be solved by the use of the perturbation theory, which enables us to solve the problem of weak interaction between the anharmonic oscillator and the field (68, 71). With computers providing a numerical solution for a particular case there is no limitation on the magnitude of oscillator-field coupling (69, 70). Furthermore, in the limiting case of a very intense field, when quantum-mechanical effects prove to be small, we may use a classical approximation, that is, give up the complicated Schrödinger equation (IV.9) and consider a simpler second-order equation for the anharmonic oscillator (63).

*Weak field-oscillator coupling.* A criterion for a weak field-oscillator interaction is that the value of level broadening (splitting) by the field is smaller than that of anharmonicity

$$\omega_R << \omega^i \left| X_{ii} \right|. \qquad (IV.10)$$

Perturbation theory can be applied when this condition is met. The most consistent treatment of this problem is given by Larsen and Bloembergen (68). Just some of the results of this work (from report (71)) are presented here. As would be expected from the simplest considerations, the excitation probability of the n-th vibrational level resonantly increases each time the condition of n-photon resonance is complied with

$$n\hbar\omega = W(n) = n\hbar\omega_n. \qquad (IV.11)$$

As a result, a succession of multiphoton resonances appears at the long-wave edge of the fundamental frequency $\omega^i$. These resonances, according to condition (IV.10) do not overlap but have their finite width determined by the Rabi frequency of their associated n-photon resonance or by the value of the effective matrix element of an n-photon transition:

$$\frac{1}{\hbar} \mu_{eff}^{(n)} E = \omega_R^{eff}(n) \qquad (IV.12)$$

The effective matrix element of transition for a two-photon resonance, for instance, has the well-known form:

$$\hbar\omega_R^{eff}(2) = \langle 0 | \mu E | 2 \rangle_{eff} = \frac{\langle 0 | \frac{1}{2}\mu E | 1 \rangle \langle 1 | \frac{1}{2}\mu E | 2 \rangle}{X_{ii} \; \hbar\omega^i} \qquad (IV.13)$$

The expression for the Rabi frequency for higher-order resonances is similar in form. For an n-photon resonance the following expression is true:

$$\omega_R^{eff}(n) = \frac{(\mu E/\hbar)^n (n!)^{\frac{1}{2}}}{2^n \left(\omega^i |X_{ii}|\right)^{n-1} \prod\limits_{n'=1}^{n-1} n'(n-n')}.$$  (IV.14)

The dynamics of an anharmonic oscillator in approximation (IV.10) remind one in essence of those of a two-level system. The anharmonic oscillator in the case of an exact n-photon resonance oscillates between the levels $v_i = 0$ and $v_i = n$ at the Rabi frequency $\omega_R^{eff}(n)$. If the laser pulse is long enough, the oscillator spends, on the average, half the time in the initial ($v_i = 0$) and half in the final ($v_i = n$) state, much as in the case of two-level systems. The dependence of the time-averaged probability $\overline{P}_R(n)$ to occupy the nth excited level on laser frequency illustrating the structure of multiphoton resonances for n = 1, 2, 3, 4 is given in Fig. 23. The width of each resonance is determined by the Rabi frequency $\omega_R^{eff}(n)$.

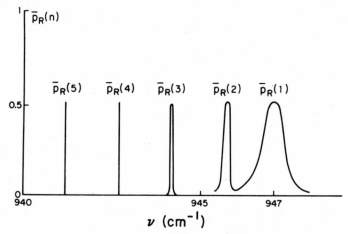

Fig. 23.  Time-averaged probability $\overline{P}_R(n)$ to occupy the n-th excited level of an isolated anharmonic vibrator by an n-photon process. The horizontal frequency scale is for the $\nu_3$ vibration of $SF_6$. Two-level perturbation theory is used with $\hbar\omega_R \simeq \frac{1}{4}\hbar\omega_3 |X_{33}|$ (71).

Consider a numerical example the parameters of which correspond to the interaction of $SF_6$ molecules with a $CO_2$-laser pulse. The parameters required for this interaction are given in Table 3 from (71). The Rabi frequency $\omega_R(1)$ in the table is evaluated for the radiation intensity corresponding to the dissociation threshold of $SF_6$ ($I_{th}$ = 23 MW/cm$^2$), the $CO_2$-laser

TABLE 3

*Relevant Energies for the Collisionless Dissociation of $SF_6$ by Infrared Radiation*

| Physical Quantity | Symbol | $E/hc$ $(cm^{-1})$ |
|---|---|---|
| Dissociation energy | D | 40,000 |
| Fundamental frequency of $\nu_3$ mode | $\hbar\omega_3$ | 948 |
| Thermal energy (300°K) | kT | 200 |
| Rotational width of P and R branch | 2B<J> | 5 |
| Coriolis splitting | — | 0.1-10 |
| Anharmonic shift | $2\lvert X_{33}\rvert\hbar\omega_3$ | 2.88 |
| Electric dipole matrix element at threshold intensity | $\hbar\omega_R = \lvert\mu\rvert\lvert E_{th}\rvert$ | 0.6 |
| Rotational constant | B | 0.0906 |
| Effective rotational constant | $\tilde{B} = B(1-\zeta)^a$ | 0.027 |
| Typical spectral width of IR radiation | $\hbar\Delta\omega_L$ | 0.03 |
| Pulse-limited width | $\hbar\tau_p^{-1}$ | $10^{-4}$ |
| Collisional width at one millitorr | $\hbar\tau_c^{-1}$ | $10^{-6}$ |
| Radiative width | $\hbar\tau_{IR}^{-1}$ | $10^{-8}$ |

a.  $\zeta$ *is the Coriolis coupling constant.*

pulse duration $\tau_p$ = 100 nsec.  For the threshold intensity, as can be seen from the table, condition (IV.10) is met, that is, anharmonicity compensation by field broadening is impossible. From (IV.13) it can be found that for the threshold intensity

$\omega_R^{eff}(4) = 1.0 \times 10^8$ sec$^{-1}$, $\omega_R^{eff}(5) = 3.2 \times 10^6$ sec$^{-1}$. Thus, throughout a pulse of duration $\tau_p = 10^{-7}$ sec the level $v_3 = 4$ must be substantially populated, the n-photon resonance width being rather small ($\sim 10^{-3}$ cm$^{-1}$). The level $v_3 = 5$ is populated slightly however ($\sim 1\%$).

*Strong oscillator-field coupling.* The Schrödinger equation ( or the density matrix equation) for an anharmonic oscillator with a finite number of levels can be numerically solved without limitations (IV.10). This was done by Hougen (69) for a nine-level oscillator the frequency of which however was tuned to the fundamental frequency rather than to that of any of the eight possible multiphoton resonances. Later Hougen carried out numerical computations for the frequency dependence of multiphoton transition probabilities of multiphoton resonances (80). The dependence computed is similar to that drawn in Fig. 23. The form of the multiphoton resonances is similar to the Lorentzian one, a multiphoton resonance of the n-th order perturbing lower-order resonances.

The frequency dependence for a four-level anharmonic oscillator was numerically computed by Larsen and presented in report (71). Figure 24 shows the dependence of the energy stored by such an oscillator which was found for two numerical values of $\hbar\omega_R(1/2\hbar\omega^i x_{ii}$ and $\hbar\omega^i x_{ii})$. As would be expected, some multiphoton resonances are materially broadened and overlap one another as the intensity increases.

In the limiting case of strong oscillator-field interaction, when

$$\omega_R \gg |x_{ii}|\omega^i \qquad (IV.15)$$

the anharmonic oscillator may be classically described (63, 81). This limiting case does not agree with the experimental conditions of molecular dissociation in fields ranging from $10^7$ to $10^9$ W/cm$^2$. Nevertheless, we consider it in order to make the picture complete. The discussion is based on the results of work (63).

The equation for classical anharmonic oscillator vibrations with the potential

$$V = \frac{1}{2} M\omega_o^2 Z^2 - \frac{b}{4} Z^4 \qquad (IV.16)$$

has the form:

$$M\ddot{Z} + M\dot{Z} \frac{1}{\tau} + M\omega_o^2 Z - bZ^3 = \ell_{eff} Ecos\omega t, \qquad (IV.17)$$

where M is the oscillator mass, $\omega_o$ is the oscillation frequency in harmonic approximation, Z is the amplitude of the small vibrations, and b is the anharmonic force constant of the

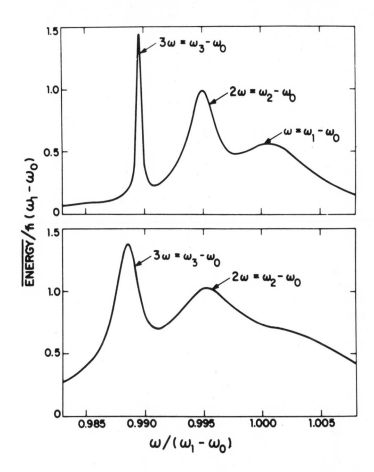

*Fig. 24. Average energy of excitation of an anharmonic vibrator as a function of frequency for an exact four-level calculation, with $\hbar\omega_R = \frac{1}{2}\,\hbar\omega_3\,|X_{33}|$ (top) and $\hbar\omega_R = \hbar\omega_3\,|X_{33}|$ (bottom) (71).*

potential. The external field $E\cos\omega t$ is an induced force the value of which depends on the effective classical charge

$$e_{eff} = |\mu|^2\,\frac{M\omega_o}{2\pi\hbar e}\;.\qquad\qquad(IV.18)$$

The steady-state amplitude of vibrations $Z$ is given by the equation

$$\left(-\omega^2+\omega_o^2\right)Z_o \;-\; \frac{3}{4}\,\left(\frac{b}{M}\right)Z_o^3 \;=\; \frac{\mu^2 E}{2\pi\hbar e}\,\omega_o \qquad(IV.19)$$

The mean vibrational level $\bar{v}$ of an anharmonic oscillator excited by the field is related to the steady-state amplitude by the relation:

$$\frac{1}{2} M\omega_o^2 \, z_o^2 = \bar{v}\hbar\omega_o \qquad (IV.20)$$

The dependence of the molecular vibrational energy $\hbar\omega_o \, \bar{v}$ on the field amplitude and its frequency $\omega$ can easily be found from relations (IV.19) and (IV.20). The anharmonic shift value is related to the parameter b by

$$2X_{ii} \, \omega_o = \frac{3}{2} \frac{b\hbar}{M^2\omega_o^2} \qquad (IV.21)$$

Equation (IV.19) may be written in the accepted notation:

$$z_o(-\Delta\omega - |X_{ii}|\omega_o \, \bar{v}) = \frac{\mu}{2\pi e} \omega_R, \quad \Delta\omega = \omega - \omega_o \qquad (IV.22)$$

Two conclusions may be drawn from equation (IV.22). First, for maximal excitation of the vibrational level $\bar{v}$ it is necessary that the field frequency be midway between the frequency of the fundamental transition $v = 0 \to v = 1$ and that of the last transition $\bar{v} - 1 \to \bar{v}$, that is, be equal to

$$\omega_{optimal} = \omega_o - |X_{ii}|\omega_o\bar{v}_i \qquad (IV.23)$$

This conclusion is quite evident since $\omega_{optimal}$ corresponds to the center frequency of the group of overlapping multiphoton resonances at frequencies ranging from $\omega_o$ to $\omega_o - 2X_{ii}\omega_o\bar{v}_i$ (Fig. 23). Secondly, considerable excitation can be realized only if the broadening $\omega_R(1)$ is comparable to the maximum anharmonic shift $X_{ii}\omega_o\bar{v}_i$.

C.    Mechanisms of Compensation for Anharmonic Shifts of

Lower Transitions

If the laser pulse frequency is tuned to that of a three- or four-photon resonance, the $SF_6$ molecule at a power of 20 MW/cm$^2$ can be excited, as it follows from (IV.13), to the levels $v = 3$ or 4. But the experiments on dissociation of the $SF_6$ molecule in a two-frequency IR field (6) show that the molecule can be excited to the lowest limit of the vibrational quasi-continuum at a lower intensity of laser pulse (about $10^4$ W/cm$^2$). It is clear that there must be a "soft" mechanism of compensation for anharmonic frequency shifts of several lower transitions which operates at intensities much lower than those required for "rougher" mechanisms of power broadening and multiphoton resonance. To explain experiments on isotopically selective collisionless dissociation of $SF_6$ molecules, two mechanisms were proposed: variations in rotational energy (33) and anharmonic splitting of degenerate vibrational energy levels (66). They are discussed below in some detail.

*1. Multistep, Vibration-Rotation Resonances*
In the simplest approximation the energy levels of a rotating anharmonic oscillator may be written as

$$W(v_i) = \hbar\omega^i \left| v_i + X_{ii} \, v_i(v_i - 1) \right| + B(v_i)J(J+1), \qquad (IV.24)$$

where $B(v_i)$ is the rotational constant in the vibrational state $v_i$. The selection rules for electric dipole vibration-rotation transitions allow changes in the rotational quantum number, J, by +1 (R-branch), -1 (P-branch) and 0 (Q-branch).
As may be seen from (IV.24), there is a unique possibility for resonance excitation of high vibrational levels by a single-frequency field if the laser frequency is tuned to any vibration-rotation transition and the anharmonic oscillator constants comply with the condition

$$B(v_i) \simeq B = - 2X_{ii}\hbar\omega_i \qquad (IV.25)$$

In this case multistep resonance excitation of the oscillator is possible by a sequence of R-transitions

$$\left| v_i{=}0, J_o \right> \xrightarrow{\ R(J_o)\ } \left| 1, J_o{+}1 \right>$$

$$\xrightarrow{\ R(J_o{+}1)\ } \left| 2, J_o{+}2 \right> \qquad (IV.26)$$

$$\xrightarrow{\quad} \ldots \ \left| n, J_o{+}n \right>$$

with precise compensation for anharmonic frequency shifts through variations in rotational energy. Unfortunately, condition (IV.25) is unique and is not met for the $\nu_3$ mode of the $SF_6$ molecule (see Table 3). For polyatomic molecules $B \ll 2|X_{ii}|$ and condition (IV.25) can be satisfied only in some very special case for a light molecule. A condition similar to (IV.25) can be met in principle for symmetric top molecules, by using changes in the rotational quantum number K. This case will be considered below. Again, it does not apply to the $SF_6$ molecule which is a spherical top.
The so-called triple vibration-rotation resonance

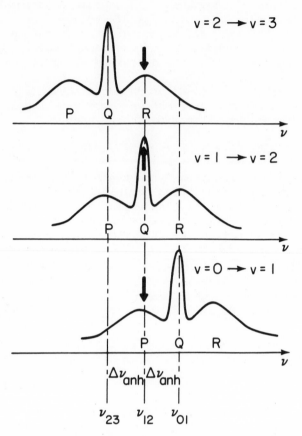

*Fig. 25. Resonance excitation of the v = 3 level due to the effect of "triple vibration-rotation resonance," when the field frequency ω is tuned to the Q-branch of the transition v = 1 → v = 2.*

illustrated in Fig. 25 is possible, however, for polyatomic molecules having a vibrational band with P, Q, R-branches. If the laser pulse frequency is tuned to the Q-branch of the transition $v_i = 1 → v_i = 2$, even though there is an anharmonic frequency shift, three successive vibration-rotation transitions are possible:

$$|v_i{=}0,J{=}J_{res}> \xrightarrow{\quad P(J_{res}) \quad} |1,J_{res}{-}1>$$

$$\xrightarrow{\quad Q(J_{res}{-}1) \quad} |2,J_{res}{-}1> \xrightarrow{\quad R(J_{res}{-}1) \quad} |3,J_{res}> \qquad \text{(IV.27)}$$

Molecules with the particular resonant rotational quantum number

$$2BJ_{res} = 2|X_{ii}|\hbar\omega_i \qquad (IV.28)$$

may be excited immediately to the vibrational level $v_i = 3$; the detuning of the field frequency with respect to the frequencies of three intermediate transitions is rather small. For instance, the $SF_6$ molecule can make three successive transitions $|0,J_{res} = 53> \longrightarrow |1,52> \longrightarrow |2,52> \longrightarrow |3,53>$. In this case the detuning of the intermediate transitions does not exceed the value $B \simeq 0.1$ cm$^{-1}$ which is 30 times less than the anharmonic shift. This process is possible, of course, if the initial state $|v_i = 0, J_0 = J_{res}>$ is sufficiently populated. In spectroscopic terms this means that the condition

$$2|X_{ii}|\hbar\omega_i < \delta\nu_P, \delta\nu_R \qquad (IV.29)$$

must be met. In this inequality $\delta\nu_P$, $\delta\nu_R$ stand for the widths of the P, R-branches of the vibrational band under excitation.

As long as the detuning of three intermediate transitions is greater than the Rabi frequency $\omega_R(1)$ for each one-photon intermediate transition, the anharmonic rotating oscillator in an electromagnetic field can be considered in the framework of perturbation theory. In a like manner to (IV.13), it is possible to introduce the Rabi frequency for the n-photon transition between the initial $|v_i=0, J_0>$ and final $|v_i=n,J_n>$ states (71):

$$\omega_R^{eff}(n,J) = \frac{<0,J_0|\frac{1}{2}\mu E|1,J_1> <1,J_1|\frac{1}{2}\mu E|2,J_2> \cdots\cdots}{\hbar^n(\omega_i+\Delta\omega_{J_0} -\omega)(2\omega_i-X_{ii}\omega_i+\Delta\omega_{J_1}-2\omega)\cdots\cdots}$$

$$(IV.30)$$

$$\frac{<n-1,J_{n-1}|\frac{1}{2}\mu E|n,J_n>}{[(n-1)\omega_i-(n-2)X_{ii}\omega_i+\Delta\omega_{J_{n-2}}-(n-1)\omega]}$$

where $\Delta\omega_{J_i}$ is the difference in rotational energy between the states $J_i$ and $J_{i+1}$. In the case above of a three-photon transition $|v_i=0,J=J_{res}> \longrightarrow |v=3,J=J_{res}>$ similar to (IV.27) a molecule can reach the vibrational level $v_i=3$ for a small detuning of the field frequency about the frequency of intermediate transitions. That is, in fields much weaker than those required by expression (IV.13) which makes no allowance for rotational states the transition takes place.

A criterion for strong coupling between a four-level rotating anharmonic oscillator and a field is the following condition:

$$\left| 2X_{ii}\hbar\omega_i + 2BJ_{res} \right| << \hbar\omega_R(1)$$    (IV.31)

In such a field the power broadening completely compensates small detunings conditioned by discrete changes (about 2B) in the frequency of successive vibration-rotation transitions (IV.27). The asymmetry of the P- and R-branches and the difference in rotational constants of vibrational levels also cause detuning which should be taken into account in (IV.30) and can be compensated for by power broadening. For a typical polyatomic molecule with B = 0.05 cm$^{-1}$ and the dipole moment of the transition $\mu \simeq 0.3$ Debye, the Rabi frequency $\omega_R(1)$ is comparable to the residual detuning, B, in laser fields with an intensity of 10$^5$ W/cm$^2$. For very weak vibrational transitions, $\mu = 0.03$ Debye, the intensity is 10$^7$ W/cm$^2$. In such fields, and especially when condition (IV.31) can be met, the maximum of $\omega_R^{eff}(3)$ approximates $\omega_R(1)$ for one-photon transitions. The number of rotational levels in the initial state involved in strong interaction with the field can be determined from the simple relation

$$2B\Delta J_o \simeq \hbar\omega_R(1)$$    (IV.32)

So, the change of rotational energy almost perfectly compensates for anharmonic detuning as the level $v_i = 3$ is excited in comparatively weak fields. This rotational compensation may also be of use in exciting the level $v_i = 4$ and even $v_i = 5$, but in these cases, of course, the anharmonic detuning can only partly be compensated for. Figure 26 shows a possible sequence of four transitions (PPRR) where small detuning may be realized in two denominators of the four for a four-photon process (68, 71). The numerical values are given for the most popular case: the $\nu_3$ mode of the SF$_6$ molecule. According to the estimates from (71), in a field intensity of 10$^8$ W/cm$^2$ a molecule can reach the level $v_3 = 4$ when the laser frequency is tuned in the range from 948 cm$^{-1}$ to 940 cm$^{-1}$. A smaller portion of molecules can reach the state $v_3 = 5$ due to the PPQRR sequence where the anharmonic detuning is also partially compensated.

In conclusion we wish to discuss the possibility of anharmonicity compensation for a large number of successive transitions, a sequence similar to (IV.26), in symmetric top molecules (35). A symmetric top has two rotational constants B ($v_i$) and A ($v_i$), and its energy levels to a first approximation may be written as (82):

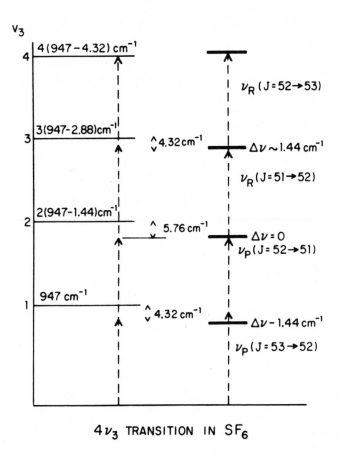

$4\nu_3$ TRANSITION IN SF$_6$

*Fig. 26. A four-photon process with partial compensation of the anharmonic energy defect by rotational energy differences (71).*

$$W(v_i) = \hbar\omega^i\left[v_i + X_{ii}v_i(v_i-1)\right]$$

(IV.33)

$$+ B(v_i)J(J+1) + \left[A(v_i) - B(v_i)\right]K^2,$$

where K is the projection of angular momentum onto the molecular axis ($-J \leqslant K \leqslant J$). The selection rules for electric dipole transitions allow changes not only in J but also in K (perpendicular bands): $\Delta K = \pm 1$, $\Delta J = 0, \pm 1$. Therefore it becomes possible to change the rotational energy in vibrational transitions. In particular, it follows from (IV.33) that under the condition

$$(A - B) = - X_{ii} \hbar \omega_i \qquad \text{(IV.34)}$$

it is possible to obtain a multistep resonance of field frequency with the following sequence of vibrational transitions (Fig. 27):

$$|v_i = 0, J_o, K_o> \xrightarrow{\quad ^PQ_{K_o} \quad} |1, J_o, K_o - 1>$$

$$\xrightarrow{\quad ^PQ_{K_o - 1} \quad} |2, J_o, K_o - 2> \xrightarrow{\quad} \ldots \qquad \text{(IV.35)}$$

$$\ldots \xrightarrow{\quad} |K_o, J_o, 0> \xrightarrow{\quad ^RQ_o \quad} |K_o + 1, J_o, -1>$$

$$\xrightarrow{\quad ^RQ_1 \quad} \ldots \xrightarrow{\quad ^RQ_{K_o} \quad} |2K_o, J_o, -K_o>$$

For such a $2K_o$-fold rotation-vibration resonance the field frequency should be tuned to the frequency of the $^PQ_{K_o}$ transition. It is the quantum number K in the initial state that determines the number of possible successive resonances. Condition (IV.34) for symmetric top molecules can be met with much higher probability than similar condition (IV.25) for spherical top molecules since the rotational constant A may be much larger than B and can be comparable to the anharmonicity constant. A similar multistep resonance may appear each time the $^PQ_K$ laser pulse frequency coincides with the center of the $^PQ_K$ band. The multiphoton absorption spectrum in the case of such a mechanism of anharmonicity compensation must have therefore a periodic structure. Such a structure in a multiple photon absorption spectrum has been disclosed for the $\perp$ band of the $\nu_7$ mode of $C_2H_4$ (35) which suggests hypotheses on the possibility of an excitation process of the type in Eq. (IV.35).

All the schemes of rotational compensation for anharmonicity considered may be realized in practice provided that the vibration-rotation energy of the anharmonic oscillator at least for the several vibrational levels under excitation can really be given in a simple form, (IV.24) or (IV.33). There

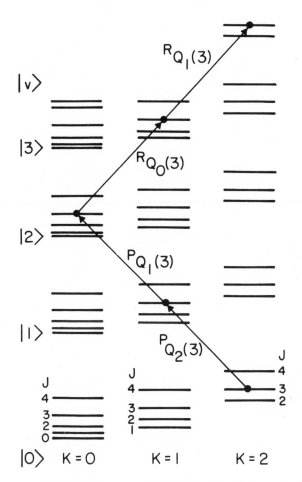

Fig. 27. *Sequence of vibration-rotation transitions on* $^{P,R}Q_K$*-branches of a perpendicular absorption band of a symmetric top, which provides exact compensation of the anharmonic frequency defect. Case with* $K_o = 2$ *(35).*

are effects, Fermi resonance in particular, which break the regular sequence of vibration-rotation levels for real molecules. So it is always necessary to analyze the applicability of the schemes considered for each particular polyatomic molecule.

## 2.    Anharmonic Splittings of Degenerate Levels

Cantrell and Galbraith directed their attention (66) to the existence of anharmonic splittings in excited, degenerate molecular vibrational states. This effect can compensate for

ANHARMONIC SPLITTING IN $\nu_3$
OVERTONE STATES OF $SF_6$

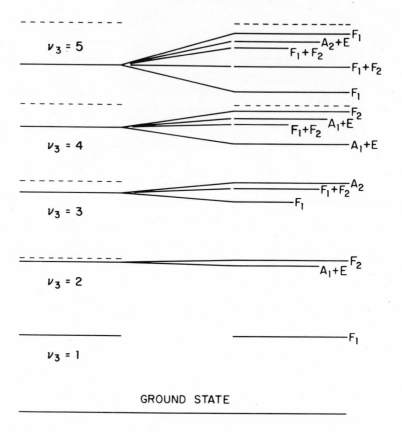

Fig. 28(a). *Compensation of the anharmonic frequency defect by means of anharmonic splitting of overtone vibrational levels in the* $\nu_3$ *mode of* $SF_6$. *On the left are shown highly idealized levels* [$v\nu_3$] *with anharmonic shift* $v(v - 1)X_{33}\nu_3$. *On the right are shown the corresponding manifolds* [$v\nu_3$]. *The positions of energy levels for a harmonic oscillator are shown by dashed lines (66).*

the anharmonicity for several lower vibrational transitions of the $SF_6$ molecule in particular. They analyzed the Hamiltonian of the vibrational mode $\nu_3$ with regard to the rotations of a tetrahedral or an octahedral spherical top. Excitation of degenerate vibrational levels always amounts to some additional angular momentum. The inclusion of anharmonicity causes degenerate level splitting. The energy levels of the vibra-

## DIPOLE-ALLOWED TRANSITIONS IN THE

## $\nu_3$ LADDER OF SF$_6$

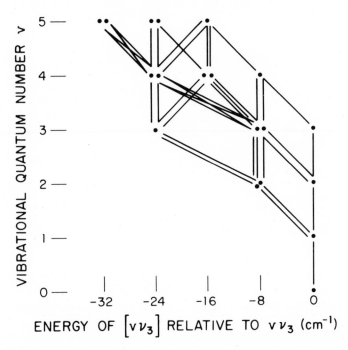

Fig. 28(b). Overtone states [$v\nu_3$] (indicated by dots) versus frequency, and dipole-allowed transitions (indicated by lines) in the $\nu_3$ mode of SF$_6$. Transitions in resonance with $\nu_3 = 948$ cm$^{-1}$ are represented by vertical lines; decreasing slope indicates a transition frequency increasingly far from resonance. For clarity, only transitions with frequency $\nu_3$, $\nu_3 \pm 8$ cm$^{-1}$ and $\nu_3 - 16$ cm$^{-1}$ have been indicated (66).

tional mode $\nu_3$, with these effects taken into account, may be written in the form:

$$W(v,\ell,J,R) = v\hbar\omega_3 + (v-1)vX_{33}\ \hbar\omega_3$$

$$+ [\ell(\ell+1)-2v]G_{33} + BJ(J+1) \qquad \text{(IV.36)}$$

$$+ B\zeta_3\ [R(R+1) - J(J+1) - \ell(\ell+1)],$$

where v is the vibrational quantum number; $\ell$ = v,v-2, ....1
or 0 is the vibrational angular momentum conditioned mainly
by excitation of degenerated vibration;  J is the total angular momentum; R is the rotational angular momentum.  Equation
(IV.36) shows that the $vv_3$ "level" is split into $[\frac{v}{2}]$ + 1 non-
degenerate vibrational levels, each of them having the
complete set of rotational sublevels.  Figure 28(a) illustrates the anharmonic splitting of the vibrational levels of
the $v_3$ mode for the $SF_6$ molecule when $2X_{33}\hbar\omega_3$ = - 2.88 cm$^{-1}$,
$G_{33}$ = 0.8 cm$^{-1}$, $T_{33}$ = 0.4 cm$^{-1}$. For simplicity, in (IV.36)
this last term was not allowed for.

Electric dipole transitions from the octahedral substate
p' of the vibrational state $[vv_3]$ to the substate p" of
$[(v+1)v_3]$ are determined by the standard selection rule:  the
group theoretical direct product p" $\textcircled{x}$ p' must contain the $F_1$
representation of the octahedral group.  These allowed transitions and their frequencies are shown on the transition "tree"
in Fig. 28(b).  The common feature of the transitions is that
they are "red"-shifted, which is quite evident.  But the
frequency of the transition nearest to the fundamental frequency $\omega_3$ is shifted by an amount much smaller than the
simple anharmonic shift $2vX_{33}\hbar\omega_3$.  The better starting point
for a chain of successive upward transitions is the upper
level of a P - branch transition from v = 0 → v = 1.  As a
molecule passes into an excited state, the largest number of
transitions may occur if the laser pulse frequency is shifted
to the region $v_3$ - 8 cm$^{-1}$.  These shifts are not beyond the
P, R-branches of the fundamental transition.

Unlike the mechanism discussed above of rotational compensation for anharmonicity, this mechanism can excite molecules from a large number of initial rotational states at the
same time.  So, one of these two mechanisms can be chosen for,
say, the $SF_6$ molecule in experiments on excitation of the
molecules in a gas phase at rather low temperatures when the
widths of the P- and R-branches do not obey condition (IV.29).
For the time being both mechanisms of "soft" compensation for
anharmonicity can be used in a consistent explanation of the
experimental data obtained.

D.   Molecular Excitation in the Vibrational Quasi-continuum

Figure 29 illustrates the position of the lower vibrational levels of the $BCl_3$ molecule.  By employing one of the
mechanisms considered, rotational energy variation and/or anharmonic splitting, the detuning of the field frequency $\omega$ from
the frequency of several lower transitions may be minimized.
As a result, quasi-resonance multiple photon excitation of the

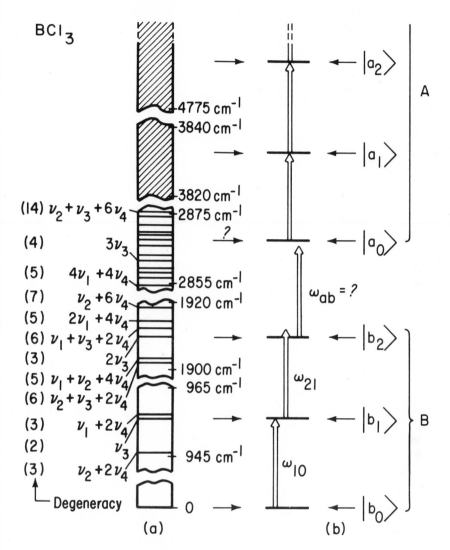

Fig. 29. a) Diagram of lower vibrational energy levels
of BCl₃; b) choice of a multilevel system as a model for
molecular excitation by an intense resonant IR laser field
which includes a sequence of discrete energy levels (subsystem
B) and ladder of equidistant energy levels within the vibrational quasi-continuum (subsystem A) (67).

energy levels $\nu_3$, $2\nu_3$, $3\nu_3$ and, probably, even $4\nu_3$ and $5\nu_3$
becomes possible. Then, as it can be felt by intuition from
Fig. 29(a), the vibrational energy level density increases so
that the field frequency may be in exact resonance with some

transition to the next highly excited vibrational state. The range of vibrational energies of a polyatomic molecule characterized by high level density is often called, for short, the "vibrational quasi-continuum."

## 1.  *The Vibrational Quasi-continuum*

The vibrational energy level density of polyatomic molecules has been considered rather comprehensively in (83). The formula deduced below for vibrational level density $\rho(E)$ in a harmonic approximation is very handy for numerical computation:

$$\rho(E) = \frac{E_o^{n-1}}{(n-1)!\prod_i \hbar\omega_i (1+\eta)} \left[ (2+\eta)(1+\frac{2}{\eta})^{\eta/2} \frac{1}{e} \right]^n$$

$$\cdot \left[ 1 - \frac{1}{(1+\eta)^2} \right]^\beta \tag{IV.37}$$

The molecular vibrational energy E is referenced to the ground state. The product is taken over all molecular vibrations; n is the number of vibrational degrees of freedom; $E_o = \frac{1}{2}\Sigma\hbar\omega_i$ is the energy of ground molecular state;

$$\eta = E/E_o, \quad \beta = (n-1)(n-2)\overline{\omega}^2/6n\overline{\omega}^2 - \frac{n}{6}.$$

Figure 30 shows the results of estimations of the $\rho(E)$ curve for a number of molecules. As seen from these curves, the vibrational level density reaches very considerable values for energies of only a few $CO_2$-laser quanta. We must also take into account the fact that each vibrational level consists of many rotational levels. In the range of the E energies, for which the condition

$$\rho(E) > \frac{1}{\Delta\omega_{las}} \tag{IV.38}$$

can be satisfied, the vibrational levels can really form a continuum due to the finite width of laser pulse spectrum

$$\Delta\omega_{las} \gtrsim \frac{1}{\tau_p}.$$

The states of the vibrational quasi-continuum are overtone and combination levels because of the admixture of different vibrations. The pure vibrational levels of the $\nu_3$ mode in resonance with the IR field are coupled with the combination vibrational levels of quasi-continuum through Coriolis or anharmonic interaction. The coupling constant between the $\nu\nu_i$ pure vibrational state with its energy $W(\nu\nu_i)$ and the frequency range in the quasi-continuum with   $E_{qc}^o = W(\nu\nu_i) + \hbar\omega$

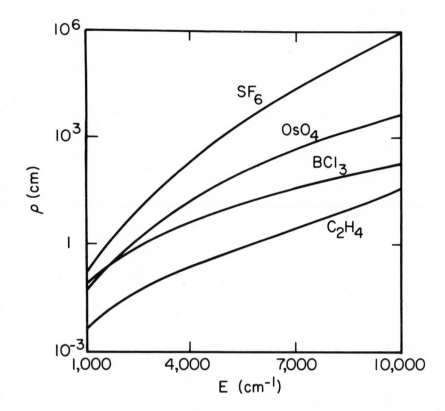

*Fig. 30.   Density of vibrational energy states of $BCl_3$, $C_2H_4$, $SF_6$, and $OsO_4$ molecules as a function of vibrational energy with respect to the ground state (73).*

that is, the matrix element of the transition from the system of discrete levels with their energy $v\nu_i$ to the quasi-continuum, depends on the admixture of $\nu_i$ vibration in the quasi-continuum Fig. 31. For estimates we may assume that the oscillator strength of the $\nu_i$ mode is distributed within the frequency range $\Delta\omega$ near the $(v + 1)\nu_i$ vibrational state the value of which may be determined, for instance, by the Coriolis interaction energy. The total admixture of the $\nu_i$ vibration in the quasi-continuum within the energy range $W(v + 1)$ may be denoted as $\left|C_i(v + 1, \nu_i)\right|^2 \Delta\omega$. For example, according to (71) for $SF_6$ $\hbar\Delta\omega \sim \left|H_{Cor}\right| \sim 10$ cm$^{-1}$ and $\left|C_3(6\nu_3)\right|^2 \Delta\omega \sim 0.01 - 0.1$. Thus, the matrix element of the transition between the system of discrete energy levels (level system "B" in Fig. 29(b)) and the quasi continuum (level system "A"), $\mu_{ba}$, is much smaller than that of the transition between

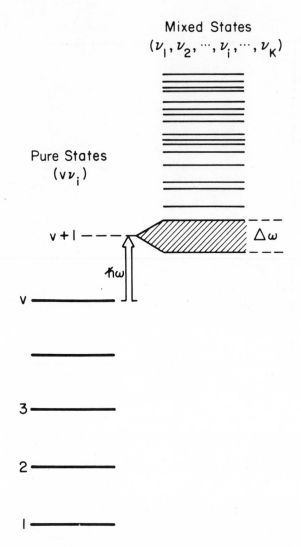

*Fig. 31. On explanation of the coupling between the discrete energy levels of the pure vibrational mode $\nu_i$ and the vibrational quasi-continuum.*

discrete levels $\mu^b_{v,v+1}$.

The transitions between the mixed vibrational states in the quasi-continuum, of course, are also far less intense than those between pure $v\nu_i$ states. Precise estimation of such transitions is a very complicated problem. If we knew the exact Hamiltonian $H(q_1, q_2, \ldots\ldots, q_N)$ for normal vibrations $q_i$, with real anharmonicity taken into account, we would try to solve the quantum-mechanical problem, that is, to estimate the position of all energy levels and the transitions between them in a strong IR field. At present this is impossible, of course. We may try to solve this problem in a classical approximation as suggested by Lamb (84). The dynamics of the vibrational quasi-continuum may be described in terms of the Cartesian coordinates of individual atoms connected with the IR field through effective charges. They interact with one another due to effective induced polarizabilities and short-range repulsive forces. Periodic shocks of such a classical system may cause one or several atoms to be ejected. The solution of this problem by the Monte-Carlo method is probably within the scope of up-to-date large computers.

## 2.   Transition to the Quasi-continuum and its Excitation

The transition of a polyatomic molecule from the system of lower vibrational levels to the quasi-continuum and the excitation of levels through this transition can be considered within the simplified model illustrated in Fig. 29(b). Such an analysis was carried out in (67, 73). Based on these results, we consider two subsystems of levels: a multilevel system consisting of discrete levels at low energies and a quasi-continuous system of levels at high energies below the dissociation limit. The lower subsystem "B" consists of several levels, the transitions between them are not necessarily in exact resonance with the field. This may be a system of n + 1 levels which are excited by one of the n-photon resonances, or specifically, a system of 4 levels excited by a triple vibration-rotation resonance, and so on. For the infinite upper subsystem "A" the field is in exact resonance at all successive transitions. The interaction of the IR field with the lower subsystem "B" may set up excitation of the uppermost state $|b_n>$. Therefore it is expected that the system "leaks" into the lower state $|a_0>$ of the upper subsystem because of its induced transition from the state $|b_n>$. The probability of such "leakage" depends on the probability amplitude of $|b_n>$, the ratio between the field frequency and the frequency $\omega_{ab}$ of transition from $|b_n>$ to $|a_0>$ for the two subsystems (the exact resonance $\omega = \omega_{ab}$ is not an optimum) and the rate of transitions between the $|a_i>$ states of the upper subsystem.

To focus our attention on the effect of leakage, we consider the case when in the lower subsystem of levels "B" we allow only two levels which weakly interact with the field. As stated above (Sec. IV, B) the anharmonic oscillator in the case of an n-photon resonance and a weak interaction with the field is similar to a two-level system with the effective Rabi frequency $\omega_R^{eff}(n)$. The lower subsystem therefore may

include several levels but in the case of an n-photon resonance
we can consider the lower subsystem oscillations between the
initial $|b_0>$ and final $|b_n>$ states. The probability of excita-
tion of the state $|b_n>$ is determined then from the expres-
sion (IV.14).

The dipole moments of transition in the lower subsystem
"B", $\mu_{k,k-1}^b$, are much larger than those for the upper subsystem    "A", $\mu_{m,m-1}^a$, which models the vibrational quasi-
continuum:

$$\mu_{k,k-1}^b \gg \mu_{m,m-1}^a \qquad (IV.39)$$

The lower subsystem "B" is coupled with the upper system "A"
by the dipole moment of the transition $|b_n> \rightarrow |a_0>$ the value
of which, according to the estimates given above, is of the
order of the dipole moments for transitions in the quasi-
continuum, i.e.

$$\mu_{ba} \sim \mu_{m,m-1}^a \qquad (IV.40)$$

Under such assumptions the transitions in the lower subsystem
"B" occur much faster than those in the upper subsystem "A."
Therefore we may consider the influence of the upper system
or of the transitions in the quasi-continuum as a small dis-
turbance.

The problem posed was solved in a rather general form
in (73) (some of the results are briefly stated in (67)).
Consideration was given to two possible dependences of the
transition dipole moment in the upper subsystem "A" on the
level number, m:

$$\mu_{m-1,m}^a = \mu_a \sqrt{m} \quad (harmonic\ oscillator) \qquad (IV.41a)$$

and

$$\mu_{m-1,m}^a = \mu_a = const. \qquad (IV.41b)$$

Under the initial condition $a_0(t = 0) = 0$ the amplitudes or
probabilities $|a_m>$ are given by the distributions (67):

$$a_m(t) = \frac{i^{m+1}\gamma_a^m}{\sqrt{m!}} \Gamma_{ba} e^{iG(t)} \int_o^t \tau^n e^{-iG\tau} e^{-\gamma_a^2\tau^2/2} d\tau \qquad (IV.42a)$$

for $\mu_{m-1,m}^a$ from (IV.41a) and

$$a_m(t) = \frac{i^{m+1}(m+1)}{\gamma_a} \Gamma_{ba} e^{iGt} \cdot$$

(IV.42b)

$$\cdot \int_o^t \frac{1}{\tau} e^{-iG\tau} J_{m+1}(2\gamma_d\tau) d\tau$$

for $\mu_{m-1,m}^a$ from (IV.41b). In (IV.42) the following notations are used:

$$\Gamma_{ba} = \frac{\mu_{ba}}{2\hbar} b_n \; , \; \gamma_a = \frac{\mu_a E}{2\hbar} \; , \; G = \omega_{ab} - \omega \pm g_n,$$

where the parameter $g_n \simeq \omega_R^{eff}(n)$ describes the power shift
because of splitting of the final n-th energy level of the
subsystem "B."

The probability of B → A "leakage" is given by the value

$$P_{B\to A}(t) = \sum_{m=o}^{\infty} |a_m(t)|^2.$$

(IV.43)

The analysis of expression (IV.42) shows that "leakage" takes
place only when the transition between the upper energy quasi-
level of the subsystem "B" ($\omega_n \pm g_n$) and the lower level of
"A" is near to exact resonance. Leakage is most efficient
when $|G| \leq \gamma_a$. Under this condition the probability of
leakage is expressed by

$$P_{B\to A}(t) \sim |b_n|^2 \gamma_a t$$

(IV.44)

where cases (IV.41a) and (IV.41b) differ only by a numerical
factor. The mean value $\bar{m}$ for distributions (IV.48), that is,
the mean energy of the upper subsystem B, is determined from
the expressions:

$$\bar{m} \sim \frac{1}{3}(\gamma_a t)^2 \quad for \; (IV.41a)$$

(IV.45a)

and

$$\bar{m} \sim \gamma_a t \quad for \; (IV.41b).$$

(IV.45b)

From relations (IV.44) and (IV.45) several important conclusions may be drawn. The threshold of strong excitation for
the levels of the upper subsystem "A" depends on the value of
$\gamma_a \tau_p$ at the end of the laser pulse, that is, it is determined
exclusively by the properties of the quasi-continuum. For
efficient excitation of high vibrational levels it is necessary that the condition $\gamma_a \tau_p \gg 1$ should be met. Therefore

full leakage ($P_{B \to A} \simeq 1$) can be realized, even though the
probability of system occupation at the upper level of the
subsystem "B" is small, that is, $|b_n|^2 \ll 1$. But, according
to (IV.44), the leakage probability is sensitive to $b_n$.
Because of this, each time the value of $b_n$ increases resonantly for instance due to an n-photon resonance, this appears as
a resonant increase in the energy of excitation of the quasi-
continuum. Resonances of this kind were studied in (67, 73).
If the lower subsystem "B" consists, for instance, of three
slightly non-equidistant levels, the excitation probability
for the levels of the upper subsystem "A" is sensitive both to
the exact resonance on the 0-1 transition and the exact two-
photon resonance 0-2. The results of calculation for such a
case are given in Fig. 32. The calculation is done for the

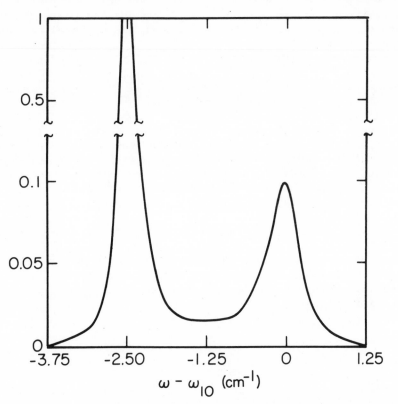

*Fig. 32.   Dependence of the "leakage" probability to the
quasi-continuum $P_{B \to A}$ for a three-level subsystem "B" with anharmonicity $\varepsilon$ on detuning of field frequency $\omega$ with respect
to exact resonance for the 0-1 transition.   This illustrates
two-photon resonance, for the following parameters:   $\tau_p = 10^{-7}$
sec, $\mu_a = 5 \times 10^{-4} D$, $\mu_{01}{}^b = 0.1\ D$, $\mu_{12}{}^b = \sqrt{2}\mu_{01}{}^b$, $\varepsilon = 5\ cm^{-1}$,
$I = 10^8\ W/cm^2$ (67).*

case when the value of "anharmonicity" $\varepsilon = 5$ cm$^{-1}$; the dipole moments of the transitions 0-1 and 1-2 are coupled by the standard relation for a harmonic oscillator $\mu_{12} = \sqrt{2}\mu_{01}$, where $\mu_{01} = 0.1$ Debye; the dipole moment of transitions in the quasi-continuum $\mu_a = 5 \times 10^{-4}$ Debye; the laser pulse duration $\tau_p$ = 100 nsec; and the pulse power $I = 10^8$ W/cm$^2$.

A similar model for excitation of high vibrational levels (a discrete set of low levels and the quasi-continuum) has been considered recently by Larsen (72). He took into account, however, two additional effects: 1) the rotational structure of lower discrete levels (as applied to the $\nu_3$ mode of the SF$_6$ molecule); 2) decay of the last (N+1)-th excited state of the quasi-continuum due to molecular transitions to the real continuum through the use of the phenomenological decay constant $\Gamma(\Gamma t \sim 1$ for $t = 10^{-9}$ sec). Several particular cases have been computed. Figure 33 gives some results of such computations for the SF$_6$ molecule at T = 193°K. There is a clearly defined three-photon resonance appearing when the laser frequency coincides with that of the Q-branch of the transition 1-2 in the lower discrete level system (PQR-resonance). The three-photon resonance has weaker P- and R-branches. In the frequency dependence of excitation and dissociation rates there is also a two-photon resonance having a clearly defined Q-branch transition, $|0,J_0\rangle \rightarrow |2,J_0\rangle$. As the decay rate of the uppermost level increases to $\hbar\Gamma = 0.05$ cm$^{-1}$, decay during 0.1 nsec, there is also a small one-photon resonance in the frequency dependence for the transition $|0,J_0\rangle \rightarrow |1,J_0\rangle$. Though the computations performed by Larsen in (72) are best applied to the $\nu_3$ mode of the SF$_6$ molecule, they nevertheless take no account of some effects which can change the shape of the calculated curves. In particular, the computations disregard the anharmonic splitting of $v \geqslant 2$ levels. There are some uncertainties in the values of $X_{33}$ and the number of the uppermost (N+1) level in the quasi-continuum where molecular dissociation occurs. It is therefore difficult for the time being to carry out quantitative comparison between calculated and experimental frequency curves.

Molecular excitation in the vibrational quasi-continuum can be considered in terms of intramolecular vibrational coupling of the mode under IR excitation with the rest of the vibrational modes. In our works (34, 35) consideration was given to this possibility. And what is more, we found the effect of intramolecular transfer of vibrational excitation in the OsO$_4$ molecule (30) (see Sec. IX below). Assume that a molecule has stored an amount of vibrational energy $E_{vib}^{(i)}$ in some vibrational mode $\nu_i$. Because of anharmonicity the normal coordinates of the molecule interact and make it impossible to localize the energy in a single mode. As a result,

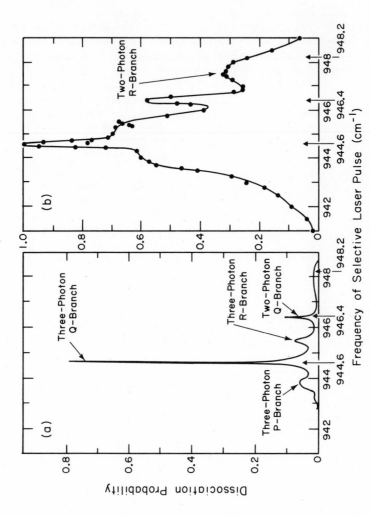

Fig. 33. Computed frequency dependence of the dissociation yield of $SF_6$. The P,Q and R branch peaks labeled refer to multiphoton Rabi oscillations between pairs of discrete states in which, if the lower level has rotational quantum number $J_o$, the upper state has a corresponding J value of $J_o - 1$, $J_o$ and $J_o + 1$, respectively. The parameter $\hbar\Gamma = 0.005$ $cm^{-1}$. In (a) $\hbar\omega_R = 0.08$ $cm^{-1}$ corresponding to 4 $MW/cm^2$ and in (b) $\hbar\omega_R = 0.25$ $cm^{-1}$ corresponding to 40 $MW/cm^2$ (72).

the energy becomes redistributed over all vibrational modes. The redistribution time depends on the interaction constants or coupling of vibrational modes and decreases as the molecule becomes more complex or the initially stored vibrational energy increases. Thus, we may assume that the molecule, excited by an intense IR field to high states far removed from the dissociation limit, transfers the excitation very quickly to other degrees of freedom. As a result, in the degree of freedom being excited, the average energy is much smaller than the dissociation energy D, but the total energy in all vibrational modes may quickly exceed the value of D and hence cause molecular dissociation (34, 35). It must be emphasized, that the intramolecular coupling of vibrational modes is just another way to describe the transitions from the vibrational mode $\nu_i$ to the quasi-continuum and those in the quasi-continuum itself. It seems to be a more convenient one if we do not know anything about the real combination vibrational states in the quasi-continuum which interact with the laser pulse.

Excitation of polyatomic molecules by an intense IR field has recently been calculated in (74) with the model of intramolecular coupling of vibrational modes. The authors of this work have considered the excitation of an anharmonic oscillator (the $\nu_3$ mode of the $SF_6$ molecule), each level being able to transfer energy to the rest of the modes of the molecule. The rate of this energy transfer upon transitions $|K> \rightarrow |K-1>$ of the anharmonic oscillator has the form

$$\gamma_{K-1,K} = (K-1)\gamma, \qquad (IV.46)$$

where $\gamma$ is a phenomenological fitting parameter. With $\gamma = 0.01$ cm$^{-1}$, transfer of vibrational energy in a characteristic time of about $10^{-9}$ sec, a molecule may absorb about 50 IR photons in $10^{-7}$ sec with the radiation intensity being $10^9$ W/cm$^2$. The excitation energy of the $\nu_3$ mode is, of course, much lower and equals several vibrational quanta. These results are rather evident if we take into account the fact that, as the calculation presupposes, the energy is to be transferred from the state $|2\nu_3>$ in $10^{-9}$ sec (!). Naturally, the inclusion of the anharmonicity of the mode being excited automatically gives rise to a sequence of multiphoton resonances in this, as in any other, model. Since the model does not allow for possible mechanisms of anharmonic compensation, the intensity required for molecular excitation proves to be too high ($10^9$ W/cm$^2$) even at an extremely high rate of intramolecular relaxation for lower vibrational levels.

*3.   Comparison Between Coherent and Incoherent Excitation*
Above we considered only coherent excitation of an isolated polyatomic molecule by an intense coherent IR field.

Collisions of the molecule being excited with other molecules
in the gas were neglected. Since vibrational energy exchange
in collisions may promote excitation of high-lying levels,
their exclusion is of primary importance in explaining the
most non-trivial effect of instantaneous collisionless dissociation of an isolated molecule in an intense IR field.
Collisions are also important in explaining retarded molecular
dissociation induced by collisions between highly excited molecules (10, 11), as well as in understanding decreased isotopic selectivity and molecular dissociation at higher pressures (12, 34, 46). To focus our attention on the difference
between coherent and incoherent excitation, we disregard level
population relaxation, excitation transfer during collisions,
etc. and restrict ourselves to consideration of the finite
time of polarization relaxation, $T_2$. This is quite reasonable
since the cross section for collisions which shift the phase
of the wave function has a larger value than that for any
other process. The transverse relaxation time, $T_2$, can be
estimated from the experimental data on collisional broadening
of narrow molecular resonances in saturated-absorption spectroscopy (85). The typical value of vibration-rotation
resonance broadening is $(1/\pi T_2) = (3 - 30)$ MHz/torr. Thus,
$pT_2 = 10^{-7} - 10^{-8}$ sec·torr.

If the cross relaxation time, $T_2$, complies with the
condition

$$\frac{1}{T_2} >> \omega_R^{eff}(n, b), \frac{\mu_{ba}E}{2\hbar}, \frac{\mu_a E}{2\hbar}, \qquad (IV.47)$$

the level excitation becomes incoherent. In this case all
three stages of the process (excitation of discrete levels,
leakage to the quasi-continuum, and excitation of the quasi-
continuum) may be considered independently in an incoherent
approximation using simple rate equations. Such comparison
between coherent and incoherent excitation was done in (73).
Here we consider the effect of level excitation coherence in
the quasi-continuum when, instead of (IV.47), a softer condition is true: $1/T_2 >> \mu_a E/2\hbar = \gamma_a$. In an incoherent approximation a multilevel system of equidistant levels (the subsystem "A") is excited by the standard stepwise mechanism.
The rate of the transition between the levels m and m-1 will
be $W_{m-1,m} = \sigma_{m-1,m}I$, where $\sigma_{m-1,m}$ is the absorption cross
section, and I is the laser radiation intensity. As in the
case of coherent excitation, we consider two physically
admissible particular cases:

$$\sigma_{m-1,m} = m \sigma_{qc} \quad \text{(harmonic oscillator)}, \qquad (IV.48a)$$

$$\sigma_{m-1,m} = \sigma_{qc} = const, \qquad (IV.48b)$$

which are analogous to relations (IV.41).

Stepwise excitation of a harmonic oscillator was analyzed in (86). The population of the m-th level $Z_m$ for the initial condition $Z_m(0) = \delta_{mo}$ has the following form for the oscillator:

$$Z_m = \frac{(Wt)^m}{(1+Wt)^{m+1}} \text{ with } W_{m-1,m} = mW. \quad (IV.49a)$$

The mean and mean-square values for the distribution (IV.49a) are given by the formulas:

$$\bar{m} = Wt, \quad (IV.50a)$$

where $W = \sigma_{qc}I$,

$$\overline{m^2} = 2W^2t^2 + Wt \quad (IV.51a)$$

The case (IV.48b), when the transition probability remains constant as the level number varies was considered in the diffusion approximation in (26). The exact solution had the form (73):

$$Z_m = e^{-2Wt} \left[ I_m(2Wt) + I_{m+1}(2Wt) \right] \quad (IV.49b)$$

with $W_{m-1,m} = W$, where $I_m(x) = (i)^{-m} J_m(ix)$ is the modified Bessel function. The mean and mean-square values for distributions (IV.49b) with $Wt \gg 1$ are given by the formulas:

$$\bar{m} \simeq 2(Wt/\pi)^{\frac{1}{2}} \quad (IV.50b)$$

$$\overline{m^2} \simeq 2Wt \quad (IV.51b)$$

The difference between incoherent and coherent excitation of the quasi-continuum by a laser pulse shows up when the corresponding distributions over levels are compared at the same stored energies, that is, at the same mean values $\bar{m}$. It follows from (IV.50) and (IV.51) that during incoherent excitation the dispersion of the distribution over levels $Z_m$ is near to the mean value $\bar{m}$. At the same time expressions (IV.42) suggest that during coherent excitation the dispersion of distribution over levels is much less than the mean value $\bar{m}$ (when $\bar{m} \gg 1$). This discrepancy is the result of flatter tails of the distributions (IV.49). From this we may conclude that during incoherent excitation the threshold for a molecule to reach high levels in the quasi-continuum may be somewhat lower than that for coherent excitation. Figure 34 shows how the probability of occupation of harmonic oscillator levels with $\bar{m} \geq 50$ depends on $\bar{m}$ for coherent and for incoherent excitation. It can be seen from the charts that for

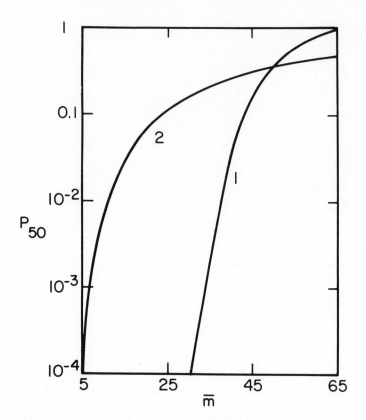

*Fig. 34. Dependence of probability $P_{50}$ of occupation of highest levels (m ⩾ 50) on the average population of harmonic oscillator: 1- coherent excitation; 2- incoherent excitation (73).*

coherent excitation the dependence has a clear threshold, while during incoherent excitation the threshold is substantially flattened. If the threshold corresponds to $P_{50} \simeq 0.1$, the laser radiation intensity required for incoherent excitation is half that for coherent excitation. Taking into account the dependence, $\bar{m} \simeq Wt$, it is possible to evaluate the cross sections of absorption for the transitions in the quasi-continuum using Fig. 34. If the threshold intensity $I_{thr} \simeq 2 \times 10^{7}$ W/cm$^2$ (the case of SF$_6$) and the laser pulse duration $\tau_p = 100$ nsec, the calculated cross section for absorption in the quasi-continuum will be $\sigma_{qc} = 5 \times 10^{-20}$ cm$^2$ (33).

Absorption by the transitions of the vibrational quasi-continuum was first considered in an incoherent approximation for model (IV.48b) by Bloembergen (63). To estimate the quantity of energy absorbed $W_{abs}$ for the transitions in the quasi-continuum, he used the simple expression:

$$W_{abs} = \sigma_{qc}(\omega) I \tau_p \qquad (IV.52)$$

It agreed exactly with the expression (IV.50a) for a harmonic oscillator provided the cross section $\sigma_{qc}(\omega)$ in the expression by Bloembergen is taken to be equal to the absorption cross section for the lower transition for a harmonic oscillator (IV.48). He also evaluated the cross section $\sigma_{qc}$ for $SF_6$ from the following considerations. The value of $\sigma_{qc}$ must be roughly hundreds of times less than the maximum absorption cross section in the P-branch of the $\nu_3$ mode at room temperature since the oscillator strength is distributed over a frequency range which is hundreds of times wider than the P-branch. From this it follows that $\sigma_{qc} \simeq 10^{-19}$ cm$^2$ which coincides with its experimental value (33).

When considering the role of coherence we must stress that the phase memory of a multilevel quantum system may be disturbed not only by collisions but also by a complex structure of the real molecular transitions in the quasi-continuum (87). Really, from a given vibration-rotation sublevel many nearly-resonant transitions may occur due to a high density of vibrational states. If the dipole moments of these transitions differ, it is expected that the excitation coherence, the phase memory of a quantum system, becomes disturbed because of the summation of a large number of coherent processes with different Rabi frequencies. Therefore the model of incoherent excitation of the quasi-continuum may be more useful for real polyatomic molecules.

The comparison between the time dependences of the energy absorbed due to the transitions in the quasi-continuum during coherent excitation (formulas (IV.45)) and incoherent excitation (formulas (IV.50)) points the way to an experimental check of the excitation regimes in each particular case. For coherent excitation the $\infty$ quantity of energy absorbed, $W_{abs}$, depends on the invariant $\int\limits_{-\infty} E(t')dt'$:

$$\frac{\mu_a}{2\hbar} \int\limits_{-\infty}^{\infty} E(t')dt' \simeq \begin{cases} (W_{abs}/\hbar\omega)^{\frac{1}{2}} & (IV.53a) \\ (W_{abs}/\hbar\omega) & (IV.53b) \end{cases}$$

for models (IV.41a) and (IV.41b) respectively. Under this regime the threshold laser pulse energy density, $I\tau_p$, must increase as the pulse duration $\tau_p$ decreases. For incoherent excitation according to (IV.50), instead of relation (IV.53), we have:

$$\sigma_{qc} \int_{-\infty}^{\infty} I(t')\,dt' \simeq \begin{cases} (W_{abs}/\hbar\omega) & \text{(IV.54a)} \\[2ex] (W_{abs}/\hbar\omega)^2 & \text{(IV.54b)} \end{cases}$$

In this case the threshold energy density, $I\tau_p$, must not depend on pulse duration. Preliminary experiments with a $CO_2$-laser pulse duration $\tau_p \simeq 1$ nsec have shown (88) that the threshold laser pulse energy density for strong excitation of the $SF_6$ molecule has the same value as that for pulses of 50 - 100 nsec. This provides support for incoherent excitation of transitions in the quasi-continuum of the $SF_6$ molecule.

The excitation coherence of the vibrational quasi-continuum probably has no specific features. Of greater importance are other quasi-continuum characteristics:  1) dependence of the transition cross section or dipole moment on level number or vibrational energy (choice between models (IV.41) and (IV.48));  2) dependence of the absorption cross section $\sigma_{qc}(\omega)$ on IR field frequency, that is, the dispersion characteristic of absorption in the quasi-continuum. The first experiment on dissociation of the $SF_6$ molecule in a two-frequency laser field (6,89) has proved the existence of the frequency dependence, $\sigma_{qc}(\omega)$. It is likely that the resonance character of the $\nu_3$ mode which is being excited through the lower levels of $SF_6$ carries over somewhat to transitions in the quasi-continuum. With allowance made for anharmonic shift, the mode $\nu_3$ may contribute to a wide resonance $\sigma_{qc}(\omega)$ shifted to the low-frequency region. This hypothesis correlates well with the experimental data from (6,89). Nevertheless, theorists have to study in more detail the properties of the vibrational quasi-continuum and its transitions for polyatomic molecules. Only after this is done, we may hope to realize the quantitative comparison between experimental and calculated data.

E.    Dissociation of Highly Excited Molecules

In the preceding sections, IV. A - IV.D, we have considered how a polyatomic molecule can absorb a certain energy $W_{abs} \gg \hbar\omega$ from an intense IR field. If the absorbed energy exceeds the dissociation energy of any molecular bond D, molecular dissociation may occur in a short time measured probably by the period of molecular vibrations. And what is more, it is expected, that due to rapid excitation of many molecular modes a molecule is able to absorb energy which is much higher than the dissociation energy of the weakest molecular bond. This follows from the experimental data of (45). Unfortunately, our present understanding of such "superexcitation" of a

polyatomic molecule and the possible channels of its dissociation is limited to the simplest remarks. Thus, the molecular
dissociation threshold can be estimated only using the simplest
relation:

$$W_{abs} = \sum_K D_K + W_{exc},$$
(IV.55)

where $\sum_K D_K$ describes the total energy for dissociation of all
bonds broken; $W_{exc}$ describes the energy (electronic, vibrational, rotational, kinetic) of the dissociation fragments.
Below we discuss possible models of some dissociation channels
which call for theoretical analysis in the future.

*Dissociation of many molecular bonds.* The breaking of
many molecular bonds, that is, dissociation not by a single
energetically most economical channel with the lowest dissociation energy $D_m$, is a well known experimental fact (11, 45).
We can imagine the following model of such a process. The
strong excitation of a polyatomic molecule by transitions in
the quasi-continuum includes excitation of many vibrational
modes at once. In order that dissociation occur when the
total energy absorbed by a molecule $W_{abs} \simeq D_m$ all the absorbed
energy must be localized in one molecular bond. The atomic
displacement in one molecular bond $\Delta X_i$ (t) is given by the
random superposition of motion in all of the vibrational modes.
It is likely that $\Delta X_i$ (t) evolves by a random interference
process in which the maximum value $\Delta X_{max}^i$ can be realized in
a characteristic time of the order of the period of lowest
frequency vibrations. But this maximum value $\Delta X_{max}^i$ is $\sqrt{N}$
times less than that which would be realized in a coherent
summation of all N absorbed vibrational quanta in a single
molecular bond. Since $N \simeq D_m/\hbar\omega \simeq 50 - 100$, we can hardly
expect collisionless dissociation of an isolated molecule with
the total store of vibrational energy of the order of $D_m$ in a
time of the order of molecular vibrations. When $W_{abs} \simeq D_m$, a
polyatomic molecule remains, probably, quite a stable system
and continues absorbing the IR field energy. On the other
hand, when $W_{abs} \simeq \sqrt{N} D_m$, collisionless dissociation of a molecular bond must come about with a probability near unity
during one period of a low-frequency molecular vibration
($10^{-13}$ sec). The real value of $W_{abs}$ seems to lie somewhere
between the lower ($D_m$) and upper ($\sqrt{N} D_m$) estimates. In this
case dissociation of many molecular bonds at once may be expected since the fluctuation of peak displacement can be
realized for any molecular bond. The probabilities of different dissociation channels with different energies must differ
at least by the factor exp ($- W_{abs}/\sqrt{N} \Delta D$), where $\Delta D$ is the
difference in dissociation energy between two different
channels. Thus, the hypothesis being proposed of a random
nature of dissociation may form the basis for explanation of

the observed "superexcitation" ($W_{abs} \gg D_m$) of polyatomic molecules in a strong IR field.

The RRKM model of unimolecular dissociation (121) gives dissociation rates and products in detail. It assumes that energy is randomly distributed among all of the vibrational coordinates on a timescale short compared to the dissociation lifetime. The research groups of Lee and Shen (119) and of Yablonovitch and Bloembergen (120) have very recently obtained data on dissociation fragment translational energies and laser energy absorption which are completely consistent with the predictions of RRKM theory. The molecular beam data indicate a lifetime of about $10^{-7}$ sec and a total $SF_6$ energy content of about six photons beyond the dissociation limit. Most of the excess energy goes into $SF_5$ vibration. Low translational energy of the fragments is also indicated by (122).

*Formation of electronically-excited fragments.* In the first experiments (22 - 24), where polyatomic molecules were acted upon by an intense IR field, visible and UV luminescence of the dissociation fragments occurred. It was rather difficult to explain this effect (24, 27). In some recent work the ratios of quantum yields of dissociation to ground state and electronically excited $NH_2$(52) and $C_2$(90) radicals were measured. This ratio for the radicals $[C_2]/[C_2^*]$ formed by dissociation of the $C_2H_4$ molecule is $10^3$. There are at least two possible mechanisms which can be proposed to explain these results. First, the dissociation giving an electronically-excited radical can be considered as an additional high-lying channel of dissociation. The considerations discussed above can then be used in discussing the breaking of many molecular bonds. A large difference in energies of dissociation by two channels with the formation of electronically-excited and unexcited radicals may be responsible for a small quantum yield of dissociation by one of the channels.

Secondly, we may assume that there exists a particular probability of molecular transition from a high vibrational level within the ground electronic state to the excited vibrational levels of an excited electronic state. After this it becomes possible to excite the molecule by virtue of transitions in the vibrational quasi-continuum of an excited electronic state up to dissociation through a new channel to yield electronically-excited fragments. Such assumptions were used in (91) to explain the formation of molecular ions by dissociation of polyatomic molecules in an intense IR field.

*Collisional dissociation of excited molecules.* Apart from collisionless dissociation in an intense IR field which is of particular interest from the physical point of view, in the experiments (10, 11) delayed dissociation is clearly observed. The discussion of its characteristics unambiguously pointed toward its collisional nature (51). When two excited

molecules collide, each having a stored vibrational energy
insufficient for its dissociation, energy may be pooled in
one of them.  As a result, after a laser pulse the collisions
of highly excited molecules give rise to delayed dissociation.
The quantum yield of collisional dissociation may be equal to,
or even greater than, that of collisionless dissociation.
When a small number of collisions is required for dissociation
of the highly excited molecules, the isotopic selectivity of
excitation and dissociation is not substantially reduced (51).
The process of collisional dissociation is therefore of great
practical interest.  The kinetics of collisional dissociation
within the framework of V-V exchange processes of highly
excited molecules has recently been discussed in (92).

## V.   MOLECULAR ISOMERIZATION BY AN INTENSE IR FIELD

Apart from dissociation, there is another unimolecular
photoprocess for polyatomic molecules, that is isomerization
of a molecule following its photoexcitation.  Photoisomerization of electronically excited molecules is rather a well-
known effect (1).  This process is all-important in photo-
biology, for instance, in the molecular mechanism of vision
(photoisomerization of rhodopsin molecules).  It is expected
that strong vibrational excitation of a polyatomic molecule by
an intense IR field may cause its atoms and (or) bonds to be
rearranged. Its stability is not disturbed.  The observation
of such a process in an intense IR field would be of particular interest since it might make possible further studies of
multiple photon absorption of IR radiation by molecules in
excited levels lower than those required for molecular dissociation.  Also, the realization of selective molecular isomerization in a strong IR field is vital to widening the scope of
isotope separation by IR laser radiation for the following
reasons.  First, isomerization gives a stable final molecule
which does not require a chemical scavenger of dissociation
fragments and thus promotes conservation of selectivity in
secondary photochemical processes.  Second, the energy of
isomerization may be rather less than that of dissociation.
This enables isotope separation with a lower consumption of
energy.  On the other hand, because of this comparatively
low isomerization energy the process is more sensitive to
thermal heating than is dissociation.  This fact makes it
possible to use the isomerization effects in investigating
the secondary thermal effects as a high-power IR pulse interacts with a molecular gas.  Studies of the isomerization of
polyatomic molecules in a strong IR field are therefore very

promising. At present primary emphasis is being given to molecular dissociation, and there has been comparatively little research on isomerization. Isomerization of molecules (trans-2-butene) by $CO_2$-laser pulses was first observed in work (28). In addition to dissociation of trans-2-butene molecules, cis-2-butene molecules were formed. The processes of dissociation and isomerization occured at approximately the same rate. Our laboratory investigated trans-cis isomerization of $C_2H_2Cl_2$ molecules throughout 1974-75. Primary attention was given to a search for the experimental conditions under which the power threshold for isomerization would be much lower than that of dissociation. However in the molecule we experimented on this objective was not accomplished. After extensive investigations we found experimental conditions under which the rate of isomerization was several times higher than that of dissociation and proved that the isomerization of $C_2H_2Cl_2$ molecules was nonthermal by nature. The results of those experiments were published in (44, 93). Finally, in (94) the isomerization of deuterated molecules of 1,5-hexadiene in an intense IR field has been investigated and the isotopic selectivity of such a process demonstrated. It must be said that the approaches to and the technique of measurements in these works differ substantially. In our work (44, 93) we used physical diagnostic techniques and the experimental conditions were sufficiently controllable, while in (28, 94) chemical methods were used.


A.    Studies of Isomerization of Polyatomic Molecules

We consider here the main results of experiments on molecular isomerization by $CO_2$-laser pulses.

1.    2-butene
In work (28) experiments were performed in which the trans- and cis-isomers of 2-butene were irradiated by focused pulses of TEA $CO_2$-laser. Both the isomers absorb $CO_2$-laser radiation with the absorption coefficients $\kappa_{trans} = 1.5 \times 10^{-3}$ torr$^{-1}$ cm$^{-1}$ and $\kappa_{cis} = 1.3 \times 10^{-4}$ torr$^{-1}$ cm$^{-1}$. The activation energy of cis-trans isomerization is 62 kcal/mole, an energy equivalent to 20 $CO_2$-laser photons. In a series of experiments pure trans-isomer, pure cis-isomer and their mixtures were irradiated. After the trans-isomer was irradiated, the following products were formed: methane, ethylene, propene, cis-2-butene and butadiene. In 3600 pulses 16% conversion of initial molecules was attained with 4 to 5% cis-2-butene formed. The product distribution and the relative rates of conversion depended only slightly on pressure (Table 4). The results were the same with the addition of

TABLE 4

*Relative Conversion of Trans-2-butene to Various Products After 3600 Pulses of $CO_2$-laser at Different Pressures (28).*

| Pressure trans-2-butene (torr) | Residual trans-2-butene (%) | % of products | | | | |
|---|---|---|---|---|---|---|
| | | meth-ane | ethyl-ene | pro-pene | cis-2-butene | buta-diene |
| 3 | 84 | 2 | 3 | 3 | 5 | 3 |
| 9 | 81 | 2 | 2 | 4 | 5 | 4 |
| 18 | 87 | 2 | 2 | 2 | 4 | 5 |

inert gases (nitrogen, helium and argon). When the cis-isomer was irradiated, the same end products were produced and the formation of trans-isomer was observed. The formation rate of trans-isomer, however, was five times smaller than in the previous experiments. The authors attribute it to the fact that the linear absorption of the cis-isomer is less than that of the trans-isomer. When the equimolar mixture of both isomers was irradiated at a total pressure of 14 torr, the concentration ratio 1:1 did not change. But at 4 torr the mixture was enriched in the cis-isomer by 15%. This result is considered by the authors as basic proof of a non-thermal character of isomerization.

The results obtained may be explained by preferential dissociation of trans-2-butene by virtue of multiple photon absorption of IR radiation. With increasing pressure the rate of V-V exchange between isomeric molecules exceeds the dissociation rate, and the effect of preferential dissociation of the trans-isomer therefore disappears.

## 2. Dichlorethylene

In the experiments (44, 93) a TEA $CO_2$-laser pulse excited trans-$C_2H_2Cl_2$ which has an absorption in the R-branch of the $v_6$ mode near 10.6 μ. The molecule of cis-$C_2H_2Cl_2$ has a much smaller linear absorption at this wavelength. In the range 700 to 1000 cm$^{-1}$ the molecules of trans $C_2H_2Cl_2$ and cis-$C_2H_2Cl_2$ have several nonoverlapping bands (Fig. 35) which may be used for quantitative analysis of the mixture composition. The original composition of cis-isomer varied from 10 to 40% and the total pressure was between 0.4 and 1.0 torr.

At a focused laser pulse intensity of over $10^9$ W/cm$^2$ and

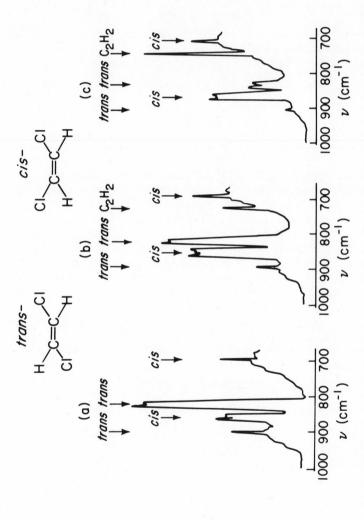

*Fig. 35. IR absorption spectrum of dichloroethylene and its dissociation products: (a) before irradiation; (b) after being irradiated by 1.5 × 10⁴ pulses; (c) irradiated by 4.5 × 10⁴ pulses ($\nu_{las}$ = 934.9 cm⁻¹, energy density in focus is 250 J/cm²) (44).*

an energy density of over 50 $J/cm^2$ there were three processes
observed at the same time:   1) trans → cis isomerization of
$C_2H_2Cl_2$;   2) dissociation of trans-$C_2H_2Cl_2$ with the formation
of ground electronic state fragments and the basic final
product of acetylene;   3) dissociation resulting in the
electronically excited radicals CH* and $C_2^*$.   The dissociation
energy for the C-C bond of the $C_2H_2Cl_2$ molecule is 5.0 eV and
the energy barrier for trans-cis isomerization is 1.8 eV (Fig.
36(a)).   Figure 35 shows the IR spectra of dichlorethylene and

## (a)

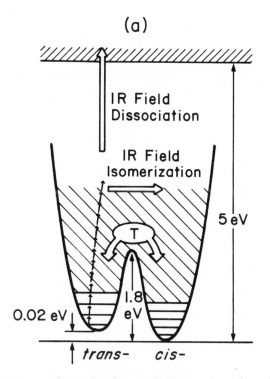

*Fig. 36(a).   Dissociation and isomerization in an intense
IR field and thermal isomerization (44).*

of its dissociation products recorded in the frequency range
700 to 950 $cm^{-1}$ at the initial concentration ratio of trans-
and cis-isomers of 0.6 : 0.4 and again after irradiation.   One
can see an increase in the amplitude of the peaks corresponding
to the cis-isomer and a decrease in the peaks belonging to the
trans-isomer.   This indicates that the trans-isomer can
convert into the cis-form.   In the region of 730 $cm^{-1}$ a new
peak appears which belongs to the main final product of IR
photolysis-acetylene.   Figures 36(b) and 36(c) show dependence
of the relative concentrations of trans- and cis-$C_2H_2Cl_2$ on the

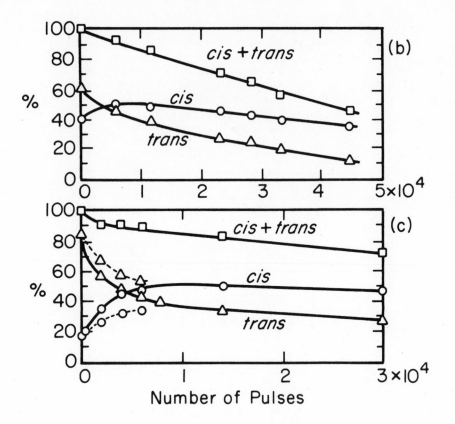

*Fig. 36(b, c). Relative concentrations of trans- and cis-dichloroethylene isomers vs number of laser pulses are shown: (b)* $\nu_{las}$ *= 934.9 cm$^{-1}$, gas cell volume V is 60 cm$^3$, initial dichloroethylene pressure is 0.4 Torr, energy density in focus* $E_f$ *= 100 J/cm$^2$, lens with f = 20 cm; (c)* $\nu_{las}$ *= 980.9 cm$^{-1}$ V = 24 cm$^3$,* $E_f$ *= 250 J/cm$^2$, f = 8 cm,* $P_O$ *= 1 Torr (solid curves) and buffer gas (Xe) pressure is 44 Torr (dashed curves) (44).*

number of radiation pulses obtained from IR absorption spectra. Although at the beginning of the irradiation the quantity of cis-isomer rises, the process of trans → cis conversion does not continue until the end. Upon reaching a maximum concentration of about 0.5, the amount of cis-isomer in the mixture begins to drop gradually. But the cis/trans concentration ratio in the mixture increases steadily and at the end of the irradiation reaches values between 2 and 3, the initial ratio being 0.67 (the curves in Fig. 36(b)), and 0.18 (the curves in Fig. 36(c)). The mixture is enriched steadily in the cis-isomer. These curves show also how the total number of dissociated

molecules depends on the number of pulses.  It can be seen
that mainly the molecules of trans-$C_2H_2Cl_2$ dissociate.

In uv photolysis of dichlorethylene the dissociation
occurs in three ways (95):

$$trans\text{-}C_2H_2Cl_2 \begin{cases} \longrightarrow C_2H_2Cl + Cl \\ \longrightarrow C_2H_2 + Cl_2 \\ \longrightarrow C_2HCl + HCl \end{cases} \qquad (V.1)$$

In all the cases only electronic ground state products are
formed, and acetylene is the main final product.  In IR photolysis of dichlorethylene the dissociation, however, is
followed by visible fluorescence, that is, as opposed to uv
photolysis, electronically excited radicals are formed.  From
the fluorescence spectrum they are identified as CH* and $C_2^*$.
The excited radicals appear in the delayed fluorescence phase,
in the tail of the laser pulse.

To see how the tail of the laser pulse affects the
processes under consideration, some experiments were conducted
with a "shortened" laser pulse without a tail (its base-
length was 250 nsec and half-length 60 nsec).  Neither fluorescence nor isomerization was observed at the same IR field
power levels, and only dissociation with the formation of
unexcited products took place.  When a pulse with a long tail
is used, all three processes occur in the same energy range
(Fig. 37).  They stop at energy densities below 50 to 100
$J/cm^2$.

As far as the choice of a particular mechanism of isomerization is concerned, experiments on buffer gases are
important.  The addition of a buffer gas has several effects.
First, when a buffer gas is added to the gas mixture, the heat
capacity is increased.  All other experimental conditions being
equal, this causes the maximum temperature to fall and must
thereby suppress thermal processes.  In this case it is only
the buffer gas heat capacity that is vital.  Secondly, a
buffer gas has an influence on vibrational, rotational and
translational relaxation, and it affects the efficiency of
multiple photon excitation and all the ensuing processes.  In
this case the effect depends significantly on the nature of
the buffer.  The behavior of various buffer gases (Xe, He, $H_2$,
$N_2$) was studied in experiments on dichlorethylene.  The
following results were obtained:

1)  A buffer gas increases the absorbed energy.  The
addition of 33 torr Xe increases the absorbed energy by more
than 10 times which almost completely compensates for the
increased heat capacity.  It is clear that in this case the

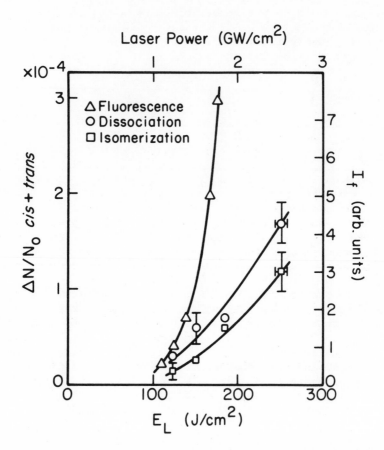

*Fig. 37. Dependence of dissociation rate, isomerization rate and intensity of fluorescence on energy density and laser radiation power ($\nu_{las} = 980.9$ cm$^{-1}$, $P_o$ ($C_2H_2Cl_2$) = 1 Torr; fluorescence is detected in the entire luminous volume) (44).*

buffer gas cannot suppress thermal processes. It must be emphasized that a buffer gas increases the absorbed energy only for multiple photon absorption. At low laser pulse intensities a buffer gas does not increase the absorbed energy.

2)  Buffer gases partly suppress the process of isomerization. The data given in Fig. 38 show that the effect of suppression varies for different gases with the same heat capacity (compare $H_2$ and $N_2$, Xe and He). It varies also for different "buffer gas + dichlorethylene" mixtures, the mixture heat capacity being the same.

3)  Buffer gases do not affect substantially the yield of dissociation (Fig. 36).

4)  At low pressures of $C_2H_2Cl_2$ (0.3 torr) all buffer

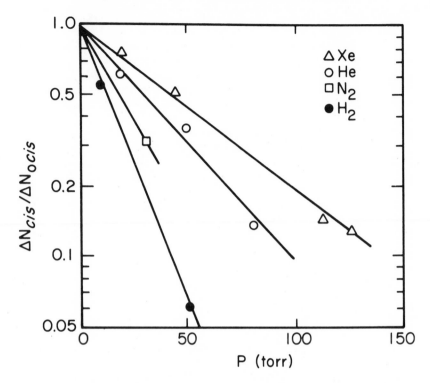

*Fig. 38. Dependence of isomerization rate on buffer gas pressure ($\nu_{las}$ = 980.9 $cm^{-1}$, $P_O$ ($C_2H_2Cl_2$) = 1 Torr) (44).*

gases increase the yield of excited dissociation products almost by an order of magnitude.

*3.    1,5-Hexadiene*
In work (94) the Cope rearrangement in the molecule of 1,5-hexadiene under $CO_2$-laser radiation has been studied recently:

$$
\begin{array}{c}
\text{(V.2)}
\end{array}
$$

Instead of hydrogen, deuterium is introduced at certain positions of this molecule. The initial and final products are chemically the same;  they have a zero-point energy difference and differ in vibrational spectrum. This enables selective excitation of one of the forms. The rearrangement rate (V.2) is (96) log K = 10.36 - 34.3/$\theta$, where $\theta$ = 2.303 RT x $10^{-3}$ and the activation energy is 34.3 kcal/mole. The IR spectrum of the deuterium-labeled 1,5-hexadiene has two absorption bands in the region of $CO_2$-laser action:  an intense band at 10.9 $\mu$ ($\kappa$ = 8.5 x $10^{-3}$ $cm^{-1}$ $torr^{-1}$) and a weak one at 10.1 $\mu$. When deuterating 1,5-hexadiene either on the external vinyl positions or on allyl ones we can observe a distinct isotope effect (10% per atom). These deuterated forms can be inter-converted using the Cope rearrangement (V.2).

One of the examples studied in (94) isomerization of deuterated compounds is:

(V.3)

(A)                          (B)

The compound A:  1,5-hexadiene - 1, 1, 2, 5, 6, 6-$d_6$ (on the left of (V.3)) was synthesized. The compound B: 1,5-hexadiene - 2, 3, 3, 4, 4, 5-$d_6$ must be formed as a result of Cope rear-rangement (V.2). By heating the initial compound a mixture of the A and B compounds was produced, their concentration ratio being 0.9. The initial A compound was quite transparent at 10.8 $\mu$, whereas the B compound displayed strong absorption ($\kappa$ = 5.8 x $10^{-3}$ $cm^{-1}$ $torr^{-1}$) at that wavelength. After the mixture of the A and B compounds was irradiated by $CO_2$-laser pulses at 10.8 $\mu$, the B compound disappeared completely, with no observable disso-ciation products formed. There was only the A compound left in the mixture. The selectivity factor for the isomerization reaction was much higher than unity, and it decreased with in-creasing pressure. The same results were obtained from experi-ments on other deuterated molecules of 1,5-hexadiene.

These experiments seem to be the first ones in which the rate of the isomerization reaction is much higher than that of the dissociation reaction. To give other proofs for the vibrational mechanism of this reaction, it is of importance to study the threshold and frequency characteristics of this reaction as a function of laser power and frequency.


B.    Discussion of Isomerization Mechanisms in an Intense IR

Field

There are several potential mechanisms which can be

responsible for isomerization (see Fig. 36(a)):

1) Multiple photon excitation of high molecular vibrational levels, with the result that a molecule can assume
another isomeric form either directly through the most excited
mode or after intramolecular energy redistribution among many
vibrational degrees of freedom with subsequent localization of
a part of energy in the degree of freedom which is responsible
for isomerization. This mechanism employs only vibrational
excitation of a molecule within the ground electron state.

2) Excitation of a molecular electronic state due to
transfer of vibrational energy, that is, due to V → E transfer,
with subsequent rearrangement of the molecule in the excited
electronic state. In this case the nature of the isomerization mechanism is the usual one (see (1), for instance); the
difference is a new method of electronic excitation.

3) Selective dissociation of the molecular isomeric form
being excited due to multiple photon absorption. A mixture
can be enriched in one isomer by dissociating another one.
But the concentration of the molecules of an unexcited isomer
(for instance, cis-$C_2H_2Cl_2$) may even be increased by virtue
of the random recombination of radicals (formed by dissociation of trans-$C_2H_2Cl_2$) to all possible isomeric forms. This
mechanism causes one isomer to burn out and other ones to
accumulate.

4) Thermal isomerization by the increased temperature in
the focal region after the absorbed energy is thermalized.
Such a process is nonselective, and an intense IR field is
required only to provide a large amount of absorbed energy.

A considerable enrichment of the mixture with one of the
isomers is a direct proof for the contribution of selective
mechanisms of isomerization 1 - 3 . For instance, in the
experiments on dichlorethylene the selectivity factor, expressed as a ratio between the initial and final ratios of
concentrations of two isomers, may be as great as 10. In the
experiments discussed on trans-dichlorethylene the reversal
of this ratio could be observed: the initial ratio was 0.18
and the final one was 2.

The thermal mechanism of dichlorethylene isomerization is
also ruled out by experiments with "shortened" laser pulses.
In both cases (a standard pulse and a shortened one) the conditions for the thermal effect are approximately the same, but
there was not isomerization with a shortened pulse. Simple
estimates of the maximum temperature in the reaction zone for
absorbed energy thermalization in the experiments on dichlorethylene give $\Delta T < 350°C$. Thus, the temperature in the reaction zone does not exceed 650°K. The time for thermal isomerization at this temperature is about $10^3$ sec (97). The
characteristic time for reaction zone cooling varies from $10^{-4}$
to $10^{-5}$ sec. Thus, we may with certainty reject any signifi-

cant contribution from the thermal mechanism to the dichloroethylene isomerization observed. Because of similar experimental conditions this conclusion may be applied to the experiments on 2-butene although the enrichment coefficient was
only 15% (28).

The choice between selective mechanisms 1)- 3), is more
difficult. In the experiments (94) with 1,5-hexadiene the
absence of significant dissociation enables us to reject the
recombination-dissociation mechanism, 3). It is difficult,
however, to choose between potential mechanisms 1) and 2) without additional experiments to ascertain the nature of the
excited states of the 1,5-hexadiene molecule in the $CO_2$-laser
field. In the experiments (44, 93) on dichloroethylene the
threshold density of energy (or power) was almost the same for
dissociation, isomerization and fluorescence (Fig. 37). This
indicates that a considerable contribution of mechanisms 2) and
3) are possible. The suppression of the dichloroethylene isomerization rate observed on increasing buffer gas pressure,
with only a slight variation in the dissociation rate, points
to a possible contribution of recombination mechanism 3. On
the other hand, the dependence of isomerization yield on
laser pulse duration may be related to the finite duration of
intramolecular relaxation of vibrational energy through both
V-V' and V-E processes. Thus, the contribution of mechanisms
1) and 2) to the dichloroethylene isomerization observed should
not be rejected.

The main purpose of experiments on isomerization of polyatomic molecules in an intense IR field is to find molecules
for which isomerization occurs without any dissociation. The
results of work (93) show, perhaps, how to realize this. The
next step is to find molecules with threshold powers of isomerization much smaller than the threshold power typical of
dissociation (from $10^7$ to $10^9$ W/cm$^2$). The possibility of
attaining this aim is far from apparent since, when the activation energy $E_{act} \simeq 1 - 2$ eV, it is necessary that one molecule should absorb several tens of IR photons. According to
the concepts of multiphoton absorption stated in Secs. II - IV,
the molecular transitions must go into the vibrational quasi-
continuum and hence rather high levels of laser pulse energy
density are required. In any case further studies into the
selective isomerization of polyatomic molecules in a strong
IR field are all-important for understanding the processes of
nonlinear interaction between high-power IR radiation and
molecules.

VI.  ISOTOPE SEPARATION BY SELECTIVE MOLECULAR DISSOCIATION BY

     IR RADIATION


The isotopically selective reactions observed during
multiple photon excitation and dissociation of $BCl_3$ (6) and
$SF_6$ (12) molecules have revealed new aspects of the application of multiple photon dissociation.  Among them are, for the
most part, laser separation of isotopes, studies of the chemical kinetics and reactivity of molecules being at vibrational
levels with $v \gg 1$, etc.

In this section we shall review some experiments on isotope separation by selective molecular dissociation in an intense IR field.  Since the process of isotope separation involves chemical transformations of the radicals or the molecules formed by dissociation, we shall try to evaluate, as far
as possible, the influence of such reactions on the enrichment
process.

By now advanced experiments have been carried out to
separate the isotopes of a large number of elements, from the
lightest ones, hydrogen and deuterium, up to those of osmium,
and there are perhaps no upper limitations on the masses of
elements to be separated.


A.  Kinetics of the Enrichment Process


In isotopically selective dissociation of molecules containing i -th isotope, the dissociation products as well as
the chemical substances formed by this dissociation will be
enriched with the i -th isotope, while the molecules left
undissociated will be enriched with another isotope, say k .
Let us briefly consider the kinetics of enrichment during
dissociation of AB molecules containing the $^iA$ and $^kA$ isotopes
(34).

In the beginning of an irradiation, when only a small
part of the AB molecules has dissociated, the enrichment in
the residual (undissociated) absorber gas is also low, while
the enrichment of the molecules formed by the dissociation of
the $^iAB$ molecules (where i is the index of the isotope under
irradiation) may be rather high.  The value of enrichment of
the resultant molecules with the $^iA$ isotope depends on dissociation selectivity of the $^iAB$ molecules with relation to
those which have a different isotopic composition, i.e. to
$^kAB$.  Under rather long irradiation, when almost all the $^iAB$
molecules become dissociated and the $^kAB$ molecules are dissociating too, though at a lower rate, the enrichment of the
resultant products with the $^iA$ isotope will cease.  At the

same time the enrichment of the very few AB molecules remaining may go on and reach extremely high values.

For quantitative analysis we consider here the simplest model of dissociation of a binary molecular mixture with the initial concentration $N_{io}$ and $N_{ko}$ by radiation pulses the frequency of which is in resonance with the molecules containing i -th isotopes. Let the dissociation rate of i -molecules be $W_i$ and that of k -molecules $W_k$. Since the i -molecules resonate with the field much better than the k -molecules, $W_i > W_k$, the dissociation selectivity of the i -molecules is determined evidently from the relation $S = W_i/W_k$; the selectivity is considered to be high when $S \gg 1$.

The concentrations of the i - and k -molecules exponentially decrease with time (or number of pulses):

$$N_i = N_{i_o} \exp(-W_i t), \quad N_k = N_{k_o} \exp(-W_k t). \qquad (VI.1)$$

For simplicity, we assume that as the starting AB molecules burn up (are dissociated) and the chemical composition and the pressure of the gas mixture change, the rates of molecular dissociation remain constant. At low gas pressures this is quite a valid first approximation for in this case it is the IR field that is responsible for dissociation.

We define the coefficient of enrichment with the i -th isotope with respect to the k -th isotope as

$$K_{res}(i/k) = \frac{\left[i_{AB}\right]^* \left[k_{AB}\right]_o}{\left[k_{AB}\right]^* \left[i_{AB}\right]_o} \qquad (VI.2)$$

where the square brackets [ ] denote the concentration of the AB molecules of an indexed isotopic composition. The signs * and $_o$ stand for the concentrations after and before irradiation respectively.

The coefficient of residual gas enrichment with the k -th isotope in this case will be

$$K_{res}(k/i) = \frac{N_k}{N_i} \cdot \frac{N_{i_o}}{N_{k_o}} = \exp\left\{\left(W_i - W_k\right)t\right\}$$

$$= \exp\left(\frac{S-1}{S}W_i t\right) \qquad (VI.3)$$

The total pressure of residual gas $N = N_i + N_k$ changes in the following way:

$$N/N_o = \delta_i \exp(-W_i t) + \delta_k \exp(-W_k t) \qquad (VI.4)$$

where

$$N_{o} = N_{i_o} + N_{k_o}, \quad \delta_i = N_{i_o}/N_o, \quad \delta_k = N_{k_o}/N_o$$

is the relative isotope abundance in the original mixture. The enrichment factor of the product formed as a result of dissociation of the i-isotope is given by the relation

$$K_{prod} (i/k) = \frac{N_{i_o} - N_i}{N_{k_o} - N_k} \cdot \frac{N_{k_o}}{N_{i_o}} = \frac{1 - exp(-W_i t)}{1 - exp(-W_k t)} \qquad (VI.5)$$

If the degree of dissociation of the original mixture is small ($W_i t$, $W_k t$ << 1),

$$K_{prod}(i/k) = W_i/W_k = S \qquad (VI.6)$$

i.e. the enrichment of the resultant molecules depends on the degree of dissociation selectivity of the i -molecules.

The relations (VI.3) to (VI.5) show that for exponentially high "burnout" of the starting i - and k -molecules, the small remaining portion of the molecules may be enriched with the k -isotope to any desired degree.

It is quite obvious that the most favorable process is that when the value $S = W_i/W_k$ of isotope separation is high. In this case the mixture is being enriched both with the i - and k -isotope.  It is also obvious that S depends on the frequency of the exciting IR field, its intensity, the initial pressure of reactants and many other parameters, some of which are considered below.

B.    Resonance Characteristics of Enrichment

The maximum value of enrichment coefficient in dissociation products (Eq. (VI.5)) attainable when there are no selectivity losses in the reactions of primary dissociation products, can be determined from the ratio of dissociation rates at a given frequency of laser radiation for molecules of different isotopic composition.  In fact, there may be various channels of selectivity loss which can cause the limiting coefficient of enrichment to be reduced.

The dependence of the dissociation rate of the $^{32}SF_6$ molecule on laser frequency was shown already in Fig. 17.

If we assume that the form of the frequency dependence of $^{34}SF_6$ dissociation rate is exactly the same as it is for $^{32}SF_6$ (the dashed line in Fig. 39, lower part) and there is only a general displacement of it by an amount equal to the isotope shift, the intersection point of these unity-normalized curves corresponds to the frequency at which no enrichment

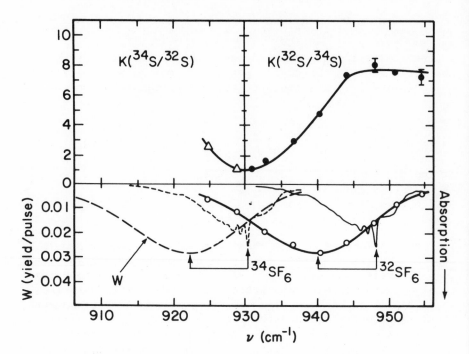

*Fig. 39. Dependence of enrichment factor K measured for the SOF₂ formed in the reaction vessel after dissociation of SF₆. From the value of K the dissociation selectivity factor S can be evaluated. The dissociation rate $W_{dis}$ curves for both isotopic species of SF₆ are shown below (46).*

must be observed. In (46) the frequency dependence of the enrichment coefficient $K_{prod}$ ($^{32}S/^{34}S$) in dissociation products has been measured. One of the products formed after the dissociation is SOF₂ which resulted from the hydrolysis on the walls of SF₄ (34). The results of such measurements are given in Fig. 39 (the top curve). One can see that in the high-frequency range, where the dissociation rate of $^{32}SF_6$ is much higher than that of $^{34}SF_6$, the enrichment coefficient is a constant and falls to 1/e as the laser frequency approaches the center of the absorption band of $^{34}SF_6$. This means that, when the mixture of isotopes is irradiated at 930 cm⁻¹, there is no isotopic enrichment. On further decrease of the laser frequency the coefficient of enrichment of the mixture with the $^{34}S$ isotope grows.

It can be seen from Fig. 39, that the maximum selectivity of dissociation is much higher than the enrichment coefficient in the reaction products, i.e. in SOF₂. The ratio $W_{32}/W_{34}$ can be estimated as high as ≈ 100 while the enrichment coefficient S* under such experimental conditions is no higher than 8 (46). Such a significant discrepancy between the results expected and obtained in (46) can be simply explained if we take into account the fact that the enrichment coefficient was measured in $SOF_2$. The real coefficient of enrichment with $^{32}S$ in (46) was much higher. The understated value of enrichment coefficient was due to the fact that the mass analysis was done using a low-resolution mass spectrometer which would not resolve the mass doublets $^{32}S^{18}OF_2$ and $^{34}S^{16}OF_2$. Thus, the limiting enrichment coefficient measured was determined mainly from the ratio of natural occurrence between the $^{18}O$ and $^{34}S$ isotopes. The abundance of the $^{18}O$ isotope taken into account, the enrichment coefficient for sulphur at 0.05 torr is approximately 50 to 70 which is quite near to the value to be expected from the dissociation rate ratio between the two isotopes S (32/34).

Similar studies on the $BCl_3$ molecule have been made in (98); their results are given in Fig. 40. This figure illustrates the dependence of the dissociation rate of $^{11}BCl_3$ on laser frequency. These data have been obtained from the dissociation of the natural mixture of the $^{10,11}BCl_3$ isotopes, the pressure of $BCl_3$ in the cell being 0.3 torr. A mixture of $O_2$ and $N_2$ in the ratio 1:4 was added into the reactor as a scavenger. The radiation lasted 30 min. at the repetition rate of 1 Hz. It was focused into the cell by a lens with f = 20 cm. As the geometric dimensions of the dissociation volume were not determined, the dissociation rate is given in relative units. At the bottom of Fig. 40 the frequency dependence of the dissociation selectivity S* is shown as measured in the reaction products. Figure 40 shows that the curve describing the dependence of S* on laser frequency is asymmetrical, i.e. the mixture is more enriched with the light isotope than with the heavy one. This asymmetry comes perhaps from the resonance nature of the quasi-continuum through which molecular dissociation is performed (42).

It follows from (VI.4) that the enrichment coefficient $K_{res}$ in the residual undissociated gas may reach very high values considerably exceeding the value of selectivity S* or S. The enrichment coefficient in the first experiments on selective dissociation of $SF_6$ (12) was more than $10^3$ and in some experiments it reached to $\gtrsim 10^4$. Some experimental results from (34) are given in Table 5.


C.  Pressure Dependence of Dissociation Selectivity

The first experiments (12) on sulphur isotope enrichment disclosed that in the pressure range from 0.2 to 2 torr the enrichment coefficient measured in the residual gas, the

TABLE 5

*Main Experimental Data on Isotope Separation by Intense IR Pulses.*

| Isotopes | Absorber gas | Scavenger | Enrichment factor measured in residual absorber gas $K_{res}$ | Enrichment in final products $S^*$ | Ref. |
|---|---|---|---|---|---|
| H, D | H$_2$CO | no | not measured | 40 | 103 |
| 10, 11$_B$ | BCl$_3$ | H$_2$, H$_2$S, HBr, NO, O$_2$ | 8[a] | from 1.2 to 10[a] | 58 31 102 |
| 12, 13$_C$ | CF$_2$Cl$_2$ | no | 1.6 | 3.6 | 102 |
|  | CCl$_4$ | no | not measured | from 1 to 7.5[b] | 57 |
| 28-30$_{Si}$ | SiF$_4$[c] | H$_2$ | 1.11 | 1.14 | 102 |
|  | SiF$_4$[d] | H$_2$ | 15 | not reported | 101 |
| 32-34$_S$ | SF$_6$ | no | $10^4$ | 70 | 46 |

| Isotopes | Absorber-gas | Scavenger | Enrichment factor measured in residual absorber gas $K_{res}$ | Enrichment final products $S^*$ | Ref. |
|---|---|---|---|---|---|
| 92-100$_{Mo}$ | MoF$_6$[e] | H$_2$ | 1.2 | not reported | 100 |
| | MoF$_6$[f] | H$_2$ | 3 | not reported | 53 |
| Os | OsO$_4$ | COS | 1.6[g] | not measured | 30 |

a. $K_{res}$ and $S^*$ are strong functions of the type of scavenger used, absorber-scavenger ratio, and irradiation time. There is no enrichment without scavenger.

b. $S^*$ is a function of laser frequency.

c. No enrichment without scavenger; reverse reaction has been seen; the intensity in focal region reached 6 GW cm$^{-2}$

d. Experiments with directly propagating beam.

e. Single frequency experiments with focusing of radiation into the reaction cell.

f. Two infrared pulse dissociation technique.

g. $K_{res}$ depends strongly on laser intensity and irradiation time. Fast reverse reactions are seen. Two IR pulse technique.

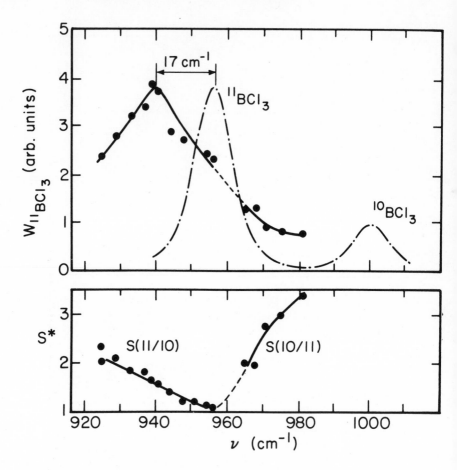

*Fig. 40. Dissociation rate $W_{dis}$ of $^{11}BCl_3$ as a function of laser frequency given in arbitrary units (the volume in which the dissociation took place was not measured precisely). The lower curve shows the frequency dependence of the selectivity S\*, enrichment in the final reaction products. The pressure of $BCl_3$ was 0.3 torr, and scavenger pressure $O_2 + N_2$ was 3 torr in all cases (98). The dashed curve gives the linear absorption spectrum.*

duration of irradiation being the same, dropped exponentially with increasing pressure of $SF_6$ in the cell (Fig. 41(a)). Since these data are obtained by measuring the residual gas undissociated by irradiation, it is obvious that such a dependence of the enrichment coefficient is conditioned by two parameters: a decrease in the dissociation rate $W_{dis}$ with increasing pressure and a decrease in the dissociation selectivity S\*. The measured selectivity S\* is given by S\* ≃ $k_{prod}$ for the first few percent of photolysis (see p. 81).

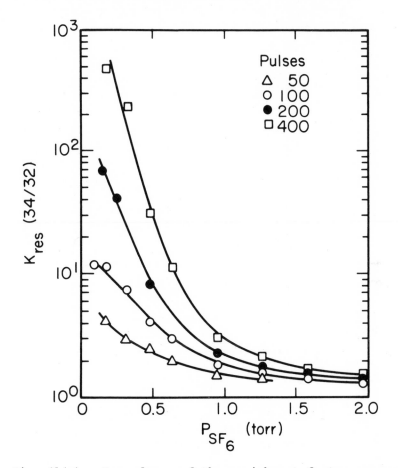

*Fig. 41(a). Dependence of the enrichment factor measured in residual undissociated $SF_6$ on initial $SF_6$ pressure. The parameter of the curves is the number of irradiation pulses (6).*

In (46) the dependence of the dissociation selectivity S* on the pressure of $SF_6$ was measured. The experiments were carried out with no gases added to the reactor as a scavenger. The results of these measurements are given in Fig. 41(b). They show that S* drops exponentially with increasing pressure of $SF_6$ and the characteristic pressure at which S* drops by e times is equal to 0.55 torr. This value coincides with that of the pressure at which the dissociation rate $W_{dis}$ also drops by a factor of e (see Sec. III, Fig. 16). This indicates that it is the V-V exchange between excited and unexcited molecules that is responsible perhaps for the drop both in selectivity and dissociation rate. According

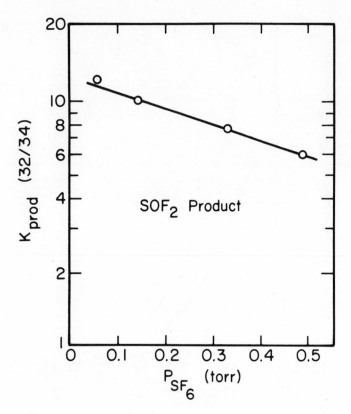

*Fig. 41(b). The dependence of enrichment factor measured in SOF$_2$ product on initial SF$_6$ pressure in the reaction vessel (46).*

to the results obtained in (49) the selectivity ceases to rise when the pressure drops below $4 \times 10^{-2}$ torr.

Figure 42(a) shows the dependence of the enrichment co-efficient in the reaction products of dissociated BCl$_3$ with the buffer gas O$_2$ + N$_2$ (in the ratio 1:4) on the pressure of BCl$_3$ in the cell. The pressure of O$_2$ + N$_2$ remained constant. Oxygen in that case acted as a radical scavenger or reaction partner for excited BCl$_3$. The selectivity value S*, as well as in the case of SF$_6$, drops with increasing pressure of BCl$_3$. If the BCl$_3$ pressure is held constant, S* rises with in-creases the pressure of the scavenger (Fig. 42(b)).

D.     Role of Radical Scavengers

The situation is more complicated when in the process of

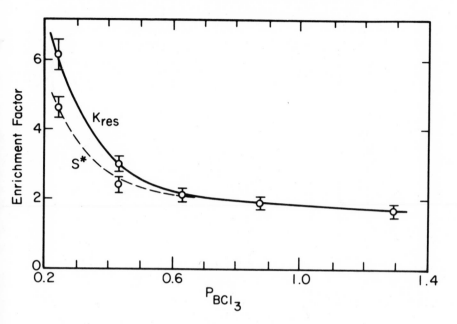

*Fig. 42(a). Dependence of the enrichment factor K for residual $BCl_3$ and of the enrichment factor S\* for final stable products formed by dissociation of $BCl_3$ on $BCl_3$ pressure, with the pressure $O_2 + N_2$, $O_2:N_2 = 1:4$, held constant at torr (58).*

isotope separation various radical scavengers are introduced into the reactor where irradiation takes place. In this case most parameters, such as $W_{dis}$, measure the disappearance rate of starting material in the cell. The isotope selectivity S\* depends on the type of scavenger, irradiation time (since a reverse reaction may take place), absorber-gas-to-scavenger ratio, and many other parameters.

As an illustration, we consider the dissociation of $BCl_3$ and $SiF_4$. In both gases (51, 58, 101) no irreversible dissociation can be observed during a long irradiation time if no scavenger is added in the cell.

The importance of choosing a scavenger can be seen from Table 6 where the experimental data on the enrichment coefficient in $BCl_3$ with various scavengers are given. The initial pressure of $BCl_3$ in all the experiments was 0.88 torr, while the pressure of the scavenger was 2.3 torr.

The enrichment coefficient in reaction products S\* when oxygen was used as a scavenger is much lower than the coefficient S determined from the data on chemiluminescence of the

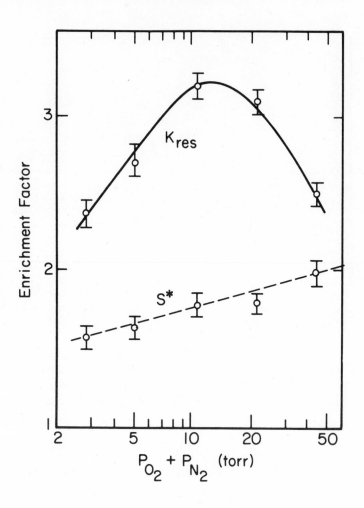

Fig. 42(b). *Dependence of the enrichment factor K for residual BCl₃ and of the enrichment factor S\* for final stable products formed by dissociation of BCl₃ on scavenger pressure ($O_2 + N_2$, $O_2:N_2 = 1:4$), with the BCl₃ pressure held constant at 0.8 torr (58).*

BO radical formed by the reaction between dissociated $BCl_3$ and oxygen (5, 51). The value of S obtained by experiments on studying the luminescence of the BO radicals under non-optimal conditions is higher than 10. The loss of selectivity is evidently due to nonselective chemical reactions.

As a consequence, it is of primary importance to choose the right ratio of molar concentrations between the gas to be dissociated and the acceptor.

TABLE 6

*Values of Enrichment Coefficient in the Reaction Products of*
*$BCl_3$ for Various Scavengers.*

| Scavenger | Enrichment coefficient in end reaction products $S*^a$ |
|-----------|:----------------------------------:|
| HBr | 2.4 |
| $O_2$ | 2.0 |
| $C_2D_2$ | 1.8 |
| NO | 1.45 |
| $H_2$ | 1.2 |

a.  *It is evident that the coefficient of enrichment S\* in*
    *final reaction products differs considerably from the*
    *parameter introduced, dissociation selectivity $S = W_i/W_k$,*
    *since, as a result of nonselective chemical reactions,*
    *the isotopic selectivity of dissociation decreases.*

In (102) the dependence of the enrichment coefficient in
the final reaction products of $BCl_3$ on irradiation time was
checked.  The results obtained (Fig. 43) show that selectivity
drops with time.  The authors attribute this to the reverse
reactions which cause the isotopes to be scrambled and the
initial concentrations to be restored.

The process of dissociation and scavenging was used to
obtain enrichment of $^{12, 13}C$ isotopes by dissociation of
$CF_2Cl_2$ molecules, $^{28, 29, 30}Si$ in $SiF_4$ (102) and $^{12, 13}C$ in
$CCl_4$ (57).  The experimental results are given in Table 5.

Analyzing the data given in Table 5 one may say that in
the cases of $SF_6$, $CCl_4$, and $CCl_2F_2$ the process of enrichment
comes about the following way:

*excitation* ⟶ *dissociation* ⟶ *scavenging.*

This follows from the high values of S\* in dissociation of
these molecules without any acceptors added.  In the case of
$BCl_3$ and $SiF_4$ the low value of selectivity S\* does not enable
us to give an unambiguous answer whether the process follows

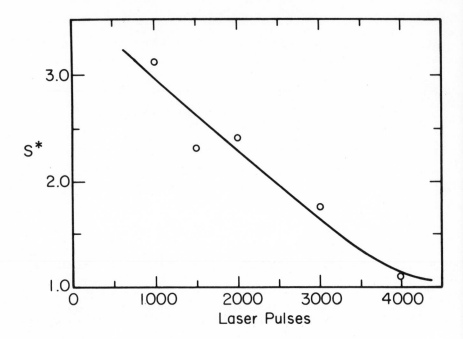

*Fig. 43. Dependence of enrichment factor measured in final products of BCl$_3$ with H$_2$ as scavenger on number of irradiation pulses - the time of irradiation (101).*

the way given above or that which is similar to the process of one quantum infrared chemistry:

*excitation* ⟶ *chemical reaction.*

An unambiguous answer to this question can be probably provided by experimenting in molecular beams with direct mass spectrometric detection of fragments formed in the dissociation process.


E.     The Influence of Radiation Intensity on Selectivity

     In the case of a large isotope shift, as in SF$_6$, where the shift between $^{32}$SF$_6$ and $^{34}$SF$_6$ equals 17 cm$^{-1}$, the large half-width of the $W_{dis}$ curve does not preclude effective isotope separation (when there are no other losses of selectivity).  As seen from Secs. III and IV, a considerable contribution to the half-width of the $W_{dis}$ curve is made by power broadening of the vibrational transition only when focused beams of intense IR radiation are used.

     Of course, the power broadening in terms of Rabi frequency is not the only factor that broadens the dissociation rate curve. One should not omit the possibility of involvement of new successive n-photon resonances as the intensity is increased.

When molecules are dissociated by a field of two infrared frequencies, the absence of power broadening results, as it was shown in Sec. III, in sharp narrowing of the half-width of $W_{dis}$ which, in principle, enables the separation of isotopes with small isotope shifts.

When dissociation is done by two IR pulses, even though the threshold of molecular dissociation may be extremely great $(0.1 - 1 \ GW/cm^2)$, narrow resonance, with half-width comparable with the value of isotope shift, can be observed in the spectrum of $W_{dis}$. Such a situation can apparently be realized in $OsO_4$ (see note added in proof, p. 315). At present isotope separation was accomplished through molecular dissociation by two IR pulses in $SF_6$ (6, 89), $MoF_6$ and $TiCl_4$ (53).

The advantage in selectivity of the method of dissociation by two IR pulses can be illustrated by the following example. The enrichment factor of $^{92}Mo$ in relation to $^{100}Mo$ in $MoF_6$ dissociation by a single-frequency pulse (100) was 20%, while in (53) where $MoF_6$ was dissociated by two pulses it equaled 300%. Comparing these results one should pay attention to the fact that in the case of $MoF_6$ the laser radiation excites a weak combination vibration $\nu_3 + \nu_5$. Therefore, the power broadening is rather small, but still greater than the isotope shift, which is $1 \ cm^{-1}/a.m.u.$ for the $\nu_3$ band of $MoF_6$.

The drastic difference between the enrichment coefficients in $SiF_4$ in (101) where, relative to (102), 15-fold enrichment was attained may be apparently attributed to the effects of transition broadening by field. Although the experimental conditions were somewhat different, they cannot be responsible for such a great difference in the enrichment coefficients. The main difference in experimental conditions between (101) and (102) was that the radiation in (101) was not focused into the cell, and this very likely made it possible to get rid of direct excitation of adjacent isotopes by laser radiation due to the effects of high-intensity fields. The importance of such potential processes was mentioned in (102) too, where no enrichment was observed in $SF_6$ when the dissociation was performed by pulses 1 nsec long with peak intensity of $10^{13} \ W/cm^2$.

In a number of cases, when there is no exact resonance between the radiation and the absorption lines in a molecular gas, the power broadening may play a positive role. In (103), for example, excitation and dissociation of the HDCO molecules took place, with the frequency of the $CO_2$-laser performing excitation and dissociation detuned by $\simeq 8 \ cm^{-1}$. The radiation intensity in the focal region (103) was 20 $GW/cm^2$.

The same work (103) reports on 40-fold enrichment with

deuterium in the dissociation products of formaldehyde.

F.    Isotope Separation by Excitation of Normal Vibrations

Without Isotope Shift.

In (43) experiments were carried out to investigate the
possibility of isotope separation during molecular dissociation by an IR field, when vibrational absorption bands without
isotope shifts are pumped.   This work disclosed isotopic selectivity during dissociation of nitromethane molecules ($CH_3NO_2$)
when the $\nu_{13}$ mode was pumped.   This effect is of particular
interest, and we shall consider the experimental data in more
detail.

In the experiment the nonlinear absorption spectra were
studied as the $\nu_{13}$ vibration of $CH_3^{14}NO_2$ and $CH_3^{15}NO_2$ was
excited   at different laser intensities.   Measurements were
also taken for the yield of electronically excited dissociation products, i.e. the intensity of visible luminescence of
the radicals in the case of excitation of each isotope modification of the $CH_3NO_2$ molecule separately and the dissociation selectivity of $CH_3^{15}NO_2$ mixed with $CH_3^{14}NO_2$ molecules.

The IR spectra of linear absorption of $CH_3^{15}NO_2$ and
$CH_3^{14}NO_2$ molecules in the range from 1000 to 1100 $cm^{-1}$ are
given in Fig. 44.   It can be seen that there is no isotope
shift within the limit of measurement error.   In an intense
field the absorption spectra of the two molecules are deformed in different ways (Fig. 44).   The authors interpret
this difference as the occurrence of an isotope shift equal
to $\approx 5$ $cm^{-1}$.   According to the experiments the value of
"isotope shift" does not substantially change as the IR field
intensity varies from $10^7$ to $10^9$ W/cm$^2$.   Even a larger differ-
rence between the spectra of luminescence excitation   for two
isotopic compounds, i.e. dissociation yield of electronically
excited fragments,was found.   The "isotope shift" in this case
was 10 $cm^{-1}$.

The different deformation of the absorption spectra in
an intense field for molecules of different isotopic composition allows successful experiments on nitrogen isotope separation to be performed by pumping with a $CO_2$-laser the vibrational band $\nu_{13}$ which does not exhibit isotope shift.   The experiments on isotope separation were conducted with the equimolar
mixture of $CH_3^{14}NO_2$ and $CH_3^{15}NO_2$ molecules irradiated by a
focused beam of $CO_2$-laser radiation, with intensity in the
focal region reaching $10^9$ W/cm$^2$.   The relative concentrations
of the molecules before and after the irradiation were measured
by IR absorption spectroscopy.

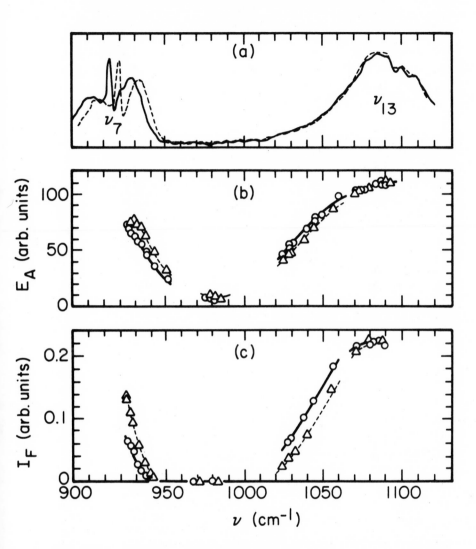

*Fig. 44. Isotopic effects in the $CH_3NO_2$ molecule in the region of the $\nu_{13}$ vibrational mode (43); a) linear absorption spectrum of $CH_3NO_2$ of $\nu_{13}$ [$\alpha(HCH)$, $\beta(NCH)$] vibrational mode, $CH_3NO_2$ pressure 20 torr. b) multiple-photon absorption spectrum of $CH_3NO_2$; $CH_3NO_2$ pressure 2 torr, $CO_2$ laser intensity $10^9$ $W/cm^2$. c) frequency dependence of the visible luminescence in a strong IR field in $CH_3NO_2$; $I_{laser} = 10^9$ $W/cm^2$, $P_{CH_3NO_2}$ = 2 torr. In all cases a), b), c) the dotted lines correspond to $CH_3{}^{14}NO_2$ and the solid lines to $CH_3{}^{15}NO_2$.*

After the mixture of isotopic molecules has been irradiated at the total pressure of 0.5 torr at 1025.3 $cm^{-1}$ by

$7 \times 10^3$ pulses with an energy of 1.2 J, the dissociation selectivity was determined from the burnout rate of each isotopic molecule

$$S^* = \frac{\Delta [CH_3^{15}NO_2]\ [CH_3^{14}NO_2]_o}{\Delta [CH_3^{14}NO_2]\ [CH_3^{15}NO_2]_o}$$

where $\Delta$ [] is the dissociated portion of molecules. The experiments gave the value of $S^*$ equal to 1.18 ± 0.08.

It follows from the ratio of luminescence intensities during excitation of $CH_3^{14}NO_2$ and $CH_3^{15}NO_2$ molecules at the frequency of 1025.3 cm$^{-1}$ that the coefficient of enrichment in primary dissociation products $S = W_i/W_k$ equals 2. The decrease in the value of $S$ in the experiments on isotope separation is very likely caused by partial losses of selectivity in secondary chemical reactions.

The experimental data from (43) cannot be interpreted unambiguously. One of the possibilities is based on assumptions of "quasi-continuum" resonance properties: since these properties are determined mainly by the absorption spectrum at several successive levels, it is quite possible that, as in the case of $SF_6$, the quasi-continuum spectrum is a wide resonance (42). Molecules of different isotopic composition in this case will have different dissociation cross sections. The dependence of the transition cross section in the quasi-continuum on frequency of exciting radiation has already been observed in $OsO_4$ (41). So the effect of "isotopic shift" occurrence observed in the absorption spectrum under intense fields may very likely relate to effects of this kind. Another possibility is that the level density per unit frequency for heavier molecules ($CH_3^{15}NO_2$) is somewhat higher than that in $CH_3^{14}NO_2$. In linear absorption of polyatomic molecules this effect is small but it can show itself more strongly when a large number of photons ($n_{real} \gg 1$) is absorbed. In any case the experiments performed (43) considerably widen the scope of applications of multiple-photon dissociation to isotope separation. (see Sec. VI.G. p. 316).

## VII. IR PHOTOLYSIS OF POLYATOMIC MOLECULES

The effect (34) of collisionless storage of energy by a molecule in an intense IR field, in an amount greater than that needed for bond breaking, puts the problem of the primary photochemical act during IR molecular dissociation in a new light.

There are no grounds for thinking that the action of an

intense IR field is equivalent to that of a single quantum
with energy equal to $\hbar\omega n_{real}$. Particular emphasis was placed
upon this fact in (104) where, perhaps, the first attempt, was
made to determine what kind of fragments are formed in the
case of dissociation in an intense IR field.

The formation of some electronically excited radicals in
IR dissociation makes it possible to determine the composition
of the radicals and the kinetics of their formation by spectroscopic methods.

It is clear that the radicals formed during a laser pulse
in the absence of collisions give information about the path-
way of molecular dissociation. The radicals formed after the
action of a laser pulse may have resulted from chemical reactions both between over-excited molecules and between fragments and molecules.

The radicals formed in the ground electronic state can be
investigated both in molecular beams setups with mass spectrometric analysis of fragments (48, 53) and through exciting
fluorescence by tunable lasers in the visible and uv (52, 90).

The spectral composition of visible luminescence in gases
which resonantly absorb $CO_2$-laser radiation ($C_2F_3Cl$, $CH_3NO_2$,
$CF_2Cl_2$, and $BCl_3$) was studied in (104). The detecting system
included a grating monochromator and a gated electronic detection system. The position of the 150 nsec long gate coincided in time with the pulse of the TEA $CO_2$-laser which
performed the dissociation of molecules. The results of these
studies are given below. During luminescence measurements the
signal was usually registered from the focal region of the
lens where the field magnitude was $\simeq 1$ $GW/cm^2$.

*1.  $C_2F_3Cl$ Molecule*

The spectrum of visible luminescence, at a $C_2F_3Cl$ pressure of 1 torr, consisted of the Swan bands of the $C_2*$ radical
($d^3\Pi_g \rightarrow a^3\Pi_u$ transition) and a wide maximum centered about
4000 Å. This maximum disappeared when the absorber gas was
pumped through the cell at a high rate. No CCl or CF radicals
were observed even at higher pressures.

*2.  $C_2H_4$ Molecule.*

The spectrum of visible luminescence, at an ethylene
pressure of 1 torr, was identical to that observed in $C_2F_3Cl$.
This was confirmed in (90) by the use of the setup shown in
Fig. 13. In addition, it was found in (90) that the concentration of $C_2$ radicals in the ground electronic state was two or
three orders of magnitude higher than that of electronically
excited radicals. The quantitative estimates of concentration
show that the main dissociation channel of $C_2H_4$ in a strong
IR field is fragmentation of ethylene into $C_2$ and hydrogen,
although the authors (35) came to the conclusion that the main

dissociation channel of $C_2H_4$ in an IR field in the same intensity region is the loss of two hydrogen atoms with subsequent formation of a triple bond.

3.   *CH₃CN Molecule*

The luminescence spectra obtained at a pressure somewhat above 1 torr confirmed the presence of the electronically excited radicals $C_2$* and CN*. The luminescence intensity of the $C_2$* radicals was much lower than that for $C_2Cl_3F$ and $C_2H_4$. A band centered at 3280 Å was not identified.

4.   *CH₃NO₂ Molecule*

The minimum pressure at which the spectral composition of visible luminescence was investigated was 5 torr, though the most intense bands of CH* could be recorded at 1 torr. The luminescence spectrum was studied when both the bands $v_7$ (921 $cm^{-1}$) and $v_{13}$ (1097 $cm^{-1}$) were pumped. In both cases the radicals CH*, CN*, $C_2$* (of very weak intensity), OH* and $O_2$* or NH* were observed, Fig. 45. The relative luminescence intensity of the radicals OH* and NH* or $O_2$*, however, in the case of pumping the $v_{13}$ band, were about 1.5 to 2 times higher than that in the case of the $v_7$ vibration.

5.   *CF₂Cl₂ Molecule*

Visible luminescence can be observed at a pressure of 10 torr only. In the spectrum the CCl* radical was identified.

6.   *BCl₃ Molecule.*

The spectrum of visible luminescence in the range from 4300 to 6000 Å was continuous at a resolution of 15 Å and was not identified. In the ultraviolet region the radical BCl* was detected ($A^1\Pi \rightarrow X^1\Sigma^+$ transition), but the luminescence of this radical is delayed relative to the laser pulse. This gives some reason  to assume that the BCl* radical is formed by secondary chemical processes or by the reaction

$$BCl_3^{**} + BCl_3 \longrightarrow BCl* + products.$$

7.   *SiF₄ Molecule*

At 1 torr the spectrum of luminescence coinciding with the laser pulse is an unidentified broad maximum at 4500 Å. In the delayed phase of luminescence, the SiF radical was seen. This enables one to consider that the origin of this radical is similar to that of BCl* (105).

8.   *HDCO Molecule*

The high enrichment factor for the D isotope in IR photolysis of a mixture of $H_2CO$ and HDCO molecules (103) suggests that the main channel of dissociation of the HDCO molecule is

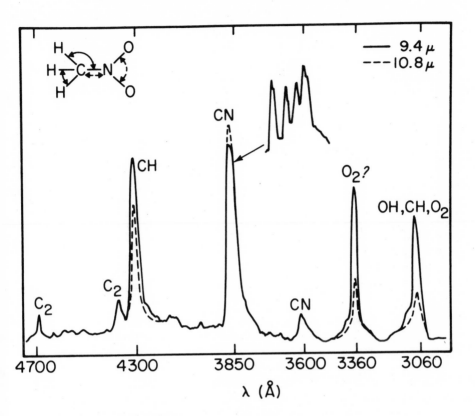

Fig. 45. A portion of the spectrum of the visible
luminescence of the dissociated products in $CH_3NO_2$. The
$CH_3NO_2$ pressure is 10 torr, resolution 20 Å. The solid line
corresponds to pumping through the deformational vibration
$\nu_{13}$, $\alpha(HCH)$ and $\beta(NCH)$, $\nu_{13} = 1097$ $cm^{-1}$. The dotted line
corresponds to the pumping of the $\nu_7$ vibration, C-N and $\gamma(ONO)$.
The structure of the band centered at 3880 Å is shown in the
circle (17 torr of $CH_3NO_2$). The structure of the band is
characteristic of CN (104).

$$HDCO \longrightarrow HD + CO,$$

even though no special studies have been done.

9. *NH₃ Molecule*

The dissociation of the $NH_3$ molecule was investigated
(52) by exciting the luminescence of the radicals formed in the
the ground electronic state using a setup similar to that
shown in Fig. 13. But, in contrast to (90), this setup
employed a cw dye laser. The $NH_2$ radicals were observed with

a time resolution of 200 nsec.  The fluorescence signal of
the laser-excited radicals is shown in Fig. 46.

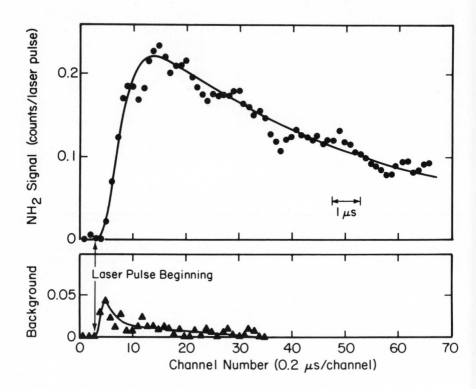

Fig. 46.  Time dependence of the laser-excited fluorescence intensity (upper curve) and the background signal (lower
curve) for 0.05 torr $NH_3$, using the P(32) $CO_2$ laser line of
energy 2 J.  The counting channel width was 200 nsec, the $CO_2$
laser fired in channel 4.  The cw dye laser was 100 mW power,
0.02 $cm^{-1}$ bandwidth, and 1.5 mm diameter; 300 laser pulses were
used to accumulate the data (52).

As Fig. 46 shows, the signal increases to its maximum
value from the point at which the $CO_2$ laser was fired with a
rise time (i.e., the time at which the signal reaches 1 - (1/e)
of its maximum) of 700 nsec.  This is considerably longer than
the width of the $CO_2$ laser pulse itself and the 200 nsec
channel width.  The rise time behavior is nevertheless consistent with prompt formation of $NH_2(^2B_1)$ radical during the 200
nsec laser pulse since the delayed formation of the excited
$NH_2(^2A_1)$ radicals due to their finite lifetime must be taken
into account.  For the pulsed formation of $NH_2$, the fluores-

cence excited by the cw laser will increase with a finite rise time which will be at least as long as the total lifetime $\tau$ of the excited state of $NH_2$. $\tau$ depends upon both radiative and collisional loss processes, and is given by $(\tau)^{-1} = (\tau_0)^{-1} + K_q[M]$, where $\tau_0$ is the radiative lifetime of the $NH_2$ $(^2A_1)$ $\Sigma(0, 9, 0)$ state, and $K_q$ its collisional deexcitation rate constant with quenching gas M. Taking M = $NH_3$ and using recently measured values of $\tau_0 = 10$ $\mu$sec and $K_q = 1 \times 10^{-9}$ $cm^3$ molecule$^{-1}$ sec$^{-1}$, $\tau$ is calculated to be 600 nsec at 0.05 torr. As this is very close to the experimentally observed risetime, one can conclude that $NH_2$ was formed virtually instantaneously, i.e., within the 100 nsec duration of the laser pulse. The discrepancy between calculated and observed pulse lengths may be caused by the formation of $NH_2$ radicals after the 200 nsec main laser pulse during the tail of the $CO_2$ laser pulse.

The degree to which collisions occured during the $NH_2$ production time can be estimated by comparing the values of pt and pt$_{coll}$, where p is the gas pressure, t - the effective time resolution, and t$_{coll}$ the time between gas kinetic collisions at room temperature. In the experiments described pt is $10^{-8}$ torr sec, which is an order of magnitude smaller than pt$_{coll}$. Therefore one can conclude that at low pressure the $NH_2(^2B_1)$ radicals were produced by the dissociation

$$NH_3 + n\hbar\omega \longrightarrow NH_2 \; X(^2B_1) + H.$$

The decay of the fluorescence signal can be seen in Fig. 46 to take place on a much longer timescale than the rise, and appears to be controlled by diffusion of ground state $NH_2$ radicals out of the cw excitation laser beam following their formation.

## 10.  SF$_6$ Molecule

The fragments formed by dissociation of $SF_6$ in the IR field were studied in molecular beams with mass-spectrometric analysis of the fragments (48, 53). The results obtained in (48) are presented in Fig. 47. From these results it follows that with increasing intensity of exciting radiation the portion of fragments with smaller masses rises. In contrast to (48), it is concluded in (53) that the only dissociation channel of $SF_6$ is

$$SF_6 + n\hbar\omega \longrightarrow SF_4 + F_2$$

In contrast to both these results the direct mass-spectrometer analysis of the fragments formed due to IR dissociation of $SF_6$ showed (118) that the only products formed are

$$SF_6 + n\hbar\omega \longrightarrow SF_5 + F.$$

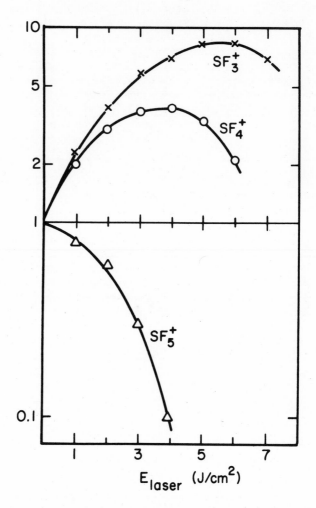

Fig. 47. *Fragment ions of dissociated $SF_6$ as a function of laser energy, showing the increasing probability for formation of lower mass fragments as the vibrational energy content in $SF_6$ increases (48).*

These measurements were also performed in molecular beam apparatus. The laser intensity reached $5 \times 10^9$ W/cm$^2$ which roughly corresponds to 20 J/cm$^2$. From the angular distribution of $SF_5$ fragment the amount of energy gone into translational motion was evaluated. It was found that the latter does not exceed 200 wavenumbers.

It seems that a sequential dissociation, $SF_6 \rightarrow SF_5 + F$ followed by $SF_5 \rightarrow SF_4 + F$, occurs (119). Thus the amount of absorbed energy per $SF_6$ molecule $\hbar\omega n_{real}$ reaches saturation after the laser intensity becomes greater than the threshold value.

### 11. Trans-$C_2H_2Cl_2$ Molecule

In work (44) the spectrum of visible luminescence in trans-$C_2H_2Cl_2$ was studied. Intense Swan bands of $C_2^*$ and $CH^*$ radical were observed. It is of interest that the dissociation of the $C_2H_2Cl_2$ molecule by a short-pulse $CO_2$-laser (the full length at the baseline was 250 nsec) did not result in electronically excited radicals, while in the case when the main pulse of the TEA $CO_2$-laser was followed by a tail a part of the radicals formed in this case was found to be electronically excited.

The summary of the results obtained from studies of the fragments formed during the dissociation of polyatomic molecules in a strong resonant IR field is given in Table 7. It was rather surprising in these studies to observe molecules dissociating to a variety of fragments at the same time.

The comparison of the primary products in IR dissociation and uv flash-photolysis (1) shows that for some molecules these two cases differ greatly. Unfortunately, we cannot draw any unambiguous conclusions at present because there are not enough experimental data available. As seen from Table 7, some experiments were conducted under conditions for which it was impossible to eliminate the collisional effects and hence chemical reactions between excited particles. Of particular interest is the dependence of fragmentation channels on intensity of pumping radiation. These observations confirm unambiguously that the energy stored by a molecule in a strong IR field is greater than its dissociation energy $D_0$.

## VIII. APPLICATIONS TO CHEMICAL SYNTHESIS AND MATERIALS
## TECHNOLOGY

The universality of multiple photon excitation, dissociation and isomerization of polyatomic molecules by intense IR radiation enables us to hope for ever increasing applications of these effects beyond the scope of isotope separation. Now it is difficult to enumerate the entire range of other potentially wider uses. Instead of this we would rather refer to several experiments which demonstrate potential applicability of the effects of multiple photon excitation and dissociation of polyatomic molecules by intense IR radiation to chemical engineering and to the technology of pure substances. Both examples that follow, however, correspond to experiments

TABLE 7

*Observed Dissociation Fragments in Intense IR Fields*

| Molecules | Radicals | Ref. | Notes. Experimental Conditions, Method |
|---|---|---|---|
| $C_2F_3Cl$ | $C_2^*$ | 104 | 1 torr; observation of visible luminescence |
| $C_2H_4$ | $C_2^*$, $C_2$ | 104,90 | 1 torr; observation of visible luminescence, excitation of fluorescence; $C_2$ is the main primary product of photolysis. |
| $CH_3CN$ | $CN^*$, $C_2^*$ | 104 | 1 torr; observation of visible luminescence. |
| $CH_3NO_2$ | $CN^*$, $CH^*$, $C_2^*$, $NH$? $O_2^*$? | 104 | CN – 1 torr, others – 10 torr; observation of visible luminescence. The role of collisions is not clear. |
| $CF_2Cl_2$ | $CCl^*$ | 104 | 10 torr; observation of visible luminescence. The role of collisions is not clear. |
| $BCl_3$ | $BCl^*$ | 104,106 | $BCl^*$ formed by collisions. |
| $SiF_4$ | $SiF^*$ | 10,105 | $SiF^*$ formed by collisions. |
| HDCO | HD, CO | 102 | Conclusion based on isotopic analysis of dissociation products. |
| $NH_3$ | $NH_2$, H | 52 | Laser excitation of fluorescence, direct observation of $NH_2$. |
| $SF_6$ | $SF_5$, $SF_4$, $SF_3$ | 48,53 119 | Dissociation in molecular beams, mass analysis of dissociation fragments. |
| trans-$C_2H_2Cl_2$ | $C_2^*$, $CH^*$, $C_2$, $CH$ | 44 | Excited radical formed if the laser pulse duration > 250 nsec, observation of visible luminescence. |

carried out at mixture pressures ranging from 10 to 100 torr. Under such conditions the effects of collisionless multiple photon excitation and dissociation, given the primary atten- tion in our present review, are not the only ones. At such pressures the best yield must occur when multiple photon absorption is combined with collisions. In practice, such a combination, as has already been mentioned, may prove to be the most economical as regards a minimum energy consumption and a maximum efficiency of the process stimulated by intense IR radiation.

A.   Chemical Synthesis

Intense IR radiation can transfer an energy of several eV to selected molecules in a mixture without giving such a high energy to other molecules. This situation differs from thermal excitation when all the molecules in a gaseous mixture have approximately the same energy content. It is therefore hoped there may be new possibilities for chemical reactions with highly excited molecules involved. It is molecular superexcitation in the transient state of reaction that before excitation thermalization may make the minor paths of chemi- cal reactions investigated under equilibrium conditions more competitive. It is known, indeed, that under equilibrium con- ditions the composition of initial and final products of reactions depends on thermodynamics only. During pulsed exci- tation in a transient, nonequilibrium regime, temporal factors become important in addition to thermodynamic factors. The ratio between the rate of some reaction and the rate of relaxation of the nonequilibrium state of a molecular mixture is the key factor.

Such an approach to chemical synthesis was demonstrated in work (107), where the chemical reaction was studied in a mixture of $BCl_3$ with $H_2$ irradiated by focused pulses of a TEA $CO_2$-laser. In the thermal reaction the compounds $BHCl_2$ and $B_2H_6$ are mainly formed in this mixture. The nonthermal reac- tion caused by laser action gives rise only to the pure compound $BHCl_2$. The total pressure of the mixture (1:1) during the experiment was from 40 to 100 torr. The maximum quantum yield of the reaction in one laser pulse was observed at the mixture pressure of 80 torr and equaled 122 IR photons/ $BHCl_2$ molecule. This quantum yield is rather high for a reaction controlled by an IR field, if we take into account the fact that the energy of the $BCl_2$-Cl bond is equivalent to about 42 $CO_2$-laser photons. This reaction comes about, perhaps, through chain reaction processes. The number of HCl molecules formed by the reaction is exactly equal to that of the resultant $BHCl_2$ molecules. Therefore the total pressure of the mixture does not materially vary in the course of the reaction. All of this supports the complete macroscopic reaction:

$$BCl_3 + H_2 + n\hbar\omega \longrightarrow BHCl_2 + HCl. \qquad (VIII.1)$$

Reaction (VIII.1) is endothermic by about 0.65 eV, or by approximately the energy of 6 photons of $CO_2$-laser. Microscopic reactions are more complex by nature, of course, because under the action of $CO_2$-laser pulses with a power $\gtrsim 10^7$ $W/cm^2$, the gas mixture fluoresces in the visible range which points to the formation of electronically excited molecules and (or) radicals.

The study of $BHCl_2$ yield with pulse power results in the dependence $(P/P_0)^{1.26}$. The fact that the reaction yield differs from the linear dependence indicates that the process is contributed to by nonlinear absorption.


B.    Materials Purification in Gas Phase

The production of highly pure substances is another promising application of chemical reactions stimulated by high-power pulses of IR radiation. The most appropriate field of application of this method is purification of a substance from those impurities  which are difficult to remove by standard methods of purification. For this purpose it is possible to combine the selective dissociation of impurity molecules in the gas phase by high-power pulses of IR radiation with conventional methods of purification to remove the resultant products. It is necessary for purification that the physical-chemical properties of the resultant products should differ from those of the original substance in a way which enables their simple separation and hence a high degree of purification.

The first experiments in this direction were performed in work (108) where the authors investigated the possibility of purification of arsenic trichloride $AsCl_3$ from 1,2-dichlorethane ($C_2H_4Cl_2$) and carbon tetrachloride ($CCl_4$). When $AsCl_3$ is purified by conventional methods, the residual content of the mentioned impurities is no less than $10^{-2}$ to $10^{-3}$%. The absorption bands of $C_2H_4Cl_2$ and $CCl_4$ fall within the lasing region of $CO_2$-laser (Fig. 49). These impurities were selectively dissociated by focused pulses from a TEA $CO_2$-laser with an energy of 2 J and a duration of 100 nsec. The basic substance $AsCl_3$ remained practically undecomposed.

The results of the experiment are listed in Table. 8. In the case of 1,2-dichlorethane the substance was also irradiated at some other frequencies of the P-branch of the 10.6 μ

TABLE 8

*Dissociation Rate of $C_2H_4Cl_2$ and $CCl_4$ Molecules as Unwanted Impurities in $AsCl_3$ (108).*

| Pressure of mixture components, (torr) | | | Radiation frequency $(cm^{-1})$ | Dissociation rate[a] $(W \cdot 10^3)$ | |
|---|---|---|---|---|---|
| $AsCl_3$ | $C_2H_4Cl_2$ | $CCl_4$ | | $C_2H_4Cl_2$ | $CCl_4$ |
| 0 | 2.5 | 0 | 938.7 | 2.5 | – |
| 2.5 | 2.5 | 0 | –"– | 2.2 | – |
| 10 | 2.5 | 0 | –"– | 2.4 | – |
| 0 | 0 | 0.4 | 980.9 | – | 0.30 |
| 10 | 0 | 0.4 | –"– | – | 0.33 |

a.   *Fractional decrease in concentration per laser pulse.*

band of the $CO_2$-laser.   The dissociation rates were approximately those given in the table.   The dissociation rate of the trans- and gauche-isomers of $C_2H_4Cl_2$ proved to be the same.

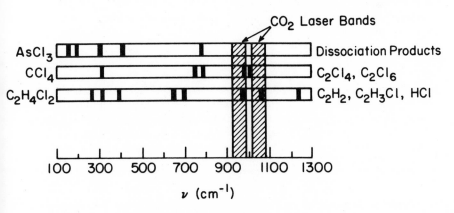

*Fig. 48.   IR absorption bands in experiment on purification of $AsCl_3$ by the multiple photon dissociation of the impurities $CCl_4$ and $C_2H_4Cl_2$ (108).*

This can be explained either by excitation of the combination vibration of the trans-isomer (probably $\nu_{17} + \nu_{18}$) or by excitation transfer from one isomer to another. The end products of dissociation of $C_2H_4Cl_2$ were identified by IR absorption spectra. Almost all the products detected were $C_2H_2$, $C_2H_3Cl$, and HCl; their composition remained constant as the laser wavelength changed and also when stored after irradiation for 24 hours. In the dissociation of $CCl_4$ the products were identified by mass spectrum (57). They turned out to be $C_2Cl_4$ and $C_2Cl_6$. In the case of 1,2-dichlorethane the resultant products differ greatly from arsenic trichloride in their physical properties which enables us to separate them easily and thereby purify $AsCl_3$.

Work (108) demonstrates a potential possibility of using high-power IR radiation in processing pure materials. This approach seems to be applicable to many other important cases and, among other things, to purifying gas mixtures from toxic and carcinogenic substances.


IX.  APPLICATIONS IN SPECTROSCOPY


Direct excitation of high vibrational levels by a high-power pulse of IR laser radiation apparently opens up new possibilities for spectroscopy of these states, states which cannot be studied by the existing experimental methods of spectroscopy. Our basic knowledge of the IR spectra of poly-atomic molecules concerns mainly the lowest vibrational transitions (see (82)). The progress of laser spectroscopy has made it possible to investigate the spectrum and the relaxation rates for the lowest energy transitions (see (110, 111)). There have not been, however, any methods for spectroscopy of vibrational transitions between highly excited states of polyatomic molecules. The approach described in this review can, in our opinion, bridge this gap. Besides, this is just a newly discovered field of research in which many improvements are yet to be realized. Therefore, as in Sec. VIII, we restrict ourselves here to considering just a few possibilities already demonstrated in studying the transitions from the ground state to highly vibrationally excited states, transitions between high vibrational states and relaxation rates of these states.


A.    Spectroscopy of Excited Vibrational States

1.    *Multiple Photon Vibrational Spectroscopy*
We use this term for spectroscopy of multiple photon vi-

brational transitions from the ground state to excited vibrational states. In the preceding sections (II - experiment,
IV - theory) we presented, in essence, such spectra for some
molecules ($SF_6$ - Fig. 8, $C_2H_4$- Fig. 9 and $CH_3NO_2$ - Fig.
44(b)), described the technique of their measurement and, in particular, the most convenient optoacoustic method (35). It
remains here only to analyze the spectroscopic potentialities
of this approach.

Since the spectrum of multiple photon absorption differs
greatly from that of linear IR absorption (Fig. 8, 9 and 44),
it gives, in principle, additional spectroscopic information
which cannot be obtained by the classical methods of linear
IR spectroscopy. The spectrum of multiphoton absorption is
simpler than that of linear absorption as, for instance, in
the case of $C_2H_4$. This seems to be attractive for molecular
analysis. But, on the other hand, it is difficult to identify
molecules if there are no fine details in the multiple photon
absorption spectrum. This is rather an essential limitation
of multiple photon spectra which results from the "looser"
selection rules. Indeed, the selection rule for the rotational quantum number J for an n-photon transition takes the
form:

$$\Delta J = 0, \pm 1, \pm 2. \ldots , \pm n \qquad (IX.1)$$

For example, there are seven overlapping vibrational branches
for a three-photon process. Besides, it must be taken into
account that the final n-th vibrational state is highly degenerate in many cases. The multiple photon vibrational
transitions in an intense IR field have therefore many possibilities for involving in the process not one but a great
number of initial rotational sublevels. This assumption is
confirmed by the recent experiments (112) where various rotational states of the $SF_6$ molecule in an intense IR field were
probed.

Thus, experimental multiple photon vibrational spectra
must be broad and hence less suitable for spectroscopic
analysis. They have some other possibilities though. First,
in studying molecular absorption at low temperatures (a cooled
gas or an overcooled molecular beam) we can decrease the molecular distribution over rotational states, and the spectrum
of multiple photon absorption in this case must be theoretically ideal (Fig. 23). Secondly, the spectrum of multiphoton
absorption is very sensitive to laser radiation intensity.
This feature may be used to discriminate resonance multiphoton
absorption against the strong background of linear continuous
absorption.

It is of great interest to investigate the isotope effect
in multiphoton absorption spectra. First experiments in this
direction were performed on $CH_3{}^{14}NO_2$ and $CH_3{}^{15}NO_2$ (43). The

isotope effect was detected on the vibrational band $\nu_{13}$. In
linear absorption this band has no isotope effect. This
phenomenon may be explained in several ways. It may be, for
example, related to the difference in the composite matrix
elements $\mu_{eff}$ for a multiple photon transition between lower
molecular levels determined by the general expressions (IV.12)
and (IV.13). The difference in composite matrix element for
two isotopic molecules may occur through the perturbation of
any intermediate level for example by Fermi resonances in the
spectrum of one isotopic modification. We must not rule out
the contribution of the difference between the absorption
cross sections in the vibrational quasi-continuum either (42).

## 2.   Spectrum of the Vibrational Quasi-continuum

Direct excitation of molecular vibrational states by
intense IR radiation makes it possible to investigate the
spectrum of molecular transitions between highly excited
states in the energy region of the vibrational quasi-continuum.
The measurement of such spectra is very necessary to gain insight into the mechanism, threshold intensity and other
characteristics of multiple photon collisionless molecular
dissociation in an intense IR field. Secondly, such spectra
are of interest in studying the structure of polyatomic molecules and, in particular, the interaction of normal vibrational modes.

The spectrum of IR absorption by molecules in the state
$v \gtrsim 3$ was first measured for the $SF_6$ molecule in (6, 89). The
results of these experiments are given in Fig. 21. They show
that the absorption spectrum is not constant and the transition cross section tends to increase as laser frequency
approaches the fundamental frequency. It is expected that the
absorption spectrum in the vibrational quasi-continuum for $SF_6$
molecules with the excited normal vibration $\nu_3$ has a wide
resonance shifted to the low-frequency region with respect
to the frequency of the fundamental transition $0 \rightarrow \nu_3$. The
measurement of the structure of such resonances for molecules
is of particular interest because it gives the answer to the
following important question: to what extent is the indivi-
duality of the excited normal mode of the vibrational quasi-
continuum transitions retained and what part of the absorption is connected with excitation of other normal modes? The
parts of IR absorption measured within a wide possible resonance and in the region of structureless background may
provide the answer to this question.

To study the spectrum of transitions between highly
excited states we have to elaborate sufficiently simple methods for such measurements. There are two important problems
arising in this case: 1) excitation of given vibrational
states and  2) detection of IR radiation absorbed by excited

molecules.

At present there exist two tested methods of pre-excitation of molecular vibrational states: by intense IR laser radiation and by thermal excitation. In (6, 89) the pre-excitation of the $\nu_3$ mode of $SF_6$ was done by a $CO_2$-laser pulse with its power ranging from $3 \times 10^4$ to $10^7$ $W/cm^2$. Such a power is not enough to dissociate the $SF_6$ molecule but it is quite sufficient to excite several lower vibrational states (with pulse power from $3 \times 10^4$ to $10^6$ $W/cm^2$) and provides multiphoton excitation of higher states (with pulse power from $10^6$ - $10^7$ $W/cm^2$). The most difficult task in this case is to control the excitation of particular vibrational states. We do not yet know exactly the frequencies of transitions between lower vibrational levels to predict the optimal position of the laser frequency for exciting a particular level. The combination of the effect of "triple rotation-vibration resonances" with excitation by a series of properly chosen laser pulses at several frequencies will probably enable us to solve this difficult problem. It must be said that its solution is more important not for spectroscopy but for chemical physics since it makes possible studies into the rate of chemical reactions of given vibrational quantum states.

In work (113) $BCl_3$ molecules were pre-excited by heating the cell. The problem was to detect the changes in absorption from thermally populated excited vibrational states at increased temperature, i.e. to detect weak "hot" bands. In such experiments the molecules in the initial state are distributed, of course, over all possible levels rather than over the levels of the $\nu_3$ mode as in the experiments in (6, 89). The absorption spectra in these two experiments must therefore differ greatly. Unfortunately, in experiment (113) the absorption spectrum was not measured, and only the temperature dependence of IR radiation absorption off the fundamental absorption bands was measured. Figure 49 illustrates such a dependence for the $BCl_3$ and $BF_3$ molecules. The slope of the curves makes it possible to determine the energy $E_i{}^*$ of the excited vibrational states from which absorption takes place.

There are two methods which may be used to detect absorption due to transitions between highly excited vibrational levels: an increase in the dissociation rate due to the increasing energy of molecular excitation and optoacoustic measurement of absorbed energy. The first one necessitates the application of an intense IR field which can, in fact, increase materially molecular vibrational energy due to transitions in the vibrational quasi-continuum. It was used in (6, 89) for selective dissociation of $SF_6$ in a two-frequency IR laser field (Sec. III). The second method does not need any intense field because absorption can be detected by observing a slight additional heating of the molecular gas

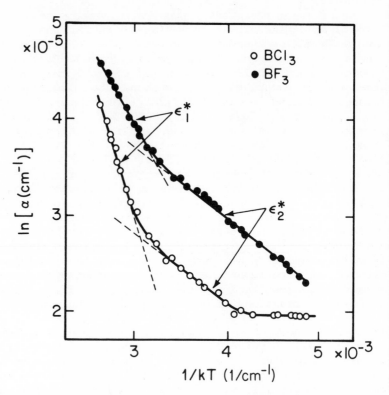

*Fig. 49. Temperature dependences of the coefficient of weak IR absorption α at a wavelength near 10 µ, outside the fundamental IR bands of BF₃ and BCl₃ molecules, which were measured by optoacoustic method (113).*

pre-excited by any method (thermal excitation (113) or laser excitation (114)). The method of transition detection in the vibrational quasi-continuum in a two-frequency pulsed field with optoacoustic measurements demonstrated in (114) is worthy of notice, too. The first laser pulse with a fixed frequency excites the molecules and gives rise to a certain optoacoustic signal. The second one with a scanning frequency induces molecular transitions between highly excited states, with the result that there arises an additional optoacoustic signal, the the magnitude of which depends materially on the frequency of the second laser pulse and the presence of the first one. By the difference between the optoacoustic signals we may evaluate the intensity of the transitions between excited vibrational levels. Because of its simplicity the optoacoustic method of measurement can be applied to many polyatomic molecules.

B.    Relaxation of Excited Vibrational States

Multiphoton excitation of high molecular vibrational states offers evidently new possibilities for studying both intramolecular and intermolecular relaxation of vibrational energy.  There have already been some advanced experiments performed in this direction.  Their difference from all previous experiments (see reviews (110, 111)) lies in their ability to study relaxation and transfer of large amounts of vibrational energy, amounts greater than the energy of a vibrational quantum.

*1.    Intramolecular Relaxation*
Our laboratory has carried out a number of experiments on the $OsO_4$ molecule, where we investigated the pressure dependence of the rate of vibrational energy exchange between two modes.  The results of these experiments are briefly presented in papers (42, 45).  The electronic absorption spectrum of $OsO_4$ shows a vibrational progression in the $\nu_1$ mode.  When the vibrational mode $\nu_3$ is pumped by a high-power $CO_2$-laser pulse, some vibrational energy can be transferred to the $\nu_1$ mode as a result of intramolecular relaxation.  The population of the vibrational levels of the $\nu_1$ mode appears as a distinct red-shift of the absorption spectrum within the vibronic progression.  Studies of the kinetics of appearance of the red-shift in uv absorption after high-rate pumping of the $\nu_3$ mode may give information about the rate of intramolecular relaxation of vibrational excitation.  This experimental technique called IR-uv double resonance is described in more detail above (Sec. II A, Fig. 4).  Direct observation of multiphoton excitation of high vibrational levels (v ≳ 10) was conducted in (30) on the $OsO_4$ molecule.

The considerable red-shift (1-2 eV) observed points to the fact that the vibrational levels $v_1$ ≳ 10 - 20 are populated due to their vibrational exchange with the excited mode $\nu_3$.  Figure 50 shows the rate of red-shift rise in uv absorption as a function of $OsO_4$ pressure.  By extrapolating the straight line in the low-pressure range we get a non-zero rate of the intramolecular vibrational exchange $\nu_3 \rightarrow \nu_1$.  Its magnitude is about $10^6$ $sec^{-1}$, and it may be interpreted as the rate of collisionless intramolecular relaxation of vibrational energy.  It is interesting to note that the rate of red-shift rise in uv absorption does not depend on the wavelength of the uv probing radiation.  From this it follows that V-V exchange occurs not in succession on vibrational levels $v_1 = 0 \rightarrow 1 \rightarrow 2 \ldots$, but that many vibrational levels of the $\nu_1$ mode are populated simultaneously.  Thus, intramolecular exchange of large portions of the vibrational energy takes place in this case.  It is quite possible to interpret physically the

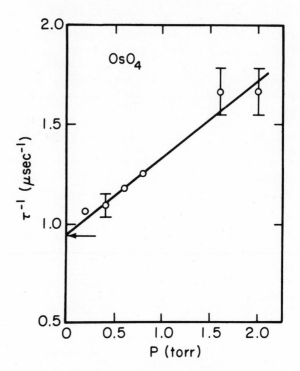

*Fig. 50. Pressure dependence of the V-V exchange rate between the $\nu_1$ and $\nu_3$ modes of $OsO_4$. The data were measured by an IR-uv double resonance method. The zero pressure intercept gives the collisionless relaxation time (42).*

collisionless intramolecular transfer of vibrational energy from the $\nu_3$ mode being excited to other modes within the ideas about multiple photon excitation described in Sec. IV. Indeed, there is an admixture of other normal modes in multiple photon absorption and in the vibrational quasi-continuum particularly. If, for example, the admixture of the $\nu_1$ vibration in multiple photon absorption of IR radiation is $|c_1(\nu\nu_3)|^2$, the vibrational energy must be distributed over the $\nu_1$ and $\nu_3$ modes in the characteristic time $1/\tau \simeq |c_1(\nu\nu_3)|^2 H_{int}/\hbar$, where $H_{int}$ is the energy of interaction (anharmonic, Coriolis, etc.) between $\nu_1$ and $\nu_3$. Such a mechanism of delocalization of vibrational energy due to excitation of the admixtures of other vibrational modes can explain why many vibrational levels of the $\nu_1$ mode can be populated simultaneously.

High-rate intramolecular V-V energy exchange in the $SF_6$ and $BCl_3$ molecules excited by a very short (2 nsec) pulse of a $CO_2$-laser has been recently investigated in (115). Probing the absorption of excited molecules with cw $CO_2$-laser radiation the author could observe rapid (> $10^{10}$ sec$^{-1}$ torr$^{-1}$) changes in the absorption at other vibrational transitions. Since the rate of changes observed in the absorption at other molecular transitions was $10^4$ times higher than the rate of gas-kinetic collisions, the author (115) interpreted the results in terms of very fast collisionless V-V excitation transfer. In essence, such a technique of double IR-IR resonance for excited molecules was used in (112) to study the "bottleneck" effect caused by the limited rate of the molecular rotational relaxation. The similar results in this work are interpreted in quite a different way, i.e. within the framework of multiphoton excitation of high vibrational levels from very many rotational sublevels of the ground vibrational state.

## 2. Intermolecular Relaxation

V-V energy exchange between highly excited and unexcited molecules is concretely demonstrated when the isotopic selectivity of molecular dissociation is measured (see Sec. VI.C). Such a process, however, may be investigated in simpler experiments, which necessitate no molecular dissociation, using not only isotopic molecules but also chemically different ones. Some experiments of the kind were carried out in work (116) where V-V exchange was studied in a mixture of highly excited $SF_6$ molecules with $NH_3$ and NO. This work, in particular, has disclosed high-rate resonance transfer of large amounts of vibrational energy as the molecules of $SF_6$ collide with NO.

The experimental technique (116) was similar to that used in experiments on $OsO_4$. A high-power collimated pulse of a $CO_2$-laser excited the $SF_6$ molecules in the gas mixture. The collisonal excitation of NO or $NH_3$ was recorded by the changes in the uv absorption spectrum of those molecules. The time of V-V energy transfer was determined from the rate of appearance of excited NO or $NH_3$ molecules, i.e. from the rate of red-shift rise in the uv absorption spectrum of the molecules. The dependences of the time of V-V energy transfer on the vibration quantum number of the level measured for $NH_3$ and NO were different. Figure 51 shows the main results on V-V exchange in the mixture $SF_6$ + $NH_3$, the intensity of $CO_2$-laser pulse being 15 MW/cm$^2$ and the frequency 942.4 cm$^{-1}$. The dependences of the reciprocal appearance time of uv absorption for a definite vibrational state of the $\nu_2$ mode of $NH_3$ on the total pressure of the mixture are straight lines. The slope of these lines gives the time of V-V transfer to the corresponding levels of the $\nu_2$ mode of $NH_3$. It is clearly seen that $p\tau_{VV}$ increases as the vibrational quantum number increases. For the level $v_2 = 1$ of $NH_3$ measurements were taken of the dependence of the time of V-V energy transfer on the

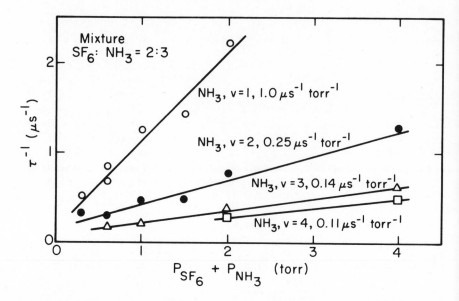

*Fig. 51. Pressure dependence of the rate of V-V energy transfer from highly excited $SF_6$ molecules to the different vibrational levels v = 1, 2, 3, 4 of $NH_3$ molecules (116).*

intensity of the $CO_2$-laser pulse exciting the $SF_6$ molecules. With a mixture pressure of 1 torr in the intensity range from 6.5 to 17 MW/cm$^2$ the time of V-V exchange was independent of intensity. When the intensity of IR radiation is 15 MW/cm$^2$ one molecule of $SF_6$ can absorb, on average, 10 $CO_2$-laser photons (34). Because of this, under the conditions of such an experiment, vibrational excitation can be transferred from many excited levels of the $\nu_3$ mode of $SF_6$. The vibrational frequency of the $\nu_3$ mode (948 cm$^{-1}$) is near to resonance with the first level of the $\nu_2$ mode of $NH_3$ ($\nu_2$ = 968 and 932 cm$^{-1}$). Therefore the process

$$SF_6(v_3) + NH_3(v=0) \longrightarrow SF_6(v_3-1) + NH_3(v_2=1) + \Delta E \qquad (IX.2)$$

is the most probable channel for the vibrational exchange. The measured time of excitation transfer to the $\nu_2$ = 1 level coincides with that of V-V exchange for the $\nu_3$ = 1 level of $SF_6$ (117). At higher levels of the $\nu_2$ mode of $NH_3$ the resonance conditions with the $\nu_3$ mode of $SF_6$ are disturbed because of the large vibrational anharmonicity of $NH_3$. The time of excitation transfer to the levels $v_2 > 1$ therefore increases (Fig. 51). In this case it is impossible to distinguish between single-quantum and multiquantum transfer of vibrational energy since both cases are nonresonant.

In the case of NO, measurements were taken only when the pressure of the mixture $SF_6$ + NO was greater than 2 torr. Some uv absorption signals from the levels v = 1, 2, 3 of the ground electronic state of NO were registered. For all three levels the appearance time of uv absorption proved to be the same, $p\tau_{vv}$ = 16 ± 4 μsec·torr. This suggests that there exists direct transfer of rather high amounts of vibrational energy (2000 - 6000 $cm^{-1}$) from the $SF_6$ molecule to NO. Really, for collisions of $SF_6$ with NO the resonance process of transfer of two vibrational quanta is as follows:

$$SF_6(v_3) + NO(v=0) \longrightarrow SF_6(v_3-2) + NO(v=1) + \Delta E. \quad (IX.3)$$

The detuning for single-quantum energy exchange is very great:

$$SF_6(v_3) + NO(V=0) \longrightarrow SF_6(v_3-1) + NO(v=1) + \Delta E', \quad (IX.4)$$

where $\Delta E' \simeq 930$ $cm^{-1}$. The typical probability of V-V exchange for processes with such detuning equals $10^{-4}$ (110), whereas the experimental value of exchange probability for this collision is 0.8 x $10^{-2}$. Such a difference in probability between two-quantum nonresonance and single-quantum resonance processes of V-V exchange indicates that the first process has a dominant role in the experiments. Processes similar to (IX.3) may bring about population of the levels v = 2, 3.

Direct high-rate multiple photon excitation of the high vibrational levels of polyatomic molecules by intense IR pulses, no doubt, provides experimentalists with both powerful and simple tools for systematic studies of intramolecular and intermolecular relaxation of vibrational excitation. This direction of research must soon become a significant part of molecular kinetics.

X.   CONCLUSION

Two years have passed since successful experiments on isotopically selective multiple photon excitation and dissociation of polyatomic molecules by intense IR radiation (5, 12). Several dozens of experimental and theoretical works in this field have followed them. We have tried to review systematically these works in the present paper.

It can be guaranteed now that the absorption of tens of IR photons is possible during the very short time of the interaction of moderate intensities with polyatomic molecules, i.e. the excitation of high-lying vibrational levels in the ground electronic state of a polyatomic molecule. It presents the basis for quite a new approach in photochemistry, in which from a practical point of view the key role is played by cheap

radiation from powerful, highly efficient, IR lasers and from
a scientific point of view, highly nonequilibrium vibrationally excited polyatomic molecules in a gas mixture. Both
aspects are entirely new for photochemistry and hence there is
every reason to think that they will have far reaching per-
spectives for science and applications.

But in spite of the extremely vast field of research
connected with multiple-photon excitation, most phenomena
described in this review are quite general in character. This
justifies our efforts in writing this review. In concluding
the review we would like to draw special attention to a number
of unsolved questions in the problem of selective interaction
of powerful IR radiation with molecules. We shall confine
ourselves to a short list of them.

1.    *The actual role of collisions in the dissociation of
highly excited polyatomic molecules*. The estimates of collision cross sections, based on existing experimental data can
be too small. One should take into account that the excited
molecules have dimensions, differing from those, which are
measured in experiments with ground state molecules. Even in
the simplest parabolic infinite potential approximation, i.e.
harmonic oscillator, the amplitude of molecular vibrations
grows proportionally to $\sqrt{v}$, where v  is the vibrational quantum number. For real potentials with the finite bond
energy of a molecule this dependence can be much steeper. If
so even very weak collisions at large distances can efficiently influence the actual dissociation mechanism of a highly
excited molecule. A careful comparison of the experimental
data obtained in the two given types of experiments - in a
molecular beam and in a low pressure gas - must give an answer
to this question.

2.    *Possible contribution of chemical reactions*. It is
impossible now to distinguish unambiguously the dissociation
process from the chemical reactions of multiple-photon excited
molecules. The first experiments on irradiation of $BCl_3 + O_2$
are a rather good example of that (the $O_2$ was a scavenger).
This type of chemical reaction leads to a disappearance of the
initial absorber gas in the cell, just as does the dissociation process. It is extremely difficult to distinguish
between them since both processes are caused by multiple-photon absorption and have quite similar features.

3.    *The primary products of collisionless dissociation
of polyatomic molecules by a strong IR field and their difference from the primary products of dissociation by uv and vuv
radiation*. The very first experiments of various groups with
a molecular beam of $SF_6$ and a $CO_2$ laser gave contradictory
results concerning the product compositions. The investigation of the distribution of absorbed energy between different
fragments and within different degrees of freedom of the

fragments translational, rotational, vibrational and electronic should be done. All of this requires careful experiments with a molecular beam employing direct detection of the
fragments and using tunable lasers for probing particular
quantum states of the fragments.

4. *Chemical synthesis of new and unusual molecular
compounds in a photochemical reactor irradiated by powerful
$CO_2$ laser pulses.* Strong molecular vibrational excitation can
increase the reactivity of molecules of one single type in a
mixture and provide fast chemical reactions through new
channels which are not competitive with a uniform, equilibrium
heating of the whole mixture. In contrast to the photodissociation approach, photochemical synthesis under the above
conditions has not been practically investigated. It has
great prospects both from the scientific and practical points
of view.

5. *Spectroscopy of excited states of polyatomic molecules and the investigation of energy transfer processes
between various vibrational modes.* It cannot be excluded that
a new approach to laser driven chemical reactions can be
found by excitation of different vibrational modes of a molecule.

We wish to thank our colleagues at the Institute of
Spectroscopy of the Soviet Academy of Sciences, Drs. N.V.
Chekalin, Yu.A. Gorokhov, N.P. Furzikov, G.N. Makarov, A.A.
Puretzkii, E.A. Ryabov, V.N. Bagratashvili, V.S. Dolzhikov,
I.N. Knyazev, and A.A. Makarov for their contribution to the
research results cited herein.

We would like to express our appreciation to our
colleagues Drs. N. Bloembergen, W. Lamb, Jr., C.B. Moore, R.
Zare, P. Kelly, C. Cantrell, A. Mooradian, P. Robinson and S.
D. Rockwood for many helpful and stimulating discussions.

XI.   REFERENCES


1.   J.G. Calvert and J.N. Pitts, Jr., Photochemistry, Wiley
     New York 1966.
2.   R.T. Hall and G.C. Pimentel, J. Chem. Phys. 38, 1889
     (1963).
3.   V.S. Letokhov and C.B. Moore, Kvant. Elektron. (Moscow)
     3, 248 - 287 and 485 - 516 (1976). [Sov. J. Quantum
     Electron. 6, 129 - 150 and 259 - 276 (1976)].
4.   V.S. Letokhov and R.V. Ambartzumian, IEEE J. Quantum
     Electron. QE-7, 305 (1971); R.V. Ambartzumian and V.S.
     Letokhov, Appl. Opt. 11, 354 (1972).
5.   R.V. Ambartzumian, V.S. Letokhov, E.A. Ryabov, and N.V.
     Chekalin, ZhETF Pis'ma Red. 20, 597 (1974) [JETP Lett.
     20, 273 (1974)].
6.   R.V. Ambartzumian, Yu. A. Gorokhov, V.S. Letokhov, G.N.
     Makarov, A.A. Puretzkii, and N.P. Furzikov, ZhETF Pis'ma
     Red., 23, 217 (1976) [JETP Lett. 23, 194 (1976)].
7.   S. Datta, R.W. Anderson, and R.N. Zare, J. Chem. Phys. 63,
     5503 (1976).
8.   E.S. Yeung and C.B. Moore, Appl. Phys. Lett. 21, 109
     (1972).
9.   V.S. Letokhov, Chem. Phys. Lett. 15, 221 (1972).
10.  N.R. Isenor, V. Merchant, R.S. Hallsworth, and M.S.
     Richardson, Can. J. Phys. 51, 1281 (1973).
11.  R.V. Ambartzumian, N.V. Chekalin, V.S. Dolzhikov, V.S.
     Letokhov, and E.A. Ryabov, Chem. Phys. Lett 25, 515
     (1974).
12.  R.V. Ambartzumian, Yu.A. Gorokhov, V.S. Letokhov, and
     G.N. Makarov, ZhETF Pis'ma Red. 21, 375 (1975) [JETP
     Lett. 21, 171 (1975)].
13.  J.I. Brauman, T.J. O'Leary, and A.L. Schawlow, Opt.
     Commun. 12, 223 (1974).
14.  J. Robieux and J.M. Auclair, French Patent No. 1391738,
     appl. October 21, 1963, publ. February 1, 1965; USA
     Patent No. 3443087, appl. September 16, 1964, publ. May
     6, 1969.
15.  S.V. Andreev, V.S. Antonov, I.N. Knyazev, and V.S. Leto-
     khov, Chem. Phys. Lett. 45, 166 (1977).
16.  V.S. Letokhov, in Optical Sciences Vol. 3, Tunable Lasers
     and Applications. Eds. A. Mooradian et al. Proceedings
     of the Loen Conference, Norway. Springer-Verlag 1976,
     p. 122.
17.  C. Bordé, A. Henry, and L. Henry, C.R. Acad. Sci. Paris,
     B 262, 1389 (1966).
18.  V.V. Losev, V.F. Papulovskii, V.P. Tichinskii, and G.A.
     Fedina, Khimiya Vysokikh Energii 8, 331 (1969) [High
     Energy Chemistry 3, 302 (1969)].

19.  N.V. Karlov, Yu.N. Petrov, A.M. Prokhorov, and O.M. Stel'-
     makh, ZhETF Pis'ma Red. 11, 220 (1970) [JETP Lett. 11,
     135 (1970)].
20.  A.J. Beaulieu, Appl. Phys. Lett. 16, 504 (1970).
21.  R. Dumanchin and J. Rocca-Serra, C.R.Acad. Sci.(Paris)
     269, 916 (1969).
22.  N.R. Isenor and M.C. Richardson, Appl. Phys. Lett. 18,
     224 (1971).
23.  V.S. Letokhov, E.A. Ryabov, and O.A. Tumanov, Opt. Commun.
     5, 168 (1972).
24.  V.S. Letokhov, E.A. Ryabov, and O.A. Tumanov, Zh. Eksp.
     Teor. Fiz. 63, 2025 (1972) [Sov. Phys. - JETP 36, 1069
     (1973)].
25.  J.L. Lyman and R.J. Jensen, Chem. Phys. Lett. 13, 421
     (1972).
26.  V.S. Letokhov and A.A. Makarov, Zh. Eksp. Teor. Fiz. 63,
     2064 (1972) [Sov. Phys. - JETP 36, 1091 (1973)].
27.  C.K. Rhodes.  Report at VII National Nonlinear Optics
     Conference, May 1974, Tashkent, U.S.S.R.
28.  A. Yogev and R.M.J. Loewenstein-Benmair, J. Am. Chem. Soc.
     95, 8487 (1973).
29.  J.L. Lyman, R.J. Jensen, J. Rink, C.P. Robinson, and S.D.
     Rockwood, Appl. Phys. Lett. 27, 87 (1975).
30.  R.V. Ambartzumian, Yu.A. Gorokhov, V.S. Letokhov, and G.N.
     Makarov, ZhETF Pis'ma Red. 22, 96 (1975) [JETP Lett. 22,
     43 (1975)].
31.  S.M. Freund and J.J. Ritter, Chem. Phys. Lett. 32, 255
     (1975).
32.  R.V. Ambartzumian, Yu.A. Gorokhov, V.S. Letokhov, G.N.
     Makarov, and A.A. Puretzkii, ZhETF Pis'ma Red. 22,
     374 (1975) [JETP Lett. 22, 177 (1975)].
33.  R.V. Ambartzumian, Yu.A. Gorokhov, V.S. Letokhov, G.N.
     Makarov, and A.A. Puretzkii, ZhETF Pis'ma Red. 23, 26
     (1976) [JETP Lett. 23, 22 (1976)].
34.  R.V. Ambartzumian, Yu.A. Gorokhov, V.S. Letokhov, and
     G.N. Makarov, Zh. Eksp. Teor. Fiz. 69, 1956 (1975)
     [Sov. Phys. - JETP 42, 993 (1975)].
35.  V.N. Bagratashvili, I.N. Knyazev, V.S. Letokhov, and V.V.
     Lobko, Opt. Commun. 18, 525 (1976).
36.  T.F. Deutsch, postdeadline paper at Tunable Lasers and
     Applications Conference, Loen, Norway 1976.
37.  R.V. Ambartzumian, V.S. Letokhov, G.N. Makarov, and A.A.
     Puretzkii, Chem. Phys. Lett. 16, 232 (1972).
38.  G.N. Makarov. Ph.D. thesis (1976).
39.  J.I. Steinfeld, in Optical Sciences Vol. 3, Tunable Lasers
     and Applications.  Eds. A. Mooradian et al.  Proceedings
     of the Loen Conference, Norway.  Springer-Verlag 1976,
     p. 190.
40.  A. Mooradian (private communication).

41. R.V. Ambartzumian, Yu.A. Gorokhov, N.P. Furzikov, G.N. Makarov, and A.A. Puretzkii, Chem. Phys. Lett. 45, 231 (1977).

42. R.V. Ambartzumian, in Optical Sciences Vol. 3, Tunable Lasers and Applications. Eds. A. Mooradian et al. Proceedings of the Loen Conference, Norway. Springer-Verlag 1976, p. 150.

43. N.V. Chekalin, V.S. Dolzhikov, Yu.R. Kolomiisky, V.N. Lokhman, V.S. Letokhov, and E.A. Ryabov, Phys. Lett. 59A, 243 (1976).

44. R.V. Ambartzumian, N.V. Chekalin, V.S. Dolzhikov, V.S. Letokhov, and V.N. Lokhman, J. Photochem. 6, 55 (1976).

45. R.V. Ambartzumian, N.V. Chekalin, Yu.A. Gorokhov, V.S. Letokhov, G.N. Makarov, and E.A. Ryabov, in Lecture Notes in Physics Vol. 43, Laser Spectroscopy. Eds. S. Haroche et al. Proceedings of the Second International Conference, Megeve, France. Springer-Verlag 1975, p. 121, and Proceedings of the Fourth Vavilov Conference on Nonlinear Optics, Novosibirsk 1975 in Kvantovaya Elektron. (Moscow) 3, 802 (1976) [Sov. J. Quantum Electron. 6, 437 (1976)].

46. R.V. Ambartzumian, Yu.A. Gorokhov, V.S. Letokhov, G.N. Makarov, and A.A. Puretzkii, Zh.Eksp. Teor. Fiz. 71, 440 (1976) [Sov. Phys. - JETP        (1976)].

47. G. Hancock, J.D. Campbell, and K.H. Welge, Opt. Commun. 16, 1977 (1976).

48. K.L. Kompa, in Optical Sciences Vol. 3, Tunable Lasers and Applications. Eds. A. Mooradian et al. Proceedings of the Loen Conference, Norway. Springer-Verlag 1976, p. 177.

49. M.C. Gower and K.W. Dillman, Opt. Commun. 20, 123 (1977).

50. T.P. Cotter and W. Fuss, Opt. Commun. 18, 31 (1976), and Appl. Phys. 12, 265 (1977).

51. R.V. Ambartzumian, V.S. Dolzhikov, V.S. Letokhov, E.A. Ryabov, and N.V. Chekalin, Zh. Eksp. Teor. Fiz. 69, 72 (1975) [Sov. Phys. - JETP 42, 36 (1975)].

52. J.D. Campbell, G. Hancock, J.B. Halpern, and K.H. Welge, Opt. Commun. 17, 38 (1976).

53. A. Kaldor (private communication).

54. S.D. Rockwood (private communication).

55. J.D. Campbell, G. Hancock, J.B. Halpern, and K.H. Welge, Chem. Phys. Lett. 44, 404 (1976).

56. J. Dupré, P. Pinson, J. Dupré-Maquaire, C. Meyer, and P. Barchewitz, C.R. Acad. Sci. (Paris) B282, 357 (1976).

57. R.V. Ambartzumian, Yu.A. Gorokhov, V.S. Letokhov, G.N. Makarov, and A.A. Puretzkii, Phys. Lett. 56A, 183 (1976)

58. R.V. Ambartzumian, Yu.A. Gorokhov, V.S. Letokhov, G.N. Makarov, E.A. Ryabov, and N.V. Chekalin, Kvantovaya Elektron. (Moscow) 2, 2197 (1975) [Sov. J. Quantum.

Electron. 5, 1196 (1975)].
59. J.I. Steinfeld, I. Burak, D.G. Sutton, and A.V. Nowak, J. Chem. Phys. 52, 5421 (1970).
60. G.A. Askarian, Zh. Eksp. Teor. Fiz. 46, 403 (1964); 48, 666 (1965) [Sov. Phys. - JETP 19, 273 (1964); 21, 439 (1965)].
61. F.V. Bunkin, R.V. Karapetian, and A.M. Prokhorov, Zh. Eksp. Teor. Fiz. 47, 216 (1964) [Sov. Phys. - JETP 20, 145 (1965)].
62. A.N. Oraevskii and V.A. Savva, Kratkie Soobschenia Po Fisike FIAN (Russian) [Short Physical Communications of the P.N. Lebedev Institute, Moscow] 7, 50 (1970).
63. N. Bloembergen, Opt. Commun. 15, 416 (1975).
64. V.M. Akulin, S.S. Alimpiev, N.V. Karlov, and L.A. Shelepin, Zh. Eksp. Teor. Fiz. 69, 836 (1975) [Sov. Phys. - JETP 42, 427 (1975)].
65. R.V. Ambartzumian and V.S. Letokhov, Comments Atom. Mol. Phys. 6, 13 (1976).
66. C.D. Contrell and H.W. Galbraith, Opt. Commun. 18, 513 (1976). See also C.C. Jensen, W.B. Person, B.J. Krohn, and J. Overend, Opt. Commun. 20, 275 (1977).
67. V.S. Letokhov and A.A. Makarov, Opt. Commun. 17, 250 (1976).
68. D.M. Larsen and N. Bloembergen, Opt. Commun. 17, 254 (1976).
69. J.T. Hougen, J. Chem. Phus. 65, 1035 (1976).
70. B.J. Feldman and C.J. Elliott, Bull. Am. Phys. Soc. 20, 1282 (1975).
71. N. Bloembergen, C.D. Cantrell, and D.M. Larsen, in Optical Sciences Vol. 3, Tunable Lasers and Applications. Eds. A. Mooradian et al. Proceedings of the Loen Conference, Norway. Springer-Verlag 1976, p. 162.
72. D.M. Larsen, Opt. Commun. 19, 404 (1976).
73. V.S. Letokhov and A.A. Makarov. Coherent Excitation of Multilevel Molecular Systems in an Intense Quasi-resonant IR Laser Field. Preprint of the Institute of Spectroscopy of the USSR Academy of Sciences, 1 - 81, February 1976.
74. D.P. Hodgkinson and J.S. Briggs, Chem. Phys. Lett. 43, 451 (1976).
75. S. Mukamel and J. Jortner, Chem. Phys. Lett. 40, 150 (1976).
76. M.F. Goodman, J. Stone, and D.A. Dows, J. Chem. Phys. 65, 5052 and 5062 (1976).
77. F.H.M. Faisal, Opt. Commun. 17, 247 (1976).
78. R.P. Feynman, Phys. Rev. 84, 108 (1951).
79. C.D. Cantrell and N. Bloembergen. Talk at Winter Colloquium on Quantum Electronics, Steamboat Springs, Colorado, February 1976.

312     R. V. AMBARTZUMIAN AND V. S. LETOKHOV

80.  J.T. Hougen (private communication, Prague, September 1976).
81.  V.I. Gorchakov and V.N. Sazonov, Zh. Eksp. Teor. Fiz. 70, 467 (1976) [Sov. Phys. - JETP 43, 241 (1976)].
82.  G. Herzberg, Molecular Spectra and Molecular Structure, Vol. 2, Infrared and Raman Spectra of Polyatomic Molecules, Van Nostrand, New York 1945.
83.  P.C. Haarhoff, Mol.Phys. 7, 101 (1963).
84.  W.E. Lamb, Jr. Talk presented at the Conference on Laser Photochemistry, Steamboat Springs, Colorado, February 1976.
85.  V.S. Letokhov and V.P. Chebotayev.   Nonlinear Laser Spectroscopy. Springer-Verlag, Heidelberg 1976.
86.  A.N. Oraevskii, A.A. Stepanov, and V.A. Scheglov, Kvantovaya Elektron. (Moscow) 1, 1117 (1974) [Sov. J. Quantum Electron. 4, 610 (1974)].
87.  V.T. Platonenko. Comments on the Session of Scientific Council "Coherent and Nonlinear Optics" of the USSR Academy of Sciences, Tzakhadzor, USSR, January 1976.
88.  P. Kolodner, C. Winterfeld, and E. Yablonovitch, Opt. Commun. 20, 119 (1977).
89.  R.V. Ambartzumian, N.P. Furzikov, Yu.A. Gorokhov, V.S. Letokhov, G.N. Makarov, and A.A. Puretzkii, Opt. Commun. 18, 517 (1976).
90.  N.V. Chekalin, V.S. Dolzhikov, V.S. Letokhov, V.N. Lokhman, and A.N. Shibonov, Appl. Phys. 12, 191 (1977).
91.  V.M. Akulin, S.S. Alimpiev, N.V. Karlov, N.A. Karpov, Yu.N. Petrov, A.M. Prokhorov, and L.A. Shelepin, ZhETF Pis'ma Red. 22, 100 (1975) [JETP Lett. 22, 45 (1975)].
92.  V.M. Akulin, S.S. Alimpiev, N.V. Karlov, B.G. Sartakov, and L.A. Shelepin, Zh. Eksp. Teor. Fiz. 71, 454 (1976).
93.  R.V. Ambartzumian, N.V. Chekalin, V.S. Dolzhikov, V.S. Letokhov, and V.N. Lokhman, Opt. Commun. 18, 400 (1976).
94.  I. Glatt and A. Yogev, J. Am. Chem. Soc. 98, 7087 (1976).
95.  R. Ausubel, M.H.J. Wijnen, J. Photochem. 4, 241 (1975).
96.  W. von E. Doering, V.G. Toscano, and G.H. Beasley, Tetrahedron, 27, 5299 (1971).
97.  V.I. Vedeneev and A.A. Kibkalo.   The Rate Constant of Monomolecular Reactions in Gases. Izd. "Nauka," Moscow, 1972.
98.  Yu.R. Kolomiisky and E.A. Ryabov, Kvantovaya Elektron. (Moscow) (to be published).
99.  A.B. Petersen, J. Tiee, C. Wittig, Opt. Commun. 17, 259 (1976).
     The recent experiments with similar experimental conditions performed by V.S. Dolzhikov in the Institute of Spectroscopy, using a spin-flip laser as a probe showed almost no absorption of excited $SF_6$ at 850 $cm^{-1}$. These

measurements confirm the conclusion that the vibrational quasi-continuum has a broad maximum near the $v_3$ band: i.e. the absorption cross section of the quasi-continuum decreases at lower frequencies.

100.  S.D. Rockwood, in Optical Sciences Vol. 3, Tunable Lasers and Applications. Eds. A. Mooradian et al. Proceedings of the Loen Conference, Norway. Springer Verlag 1976, p. 140, and S.M. Freund and J.L. Lyman, postdeadline paper at Tunable Lasers and Applications Conference, Loen, Norway 1976.

101.  A. Kaldor, D. Cox, and P. Rabinowitz, postdeadline paper at IX Quantum Electronics International Conference, Amsterdam, June 1976. (no abstract).

102.  J.L. Lyman, S.D. Rockwood, J. Appl. Phys. 47, 595 (1976).

103.  G. Koren, U.P. Oppenhein, D. Tal, M. Okon, and R. Weil, Appl. Phys. Lett. 29, 40 (1976).

104.  R.V. Ambartzumian, N.V. Chekalin, V.S. Letokhov, and E.A. Ryabov, Chem. Phys. Lett. 36, 301 (1975).

105.  N.V. Chekalin, V.S. Dolzhikov, and V.N. Lokhman, Kvanto-vaya Elektron. (Moscow) (to be published).

106.  V.N. Bourimov, V.S. Letokhov, and E.A. Ryabov, J. Photochem. 5 , 49 (1976).

107.  S.D. Rockwood and J.W. Hudson, Chem. Phys. Lett. 34, 542 (1975).

108.  R.V. Ambartzumian, Yu.A. Gorokhov, S.L. Grigorovich, V.S. Letokhov, G.N. Makarov, Yu.A. Malinin, A.A. Puretzkii, E.P. Filippov, and M.P. Furzikov, Kvanto-vaya Elektron. (Moscow) 4, 171 (1977) [Sov. J. Quantum Electron. 7,     (1977)].

109.  N.V. Chekalin, V.S. Dolzhikov, V.S. Letokhov, V.N. Lokhman, and A. Shibanov, (to be published).

110.  C.B. Moore, Adv. Chem. Phys. 23, 41 (1973).

111.  G.W. Flynn, in Chemical and Biochemical Applications of Lasers, Vol. I. Ed. C.B. Moore. Academic Press, New York 1974, p. 163.

112.  S.S. Alimpiev, V.N. Bagratashvili, E.M. Khokhlov, N.V. Karlov, V.S. Letokhov, V.V. Lobko, A.A. Makarov, and B.G. Sartakov, Zh. Eksp. Teor. Fiz. (to be published).

113.  V.P. Zharov, V.S. Letokhov, and E.A. Ryabov, Appl. Phys. 12, 15 (1977).

114.  V.N. Bagratashvili, V.S. Letokhov, and V.V. Lobko, (in press).

115.  D.S. Frankel, Jr., J. Chem. Phys. 65, 1696 (1976).

116.  R.V. Ambartzumian, Yu.A. Gorokhov, N.P. Furzikov, G.N. Makarov, and A.A. Puretzkii, (in press).

117.  T.A. Dillon and J.C. Stephenson, Phys. Rev. A 6, 1460 (1972), and J. Chem. Phys. 58. 2056 (1973).

118.  M.J. Goggiola, P.A. Schulz, Y.T. Lee, and Y.R. Shen, Phys. Rev. Lett. 38, 17 (1977).

119. E.R. Grant, M.J. Coggiola, Y.T. Lee, P.A. Schulz, and
     Y.R. Shen, Chem. Phys. Lett., (in press).
120. J.G. Black, E. Yablonovitch, N. Bloembergen, and S.
     Mukamel, (to be published).
121. P.J. Robinson and K.A. Holbrook, Unimolecular Reactions,
     Wiley-Interscience 1972.
122. J.M. Preses, R.E. Weston, Jr., and G.W. Flynn, Chem.
     Phys. Lett. $\underline{46}$, 69 (1977).

Such a situation was realized in our Os isotope separation experiments. Two laser beams with 10 MW/cm$^2$ were used to dissociate the OsO$_4$ molecule. The intensity of both beams was significantly lower than the dissociation threshold for single frequency dissociation (25 - 28 MW/cm$^2$). Both beams propagated unfocused through the cell with a one microsecond delay. The frequency of the dissociating field $\nu_2$ was fixed on the P(20) line of the 10.6 $\mu$ band of the CO$_2$ laser. Frequency $\nu_1$ was scanned.

The resulting Os isotope separations are given in the Table. The starting pressure of OsO$_4$ in each case was 0.3 torr with 2 torr of OCS added as a scavenger. Ninety percent of the gas was dissociated and the enrichment was measured in the residual OsO$_4$ gas.

*Enrichment of Os Isotopes by Two-Frequency IR Dissociation of OsO$_4$.*

| Laser Line | Enrichment Factor K ($\pm$ 0.02) | | | | |
| --- | --- | --- | --- | --- | --- |
| | $\dfrac{192_{Os}}{190_{Os}}$ | $\dfrac{192_{Os}}{189_{Os}}$ | $\dfrac{192_{Os}}{188_{Os}}$ | $\dfrac{192_{Os}}{187_{Os}}$ | $\dfrac{192_{Os}}{186_{Os}}$ |
| R(2), P(20) | 1.02 | 1.02 | 1.04 | 1.08 | 1.17 |
| P(6), P(20) | 1.11 | 1.13 | 1.24 | 1.48 | 1.60 |
| P(12), P(20) | 1.05 | 1.07 | 1.13 | 1.14 | 1.19 |
| P(20), P(20) | 1.02 | 1.02 | 1.08 | 1.00 | 1.01 |

One can easily see from the data presented that when two successive pulses at P(20) are applied only one isotope is involved in enrichment process. The enrichment, in our opinion, can be made much higher with the laser at $\nu_1$ more finely tunable and with a better choice of scavenger (from p.281).

VI.G.   Optimal Regime For Isotope Separation

A sufficient number of experiments on isotopically selective dissociation of polyatomic molecules by intense IR radiation have been performed to permit formulation of principles for an optimal regime of isotope separation. As shown in Sec. III dissociation in a two-frequency IR field has the following two advantageous properties. First, a tunable laser with a relatively low power level may be used for isotopically selective excitation of a molecule by frequency $\nu_1$. This frequency is tuned to maximize the isotopic selectivity of multiple photon excitation and dissociation.

Second, a more powerful but less finely tuned laser may be used for dissociation of the selectively excited molecules. Its frequency $\nu_2$ is coarsely tuned to the maximum of the wide multiple photon absorption resonance in the vibrational quasi-continuum which is shifted to the red from the fundamental vibrational frequency. The correct tuning of the powerful pulse increases the molecule's capacity to absorb IR radiation, i.e. reduces the dissociation threshold. For instance, in our experiments on $OsO_4$ a nonlinear dependence of absorption on the intensity $I_2$, between $I_2{}^3$ and $I_2{}^4$ is observed when tuning frequency $\nu_2$ to the blue side of the absorption band. At the same time when the frequency $\nu_2$ was tuned to the red from the absorption band, the dependence on intensity became linear and the absolute value of the dissociation rate increased by several orders of magnitude compared to the blue-shifted case.

The dissociation of molecules in a two-frequency field looks to be the most practical from the point of view of a scalable isotope separation process. For a proper selection of frequency $\nu_1$ the high isotopic selectivity of such a process permits isotope separation to be realized for heavy elements with isotopic shifts $\Delta\nu_{is} \simeq 10^{-3} \nu_{vib}$, as has been shown in our experiments on $OsO_4$. With single-frequency excitation it had not been possible to detect a particular enrichment from natural isotope abundance. However, with two-frequency excitation the enrichment coefficient reached 50 - 60% in spite of the presence of several isotopes in the primary mixture. With the correct selection of frequency $\nu_2$ the low threshold of dissociation permits experiments on isotope separation in unfocused beams to be performed with an intensity $I_2 \gtrsim 10^6$ W/cm$^2$. This is of great importance from the point of view of process scalability and maximally efficient use of the IR radiation.


These notes are parts of the original manuscript from pp. 281 and 284 which were lost in the mail between Moscow and Berkeley. The editor apologizes for this consequently awkward format.

# Index

# RETURN TO: PHYSICS-ASTRONOMY LIBRARY
### 351 LeConte Hall

| LOAN PERIOD 1 | 2 | 3 |
|---|---|---|
| **1-MONTH** | | |
| 4 | 5 | 6 |

## ALL BOOKS MAY BE RECALLED AFTER 7 DAYS
Books may be renewed by calling 510-642-3122

## DUE AS STAMPED BELOW

|  |  |  |
|---|---|---|
|  |  |  |
|  |  |  |
|  |  |  |
|  |  |  |
|  |  |  |
|  |  |  |
|  |  |  |
|  |  |  |
|  |  |  |
|  |  |  |
|  |  |  |

FORM NO. DD 22
2M 7-10

UNIVERSITY OF CALIFORNIA, BERKELEY
Berkeley, California 94720–6000